Circular Economy, Ethical Funds, and Engineering Projects

Circular Economy, Ethical Funds, and Engineering Projects

Special Issue Editors
Konstantinos P. Tsagarakis
Ioannis Nikolaou
Foteini Konstantakopoulou

MDPI • Basel • Beijing • Wuhan • Barcelona • Belgrade

Special Issue Editors
Konstantinos P. Tsagarakis
Democritus University of Thrace
Greece

Ioannis Nikolaou
Democritus University of Thrace
Greece

Foteini Konstantakopoulou
Hellenic Open University
Greece

Editorial Office
MDPI
St. Alban-Anlage 66
4052 Basel, Switzerland

This is a reprint of articles from the Special Issue published online in the open access journal *Sustainability* (ISSN 2071-1050) from 2018 to 2019 (available at: https://www.mdpi.com/journal/sustainability/special_issues/circular_economy_ethical_funds_engineering_projects).

For citation purposes, cite each article independently as indicated on the article page online and as indicated below:

LastName, A.A.; LastName, B.B.; LastName, C.C. Article Title. *Journal Name* **Year**, *Article Number*, Page Range.

ISBN 978-3-03928-252-4 (Pbk)
ISBN 978-3-03928-253-1 (PDF)

© 2020 by the authors. Articles in this book are Open Access and distributed under the Creative Commons Attribution (CC BY) license, which allows users to download, copy and build upon published articles, as long as the author and publisher are properly credited, which ensures maximum dissemination and a wider impact of our publications.

The book as a whole is distributed by MDPI under the terms and conditions of the Creative Commons license CC BY-NC-ND.

Contents

About the Special Issue Editors vii

Preface to "Circular Economy, Ethical Funds, and Engineering Projects" ix

Panagiotis Marhavilas, Dimitrios Koulouriotis, Ioannis Nikolaou and Sotiria Tsotoulidou
International Occupational Health and Safety Management-Systems Standards as a Frame for the Sustainability: Mapping the Territory
Reprinted from: *Sustainability* **2018**, *10*, 3663, doi:10.3390/su10103663 1

Ionica Oncioiu, Sorinel Căpușneanu, Mirela Cătălina Türkeș, Dan Ioan Topor, Dana-Maria Oprea Constantin, Andreea Marin-Pantelescu and Mihaela Ștefan Hint
The Sustainability of Romanian SMEs and Their Involvement in the Circular Economy
Reprinted from: *Sustainability* **2018**, *10*, 2761, doi:10.3390/su10082761 27

Wenqing Wu, Kexin Yu, Chien-Chi Chu, Jie Zhou, Hong Xu and Sang-Bing Tsai
Diffusion of Corporate Philanthropy in Social and Political Network Environments: Evidence from China
Reprinted from: *Sustainability* **2018**, *10*, 1897, doi:10.3390/su10061897 46

Ulrike Meinel and Ralf Schüle
The Difficulty of Climate Change Adaptation in Manufacturing Firms: Developing an Action-Theoretical Perspective on the Causality of Adaptive Inaction
Reprinted from: *Sustainability* **2018**, *10*, 569, doi:10.3390/su10020569 63

Amtul Samie Maqbool, Francisco Mendez Alva and Greet Van Eetvelde
An Assessment of European Information Technology Tools to Support Industrial Symbiosis
Reprinted from: *Sustainability* **2019**, *11*, 131, doi:10.3390/su11010131 79

António Cavaleiro de Ferreira and Francesco Fuso-Nerini
A Framework for Implementing and Tracking Circular Economy in Cities: The Case of Porto
Reprinted from: *Sustainability* **2019**, *11*, 1813, doi:10.3390/su11061813 94

André Luis Azevedo Guedes, Jeferson Carvalho Alvarenga, Maurício dos Santos Sgarbi Goulart, Martius Vicente Rodriguez y Rodriguez and Carlos Alberto Pereira Soares
Smart Cities: The Main Drivers for Increasing the Intelligence of Cities
Reprinted from: *Sustainability* **2018**, *10*, 3121, doi:10.3390/su10093121 117

Maria Milousi, Manolis Souliotis, George Arampatzis and Spiros Papaefthimiou
Evaluating the Environmental Performance of Solar Energy Systems Through a Combined Life Cycle Assessment and Cost Analysis
Reprinted from: *Sustainability* **2019**, *11*, 2539, doi:10.3390/su11092539 136

Fengjiao Ma, A. Egrinya Eneji and Yanbin Wu
An Evaluation of Input–Output Value for Sustainability in a Chinese Steel Production System Based on Emergy Analysis
Reprinted from: *Sustainability* **2018**, *10*, 4749, doi:10.3390/su10124749 159

G.K. Koulinas, O.E. Demesouka, P.K. Marhavilas, A.P. Vavatsikos and D.E. Koulouriotis
Risk Assessment Using Fuzzy TOPSIS and PRAT for Sustainable Engineering Projects
Reprinted from: *Sustainability* **2019**, *11*, 615, doi:10.3390/su11030615 178

Long Li, Zhongfu Li, Guangdong Wu and Xiaodan Li
Critical Success Factors for Project Planning and Control in Prefabrication Housing Production:
A China Study
Reprinted from: *Sustainability* **2018**, *10*, 836, doi:10.3390/su10030836 193

Dongxiao Niu, Weibo Zhao, Si Li and Rongjun Chen
Cost Forecasting of Substation Projects Based on Cuckoo Search Algorithm and Support
Vector Machines
Reprinted from: *Sustainability* **2018**, *10*, 118, doi:10.3390/su10010118 210

Sung-Hwan Jo, Eul-Bum Lee and Kyoung-Youl Pyo
Integrating a Procurement Management Process into Critical Chain Project Management
(CCPM): A Case-Study on Oil and Gas Projects, the Piping Process
Reprinted from: *Sustainability* **2018**, *10*, 1817, doi:10.3390/su10061817 221

Yonggu Kim and Eul-Bum Lee
A Probabilistic Alternative Approach to Optimal Project Profitability Based on the Value-at-Risk
Reprinted from: *Sustainability* **2018**, *10*, 747, doi:10.3390/su10030747 243

Kingsley Adjenughwure and Basil Papadopoulos
Towards a Fair and More Transparent Rule-Based Valuation of Travel Time Savings
Reprinted from: *Sustainability* **2019**, *11*, 962, doi:10.3390/su11040962 267

About the Special Issue Editors

Konstantinos P. Tsagarakis is a Professor of Economics of Environmental Science and Technology in the Department of Environmental Engineering at the Democritus University of Thrace. He holds a degree from the department of Civil Engineering of the Democritus University of Thrace, a BA degree from the Department of Economics of the University of Crete, and a Ph.D. Degree in Public Health from the School of Civil Engineering from the University of Leeds, UK. His research interests include: technical–economic project evaluation; environmental and energy economics; public health economics; environmental and energy behavior; big data; online behavior; environmental performance of firms; and quantitative methods. His research work has been published in more than 80 papers in refereed journals. He is an Associate Editor in the Water Policy Journal and has served as Guest Editor in several others.

Ioannis Nikolaou is an Associate Professor of Corporate Environmental Management and Environmental Management Systems in the Department of Environmental Engineering at the Democritus University of Thrace. His research interests include: corporate environmental management; corporate sustainability; corporate social responsibility; business circular economy models; and environmental economics. His research work has been published in more than 60 papers in peer-reviewed journals. He has served as Guest Editor in many journals.

Foteini Konstantakopoulou is an Adjunct Professor of Construction Law and Construction Safety in the School of Science and Technology at the Hellenic Open University. She holds a degree from the Department of Chemistry and a Ph.D. Degree in Applied Chemistry, both from the University of Patras, Greece. Her research interests include: engineering project management; waste management; sustainability; applied chemistry; and environmental engineering. Her research work has been published in more than 30 papers in refereed journals and conference proceedings. She has served as a reviewer in numerous journals.

Preface to "Circular Economy, Ethical Funds, and Engineering Projects"

While engineering projects are designed to meet human needs, they bear, apart from benefits to society or humans, environmental impacts and resource limitations. Technology progress, resources, quality, impacts, and awareness are interrelated variables in the design, construction, operation, and end-of-life management of such projects. There are many agents involved in critical roles in engineering projects, such as governments, financial institutions, construction industries, local authorities, and communities. For this purpose, policy makers and entrepreneurs need to make the most suitable decisions to meet the needs of local societies and individuals under sustainability principles, having, above all else, safeguarded the prosperity of generations to come. In particular, each agent should integrate sustainability principles in every stage of the management and design of engineering projects. For example, local authorities have to be open-minded about engineering projects with better social and corporate performance, while construction industries should take into account Circular Economy principles to minimize environmental impacts, while sustainably utilizing natural resources. Similarly, local entrepreneurs and consumers should contribute to sustainable engineering by developing green entrepreneurship and green consumerism. Financial institutions should also play a critical intermediary role in engineering by incorporating sustainable criteria in lending procedures, such as Equator principles. To build an engineering program, there are several classical methodologies and tools that can be employed. Information technology has facilitated the evaluation process by improving classical project evaluation approaches and assisting with the development of new technology-based ones. This Special Issue provides a collection of 15 papers with modern theories and applications for circular economy, engineering projects, entrepreneurship models, and investor decisions. After the commencing review on Occupational Health and Safety Management-System Standards follow papers which can be classified into four categories which cover the overall scope of the Special Issue.

The first category includes papers regarding the microlevel of the circular economy. This means case studies in firm-level which implement different techniques to achieve sustainable development and circular economy goals. The findings reveal interesting achievements which are associated with cultural characteristics of the countries in which these case studies have been conducted. The second category of papers refers to the mesolevel of the circular economy where firms cooperate with each other by exchanging byproducts and organizing common operational procedures and routines to face environmental problems. The findings suggest assessment information technology tools to support industrial symbiosis among European firms. The next body of literature encompasses the macrolevel, where circular economy techniques are implemented at a country level. Findings suggest many methodologies for implementing and tracking circular economy in cities. Finally, a number of papers are included that focus on advanced engineering techniques. These techniques are useful tools for achieving circular economy and sustainability.

Konstantinos P. Tsagarakis, Ioannis Nikolaou, Foteini Konstantakopoulou
Special Issue Editors

Review

International Occupational Health and Safety Management-Systems Standards as a Frame for the Sustainability: Mapping the Territory

Panagiotis Marhavilas [1,*], Dimitrios Koulouriotis [1], Ioannis Nikolaou [2] and Sotiria Tsotoulidou [3]

1. Department of Production & Management Engineering, Democritus University of Thrace, Vas. Sofias 12 St., 67132 Xanthi, Greece; jimk@pme.duth.gr
2. Department of Environmental Engineering, Democritus University of Thrace, Vas. Sofias 12 St., 67132 Xanthi, Greece; inikol@env.duth.gr
3. Department of Engineering Project Management, Faculty of Science & Technology, Hellenic Open University, Parodos Aristotelous 18 St., 26335 Patra, Greece; riatsotoulidou@gmail.com
* Correspondence: marhavil@ee.duth.gr; Tel.: +30-2541-079-640

Received: 29 August 2018; Accepted: 10 October 2018; Published: 12 October 2018

Abstract: A significant part of literature has shown that the adoption of Sustainability and Health-Safety management systems from organizations bears some substantial benefits since such systems (i) create a suitable frame for the sustainable development, implementation and review of the plans and/or processes, necessary to manage occupational health-safety (OHS) in their workplaces and (ii) imply innovative thinking and practices in fields of economics, policy-making, legislation, health and education. To this context, the paper targets at analysing current sustainability and OHSMSs in order to make these issues more comprehend, clear and functional for scholars and practitioners. Therefore, a literature survey has been conducted to map the territory by focusing on two interrelated tasks. The first one includes the presentation of the main International Management Systems (IMS) with focus on Sustainability and OHS (S_OHSMS) topics and the second task depicts a statistical analysis of the literature-review findings (for the years 2006–2017). In particular, the main purposes of the literature research were: (i) the description of key points of OHSMS and sustainability standards, (ii) the comparative analysis of their characteristics, taking into account several settled evaluation-criteria and (iii) the statistical analysis of the survey's findings, while our study's primary aim is the reinforcement of OHMSs' application in any organization. The results evince, that the field of industry (with 28%) and also of the constructions (with 16%), concentrate the highest percentage of OHSMS use. In general, there were only few publications including OHSMSs (referred to various occupational fields) available in the scientific literature (during 2006–2017) but on the other hand, there was a gradually increasing scientific interest for these standards (especially during 2009–2012).

Keywords: Occupational Health and Safety (OHS); sustainability; Management Standards

1. Introduction

Occupational accidents have a key impact upon human probity, create high expenses for the social health/insurance system of any country and deteriorate the sustainability of societies. Moreover, occupational "health and safety" is one of the most vital issues in any organization (or part thereof) because it assures its continual operation, productivity and efficiency. It is known that any occupational accident or illness can affect both the employee, business operation and overall sustainability performance of firms. These disturbances, which can be valued mainly through the lost working-hours and the production-delays, can affect the quality of the enterprise's product [1] and the reputation of firms.

To overcome such problems, many organizations have adopted health/safety and sustainability management systems (with sufficient documentation/certification). Any organization is gradually more concerned with improving sustainability and occupational health and safety (OHS) performance and this is achieved by controlling sustainability and OHS risks, in accordance with their sustainability and OHS policy and in the context of strict legislation. There are plenty of organizations that apply sustainability and OHS reviews (or audits) to assess their sustainability and OHS performance. Nevertheless, these reviews and/or audits may not be sufficient to afford an organization with the assurance that its performance will maintain to fulfill the specific legal and policy requirements of this organization. To be efficient, they must be carried out within a structured management system that is embedded in the organization [2].

Moreover, Occupational Health and Safety Management Systems were created after a lot of well-documented and severe industrial-accidents, during the decades of 1970 and 1980 (e.g., the Flixborough Accident in 1974, the Seveso incident in 1976 and the Piper Alpha disaster in 1987). Studies and research applied on these incidents, unveiled deficiencies in main approaches concerning S_OHMS and regulation and revealed the need to approve approaches which thoroughly addressed both education and engineering responses. The propagation of OHS management systems that have been observed globally since the decade of 1990 [3], has noticeably increased the focus on techniques (and/or tools) concerning performance measurement [4].

A Health and Safety Management System provides a framework for managing health and safety risk. Generally speaking, we can consider the term "risk" as the likelihood that someone (or something) will be harmfully affected by the hazard, while "hazard" is any insecure condition (or source of undesirable/adverse events) with strong potential for creating harm or damage. Alternatively, "risk" would be defined as a measure (under ambiguity) of the hazard severity or a measure of the likelihood and consequence of injurious/adverse effects [5–8].

Public interest in the field of risk analysis and assessment (RAA) has been expanded during the last four decades, so that risk analysis constitutes an efficient and widespread procedure that completes the whole management of nearly all aspects of our life. Thus, almost all managers (e.g., of health care, environment, physical infrastructure systems, etc.) incorporate RAA techniques in their decision-making process. In addition, the universal adjustments of risk analysis by many disciplines (like industry, government agencies) in decision-making, have led to a unique development of theory and methodology and also of practical tools [8].

According to P. Marhavilas [9], risk analysis is a vital process for the safety strategy of any firm, having as main objective the elimination of any potential of damage or harm in its production, while the quantified risk evaluation apparently is the most critical part of the entire procedure of assessing occupational hazards and/or unsafe situations in the workplaces. Furthermore, a complex human-machine system that is composed of humans, machines and their interaction, could suitably be expressed by a system model. Therefore, RAA constitutes a substantial tool for the safety strategy of an organization and also for the assessment process of the occurrence, the consequences and the impact of human activities on systems with hazardous features.

The introduction of a management system in any organization provides a frame for the sustainable development, implementation, sustainability and review of the plans and/or processes which are essential for the occupational health-safety (OHS) management in the workplaces. Since the appearance of such systems during the decade of 1970, significant growth of the approach has occurred, driven by the following factors: (i) OHS is affected by all aspects of the design and functioning of an organization, (ii) the design and management of health and safety systems must associate people, environment and also technical systems in extent that reveal an organization's unique features, (iii) health and safety is a management function and requires broad management involvement, (iv) accidents, injuries and diseases are an indication of a problem in the system and are not coming from a human error and (vi) performance goals must illustrate management objectives [10].

The international management systems (IMS) standards, covering the field of occupational health and safety (OHS) in worksites, are intended to provide organizations and enterprises with elements of an effective occupational health and safety management system (OHSMS) that can be associated (or integrated) with other management requirements and help organizations achieve OHS and economic objectives.

The S_OHSMS standards specify requirements for an OHS management system, in order to allow an organization to develop and implement a strategy which take into account legal requirements about OHS risks. These are intended to apply to all types of corporations and to establish various geographical, cultural and social conditions. Such a system enables a corporation to create an OHS strategy, develop objectives, scopes and processes to achieve the policy obligations, take action as needed to improve its performance and demonstrate the compliance of the system to the requirements of this OHSMS standard. Moreover, the general aim of OHSMS standards is to support capable OHS practices, in the framework of socio-economic needs [2].

The British Safety Council (BSC) and the International Labour Organisation (ILO) made a research in which valued the rewards of the prevention of accidents and/or diseases in enterprises within a period of 2 years. This study shows that the corporation which had adopted such a safety management system had the following results [11]: (i) productivity improvement, (ii) significant reduction of the frequency of cases of absence, (iii) significant reduction of compensation claims and insurance costs, (iv) improvement of the psychology of labor in addition with the increase of morale and concentration at work and (v) improving the company image to customers and suppliers.

In this work, the foremost IMS standards of promoting sustainability and OHS are presented, on the one hand and on the other side, the statistical results of a research (literature survey), reviewing vicarious scientific journals (for years 2006–2017). Thus, the main aim of our study is the strengthening of OHSM standards' application, in any organization (i.e., of any type and size).

The rest of the paper consists of five sections including (i) an overview of the OHSMS standards, (ii) a methodology (iii) the results of a statistical analysis, (v) the discussion and (vi) the conclusions.

2. Theoretical Background

A significant part of literature focuses on RAA of sustainability and health/safety accidents. The growing complexity of services, processes and products, entering the market, requires that the safety aspects must be considered with high priority. Undoubtedly, there is no absolute safety, so that some risk always remains in a specific worksite, constituting the "residual risk." Thus, any service, process and/or product can only be relatively safe. To continue, relative safety is achieved by risk degradation to a tolerable level, which is called as "tolerable risk," which is defined by the exploration of the finest balance between the ideal safety and the demands to be met by a service/process/product and factors such as profit for the user and cost effectiveness. Tolerable risk is succeeded by the procedure of risk assessment (risk analysis and risk evaluation) and risk reduction [12], while "risk management" can be considered as the entire methodology that includes both "qualitative" and "quantitative analysis techniques" [13–15].

In the scientific literature, four phases are prominent, as far as quantitative risk assessment is concerned (see for example the works [9,16–18]) depicted as follows: (a) Qualitative analysis, that incorporates the system definition and its scope, the hazards identification/description and the failure scenarios as well. (ii) Quantitative analysis, which incorporates the probabilities determination and the consequences of the defined undesirable events and also the risk quantification by a number (i.e., the risk quantity) or by a graph as a function of probabilities and consequences. (iii) Risk evaluation, which incorporates the evaluation process, on the base of the results of the former analysis. (iv) Risk control and reduction phase, which includes the step of taking measures (in order to be reduced the risk) and taking into account how the risks can be controlled (for example by inspection, maintenance or warning systems).

According to the IEC [15] the concept of risk presents two components that is, the frequency (or probability) that a harmful event (or an unsafe situation) is expected to occur and the consequences of this event. Moreover, CCPS [19] determines the risk as a measure of economic loss or human hurt in the frame of the likelihood and the magnitude of the loss (or damage).

To eliminate risks from sustainability and Health and Safety problems, a number of management systems has been proposed. In general, a management system is the methodology or the way by which an organization manages its internal procedures (or subjects) in order to achieve its objectives, which are associated with a number of different topics (including service quality or product quality, operational capability, environmental accomplishment, health and safety in the workplaces, et cetera). The level of the system's intricacy will depend on each organization's specific context. In small organisations, there is no (or less) need for extensive documentation because it is transparent how the employees contribute to the organization's overall aims. On the other side, more complicated corporations operating, may need extensive documentation in order to accomplish their organizational goals. Moreover, international management system (IMS) standards help organizations improve their performance by specifying repeatable steps that organizations consciously implement to accomplish their aims and to develop an organizational culture [20].

According to Gallagher [21], OHSM systems have been defined as "a combination of the planning and review, the management organizational arrangements, the consultative arrangements and the specific program elements that work together in an integrated way to improve health and safety performance."

Table 1 presents an overview of the most important worldwide OHSMS standards, based on selected information that has been collected from various sources and from the review of scientific literature as well.

Table 1. An overview of OHSMS standards.

Codes	Edition Year	Institutions	Description	Focus	Reference
BS 8800	1996 (as BS 8800:1996) and revised in 2004 (as BS 8800:2004) and in 2008 (as BS 18004:2008).	BSI	"It gives guidance on OHS management systems for assisting compliance with stated OHS policies and objectives and on how OHS should be integrated within the organization's overall management system"	Social dimension	[22–24]
HSG 65	1991 and revised in the years 1997 and 2013.	HSE	"A useful guide for directors, managers, health/safety professionals and employee representatives who wanted to improve health and safety in their organizations"	Social dimension	[25,26]
OHSAS 18001	The first edition (OHSAS 18001:1999) has been technically revised and replaced by the OHSAS 18001:2007 edition (second one).	44 cooperating organizations (constituting OHSAS Project Group)	"It is based on (i) "Plan": establish the aims and processes which are essential for the achievement in accordance with the organization's OHS policy, (ii) "Do": implement the processes, (iii) "Check": monitor and measure processes against OHS policy, objectives, legal and other requirements and report the results, (iii) "Act": take actions to continually improve OHS performance"	Social dimension	[2]
ILO-OSH 2001	2001 and revised in 2009.	ILO	"It provides a unique and powerful instrument for the development of a sustainable safety culture within organizations. The practical recommendations of these guidelines are intended for use by all those who have responsibility for occupational safety and health management"	Social dimension	[27]

Table 1. Cont.

Codes	Edition Year	Institutions	Description	Focus	Reference
AS/NZS 4801:2001	2001	AS/NZS	"The scope of this standard is to set auditable criteria for an OHSMS. It is a specification that aims to cover the best elements of such systems already widely used in New Zealand and Australia. It incorporates guidance on how those criteria may be accomplished"	Social dimension	[28]
ANSI/ AIHA Z10-2005	2005 and revised in 2012	ANSI	"Significant features that define Z10 include focus on management leadership roles, efficient employee participation, design review and change. It provides a tool to help organizations create and develop OHS performance"	Social dimension	[29,30]
SS 506	2004 (as SS 506:2004) and revised in 2009 (as SS 506:2009).	SSC	"It is consisted of three parts: (i) Requirements, (ii) Guidelines for the implementation of SS 506, (iii) Requirements for the chemical industry. It designates requirements for an OSH management system to activate a company to develop and implement a strategy and scopes which take into account legal requirements and information about OSH risks"	Social dimension	[31,32]
Une 81900:1996 EX	1996	AENOR	"- UNE 81900:1996 EX: Prevention of Occupational Hazards. Rules for the implementation of a SGPRL. - UNE 81901:1996 EX: Prevention of Occupational Hazards. General Rules for the Evaluation of SGPRLs. - UNE 81902:1996 EX: Prevention of Occupational Hazards. Vocabulary. - BUNE 81903:1997 EX: Prevention of Occupational Hazards. General Rules for the Evaluation of a SGPRL. Criteria for the qualification of the Auditors of Prevention. - UNE 81904:1997 EX: Prevention of Occupational Hazards. General Rules for the Evaluation of SGPRLs. Management of audit programs. - UNE 81905:1997 EX: Prevention of Occupational Hazards. Guide for the implementation of a SGPRL"	Social dimension	[33]
Uni 10616	1997 (and withdrawn in 2012	UNI	"Some of the major qualifying points are: (i) Espousal of inherent safety principles. (ii) Espousal of matrices or risk charts for assessing the acceptability/tolerability of risks. (iii) Definition of inspection activities and periodic checks of critical lines and equipment. (iv) The assessment of the external domino effect between neighbouring plants, (v) The adoption of a work-permission system, (vi) Selection of suppliers of goods and services such as companies, companies, builders, consortia, (vii) Adopting procedures for periodic internal auditing with internal or external auditors"	Social dimension	[34]

Table 1. Cont.

Codes	Edition Year	Institutions	Description	Focus	Reference
ISO 14000	ISO 14001:2004	ISO	"The ISO 14000 family of standards emphasize on manage their environmental responsibilities. In particular, ISO 14001:2015 and its accompanying standards such as ISO 14006:2011 concentrate on environmental systems to achieve this"	Environmental dimension	[35]
ISO 45001	2018	ISO	"ISO 45001 is intended for use by any organization, regardless of its size or the nature of its work and can be integrated into other health and safety programmes such as worker wellness and wellbeing. It can assist an organization to conform its legal requirements"	Social dimension	[36,37]

Over the past three decades, the use of OHSMS has become common in worksites in the developed economies, noting the fundamental elements of a OHSMS [38]. It is worth noting that many international management system standards have contiguous structure, containing a large number of the same terms and definitions. These characteristics are helpful for those organizations that operate an "integrated" management system [20] which can merge the requirements of two or more management system standards simultaneously (for example OHS with Environmental, or OHS with Quality management systems).

In Table 2 we depict the evolution of the OHSMS standards throughout the years 1990–2018. More specifically, the symbols "−" denote (in association with the year) the nonexistence of a standard, while the symbols "+" the appearance of the standard. In addition, the symbols "++," "+++," "++++" and "+++++," denote the 1st, the 2nd, the 3rd and the 4th update of it, correspondingly.

Table 2. The evolution of the OHSMS standards throughout the years 1990–2018.

Year	ISO 14001	ILOOSH 2001	BS 8800	OHSAS 18001	HSG65	ANSI/AIHA Z10	AS/NZS 4801	SS 506	Une 81900	Uni 10616	ISO 45001
1990	−	−	−	−	−	−	−	−	−	−	−
1991	−	−	−	−	+	−	−	−	−	−	−
1992	+	−	−	−	+	−	−	−	−	−	−
1993	+	−	−	−	+	−	−	−	−	−	−
1994	+	−	−	−	+	−	−	−	−	−	−
1995	++	−	−	−	+	−	−	−	−	−	−
1996	+++	−	+	−	+	−	−	−	+	−	−
1997	+++	−	+	−	++	−	−	−	+	+	−
1998	+++	−	+	−	++	−	−	−	+	+	−
1999	+++	−	+	+	++	−	+	−	+	+	−
2000	+++	−	+	+	++	−	++	−	+	+	−
2001	+++	+	+	+	++	−	+++	−	+	+	−
2002	+++	+	+	+	++	−	+++	−	−	+	−
2003	+++	+	+	+	++	−	+++	−	−	+	−
2004	++++	+	++	+	++	−	+++	+	−	+	−
2005	++++	+	++	+	++	+	+++	+	−	+	−
2006	++++	+	++	+	++	+	+++	+	−	+	−
2007	++++	+	++	++	++	+	+++	+	−	+	−
2008	++++	+	+++	++	++	+	+++	+	−	+	−
2009	++++	++	+++	++	++	+	+++	++	−	+	−
2010	++++	++	+++	++	++	+	+++	++	−	+	−
2011	++++	++	+++	++	++	+	+++	++	−	+	−
2012	++++	++	+++	++	++	++	+++	++	−	−	−
2013	++++	++	+++	++	+++	++	+++	++	−	−	−
2014	++++	++	+++	++	+++	++	+++	++	−	−	−
2015	+++++	++	+++	++	+++	++	+++	++	−	−	−
2016	+++++	++	+++	++	+++	++	+++	++	−	−	−
2017	+++++	++	+++	++	+++	++	+++	++	−	−	−
2018	+++++	++	+++	++	+++	++	+++	++	−	−	+

Annotations: the symbols "−," "+," "++," "+++," "++++" and "+++++," denote the nonexistence, the appearance, the 1st, the 2nd, the 3rd and the 4th update of a standard, respectively.

3. Methodology

Today, OHS issues are considered very important for organizations for economic (e.g., decrease lost working days), environmental (e.g., environmental hazards for employees) and social issues (e.g., ethical working conditions). It is well known that the major body of relevant literature focuses on the regulatory requirements of organizations regarding OHS issues, while a smaller part of the literature emphasizes on voluntary initiatives of organization on OHS issues. However, the voluntary trend of organizations has lately gained ground under the context of social responsibility of organizations to contribute to sustainable development [39]. This is integrated into the context of organizations as a commitment to OHS issues beyond the law which should be achieved through voluntary implementation of OHS standards (e.g., OSHAS 18001, ISO 45001). To this end, the suggested research methodology recommends sustainability concept as a frame to examine a set of current OHS standards (Table 1) through: (a) environmental dimension of sustainability (ISO 14001) and (b) social dimension of sustainability (BS 8800; HSG 65; OHSAS 18001; ILO-OSH 2001; ASINZS 4801; SS506; Une 81900: 1996 EX; Uni10616 and ISO 45001). It is necessary to clarify that the sustainability concept is only utilized as a frame of analysis and classification of OHS standards and none effort has been made to explain how the OHS examined standards contribute to aspects of sustainable development.

Additionally, the suggested research methodology is based on three sequential steps. The first step pertains the selection of journals, the second includes the appropriate keywords for addressing research questions and the third shows the coding method. In particular, the review of the scientific literature was accomplished by the research of ten representative scientific journals which focus on sustainability and health and safety issues (Table 3). The selection of these papers was based on the following two criteria: (a) the focusing on health and safety issues and sustainability and (b) the existence of high impact among scholars (Q1 and high impact factor). More specifically, taking into account that few other systematic (e.g., [3]) or narrative (e.g., [40–43]) literature reviews exist, on the topic of OHSMS standards, the time period before the year 2006, we investigated and studied published articles of the previous referred journals, collecting a large number of around N = 9822 papers, throughout the years 2006–2017.

Table 3. The ten investigated journals/sources (throughout the years 2006–2017).

Nr	Source	Publisher
1	"Applied Ergonomics"	
2	"Accident Analysis and Prevention"	
3	"Journal of Cleaner Production"	
4	"Journal of Operations Management"	"Elsevier B.V."
5	"Safety Science"	
6	"Journal of Loss Prevention in the Process Industries"	
7	"International Journal of Industrial Ergonomics"	
8	"Journal of Safety Research"	
9	"Architectural Science Review"	"Taylor & Francis"
10	"Professional Safety"	"American Society of Safety Engineers"

In particular, the methodological steps of the survey included: (i) investigation of the literature (e.g., through SCOPUS); (ii) screening the journals with the highest number of articles and the most important studies on S_OHSMS standards; (iii) selection of relevant studies; (iv) appraisal of the quality of the research evidence in the studies. The appropriate keywords we used in the survey were "Occupational Health and Safety", "Sustainability", "Management Standards".

4. Results

The procedure of reviewing the scientific literature, unveiled only a few published papers on OHSMS standards referred to many different fields (like construction, industry, engineering, transportation, chemistry, oil and refinery, food sector, et cetera). These papers address concepts, tools and methodologies that have been created and practiced in such areas as design, development, quality-control and maintenance, in association with occupational risk assessment.

The different OHSMS standards follow, in general, similar paths between "start" and "finish." Taking into account the results of our literature review, we present in the following Table 4, the comparison between the above referred OHSMS standards. This table depicts an overview of their features, comparatively with several settled evaluation-criteria.

Table 4. Comparison of OHSMS Standards.

Characteristics	OHSMS Standards							
	ISO 14001	ILOOSH 2001	BS 8800	OHSAS 18001	HSG65	ANSI/ AIHA Z10	AS/NZS 4801	SS 506
General requirements	YES	NO	NO	YES	NO	NO	NO	NO
Initial or periodic status review	NO	NO	YES	NO	YES	NO	NO	NO
Management leadership and Labour participation	NO	NO	NO	NO	NO	YES	NO	NO
Policy	YES	YES	YES	YES	YES	NO	YES	YES
Organising	NO	YES	YES	NO	YES	NO	NO	NO
Planning	YES	YES	YES	YES	YES	YES	YES	YES
Implementation/ Operation	YES	YES	YES	YES	YES	YES	YES	YES
Inspection and Evaluation	YES	YES	NO	YES	NO	YES	YES	YES
Performance measurement	NO	NO	YES	NO	YES	NO	YES	NO
Improvement actions	NO	YES	NO	NO	NO	NO	NO	NO
Corrective actions	YES	NO	NO	YES	NO	YES	NO	YES
Management Review	YES	NO	NO	YES	YES	YES	YES	YES
Audit	NO	NO	YES	NO	YES	NO	NO	NO
Continuous Improvement	YES	NO	NO	YES	NO	YES	YES	YES
Performance inspection	YES	YES	YES	YES	YES	YES	YES	YES
Origin	International	International	UK	International	UK	USA	Australia/ New Zealand	Singapore
Year of establishment	1992	2001	1996	1999	1991	2005	1999	2004
Update	YES	YES	YES	YES	YES	YES	YES	YES
Good embedded OHS practices	NO	YES	YES	YES	NO	YES	NO	YES
Weak issues	YES	YES	YES	YES	YES	YES	YES	YES
Glossary Terms and definitions	YES	YES	YES	YES	NO	NO	NO	NO
Workers participation	NO	YES	NO	NO	NO	NO	NO	NO
Documentation	YES	YES	YES	YES	YES	YES	YES	YES
Free of Charge Manual	NO	YES	NO	NO	NO	NO	NO	NO
Risk assessment technique	NO	NO	YES	YES	NO	YES	NO	NO
Application on Organisations of any type and size	NO	YES	YES	YES	NO	YES	NO	YES

Table 4. Cont.

Characteristics	OHSMS Standards							
	ISO 14001	ILOOSH 2001	BS 8800	OHSAS 18001	HSG65	ANSI/ AIHA Z10	AS/NZS 4801	SS 506
Accompanied with extra guidelines series	NO	NO	YES	YES	NO	NO	YES	YES
Incorporated examples	NO	NO	YES	NO	NO	YES	NO	NO
Annexes	YES	YES	YES	YES	YES	YES	YES	YES
Embedded comparisons with other OHSMS standards	NO	NO	YES	YES	NO	YES	NO	NO
Possible combination with other OHSMS standards	NO	NO	YES	YES	NO	YES	NO	NO
Compatibility with other OHSMS standards	YES	YES	YES	YES	NO	YES	NO	NO
Compatibility with Quality and/or Environmental IMS standards	YES	NO	YES	YES	YES	YES	YES	YES
Future improvement	YES	YES	YES	YES	YES	YES	YES	YES
Taking into account recent legislative changes	NO	NO	YES	NO	NO	NO	NO	NO

The investigation of the scientific literature revealed S = 75 published technical articles which were associated with OHSMS Standards in the worksites and concerned many different fields (like construction, chemistry, engineering, transportation, medicine, etc.).

In Appendix A (Table A1) we show the classification results of the 75 articles incorporating OHSMS standards which were defined by the investigation of N = 9822 papers of ten sources covering the time period 2006–2017. Table A1 uses eight columns for example, the number (or numerical code) of the paper (A), the paper's citation information (columns B, C, D), the OHSMS standard's name (E), the kind of the paper's data or material (F), the field of application (G) and the source (journal's name) [column H].

In Appendix B (Table A2) we illustrate the statistical results of the survey including the following:

(i) the absolute frequency N_i that is, the number of investigated papers per journal (col. C, that is, **JAE:** 886; **AAP:** 1522; **JCP:** 1005; **JOM:** 995; **ASR:** 1007; **JSS:** 945; **JPS:** 998; **JLPPI:** 881; **IJIE:** 955; **JSR:** 628),

(ii) the relative frequency $F_i = N_i/N$ of the ten journals, concerning the total amount of the published papers during 2006–2017 (column D, that is, **JAE:** 9.02%; **AAP:** 15.50%; **JCP:** 10.23%; **JOM:** 10.13%; **ASR:** 10.25%; **JSS:** 9.62%; **JPS:** 10.16%; **JLPPI:** 8.97%; **IJIE:** 9.72%; **JSR:** 6.39%),

(iii) the number of papers $n_{ST(i)}$ concerning OHS which include or use or refer to OHSMS standards (column E, that is, **JAE:** 3; **AAP:** 8; **JCP:** 3; **JOM:** 1; **ASR:** 1; **JSS:** 40; **JPS:** 4; **JLPPI:** 8; **IJIE:** 1; **JSR:** 6),

(iv) the relative occurrence frequency of papers (referred to N) which include (or use or refer to) OHSMS-standards $f_{ST(i)} = n_{STS(i)}/N$ (column F, that is, **JAE:** 0.03%; **AAP:** 0.08%; **JCP:** 0.03%; **JOM:** 0.01%; **ASR:** 0.01%; **JSS:** 0.41%; **JPS:** 0.04%; **JLPPI:** 0.08%; **IJIE:** 0.01%; **JSR:** 0.06%),

(v) the normalized (per journal) occurrence frequency of papers which include OHSMS standards $f_i^* = n_{ST(i)}/N_i$ (column F, that is, **JAE:** 0.34%; **AAP:** 0.53%; **JCP:** 0.30%; **JOM:** 0.10%; **ASR:** 0.10%; **JSS:** 4.23%; **JPS:** 0.40%; **JLPPI:** 0.91%; **IJIE:** 0.10%; **JSR:** 0.96%),

(vi) the relative occurrence frequency of papers (referred to S) which include OHSMS standards $f^{**}_{ST(i)} = n_{ST(i)}/S$ (column F, that is, **JAE:** 4.00%; **AAP:** 10.67%; **JCP:** 4.00%; **JOM:** 1.33%; **ASR:** 1.33%; **JSS:** 53.33%; **JPS:** 5.33%; **JLPPI:** 10.67%; **IJIE:** 1.33%; **JSR:** 8.00%).

Moreover, Figure 1 illustrates for the time period of 2006–2017 the following: (a) the relative frequency $F_i = N_i/N$ of the ten journals, concerning the total amount of their published articles, (b) the relative occurrence frequency of papers (referred to N) with OHSMS-standards $f_{ST(i)} = n_{ST(i)}/N$, (c) the normalized (per journal) occurrence frequency of papers concerning OHS which include OHSMS standards $f_i^* = n_{ST(i)}/N_i$ (d) the relative occurrence frequency of papers (referred to S) which include OHSMS standards $f^{**}_{ST(i)} = n_{ST(i)}/S$, (e) the relative occurrence-frequency of the various OHSMS standards (which are included in the above referred S = 75 papers) and (f) the percentage distribution of the papers with OHSMS standards in association with different types of data.

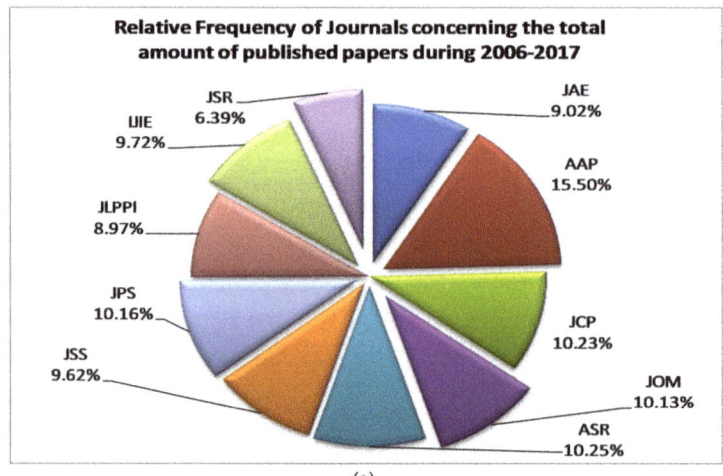

(a)

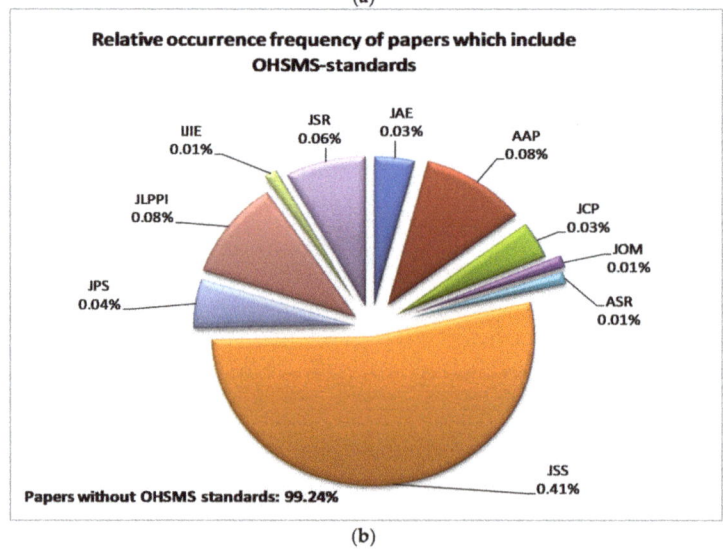

(b)

Figure 1. *Cont.*

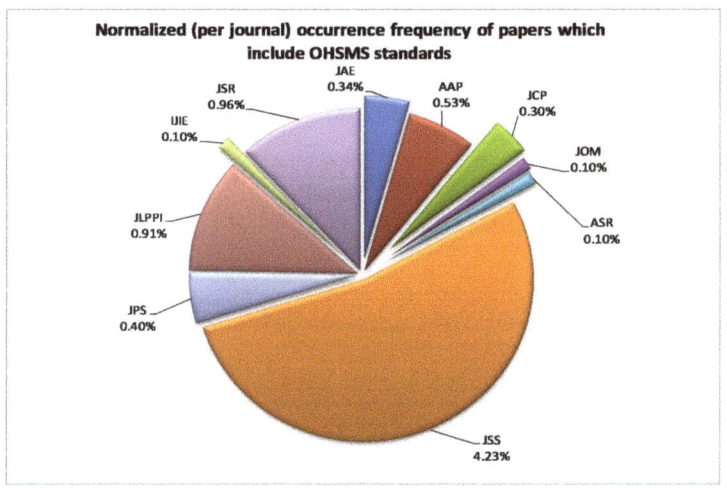

(c)

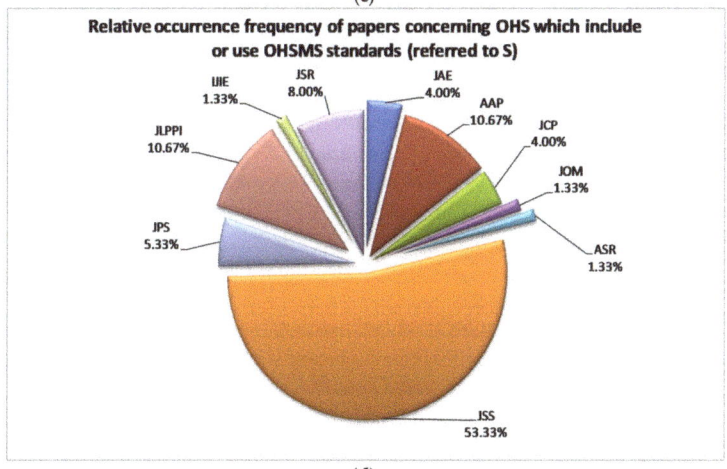

(d)

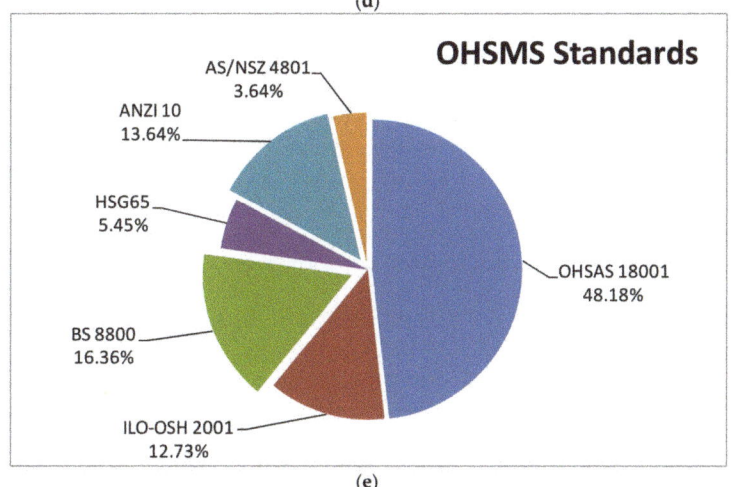

(e)

Figure 1. *Cont.*

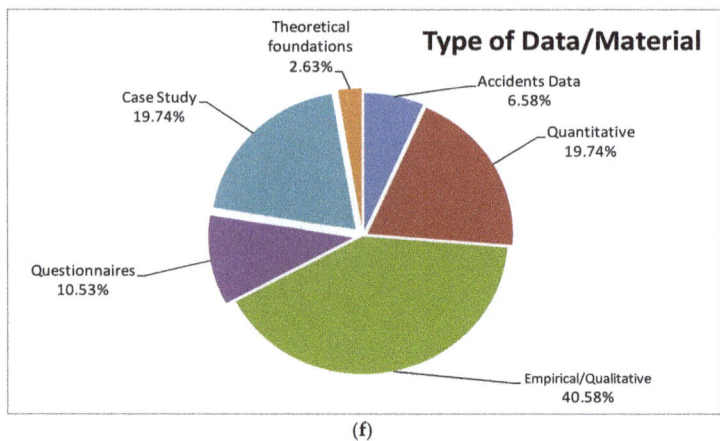

(f)

Figure 1. It depicts for the period of 2006–2017 the various relative occurrence frequencies of papers with OHSMS-standards.

The survey relating to the above 10 journals revealed (according to Appendix B, Table A2) that the papers concerning OHS and including OHSMS standards were very few (i.e., for JAE the maximum percentage is only 0.34% taking into account the total number N of the investigated papers and only ~4% as far as the normalized per journal percentage is concerned), while the majority (i.e., 99.24%) is represented by papers without OHSMS standards. Taking into account the graphs of Figure 1, we note that although AAP is the journal with the most publications for 2006–2017 (Figure 1a), JSS is the journal with the greatest number of publications, concerning OHSMS standards (Figure 1b–d).

Besides, the pie-chart of Figure 1e reveals that the OHSAS 18001 standard, presents the higher relative occurrence-frequency (47.06%) in comparison with the other OHSMS standards (of Table A2 in Appendix B), that is, OHSAS 18001 is the most frequent OHSMS standard according to scientific literature review. This can be related with the fact that this international standard was the result of the co-operation of 13 different international organizations which represent 80% of the certification bodies.

In addition, the graph 1f which depicts the percentage distribution of the articles (with OHSMS standards) shows that "Empirical/Qualititative" is the most frequent type compared with various types of data.

In Figure 2 we illustrate the yearly variation of the number (nST) of papers with OHSMS standards, published by the previous referred 10 journals, during 2006–2017 (panel a) and the corresponding percentage distribution of papers in association with the year of publication (panel b). The inserted bar-graph in panel a, depicts the relative occurrence frequency of papers concerning OHS which include OHSMS standards in association with the title of the 10 journals (horizontal axis).

The curve of the graph of Figure 2a shows the existence of a long-term trend factor with positive inclination (throughout the period 2007–2012), with negative inclination (for the period 2013–2015) and also with a positive inclination (throughout the 2016–2017). In particular, there is a gradual increasing for the period 2006–2012 (with a maximum in years 2011 and 2012), while for the years 2013–2015, an abrupt decreasing with an intensive negative slope. The second graph (pie-chart) shows that 2011 and 2012 are the years with the greatest percentage of papers referring to OHSMS systems.

To continue, the pie-chart of Figure 3 displays the distribution of papers with OHSMS standards (published by the 10 journals during 2006–2017) in association with various fields of application (Industrial Sector: 28%; Construction Sector: 16%; Chemical Sector: 6.67%; Oil and Refinery: 4%; Mining: 4%; Shipbuilding Sector: 4%; Food Sector: 4%; Railways Sector: 2.67%; Transportations Sector: 1.33%; Telecommuncations: 1.33%; Other: 28%). The major evident feature of this pie-chart is that the field of "Industry" concentrates the maximum number of the papers with OHSMS standards.

Apparently, one reason is that the industrial organizations present more dangerous working conditions in comparison with other corporations (for example due to the existence of heavy machines in the production procedure) [9,44]. Moreover, it is apparent that the construction sector is following, due to the greatest number of occurring incidents.

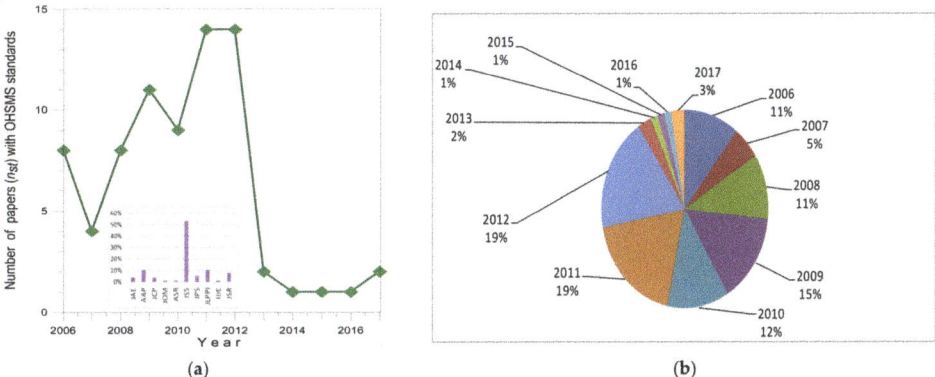

Figure 2. (a) The curve illustrates the variation of the number of papers with OHSMS standards published by the 10 journals during 2006–2017. (b) Distribution of papers with OHSMS standards in association with the year of publication.

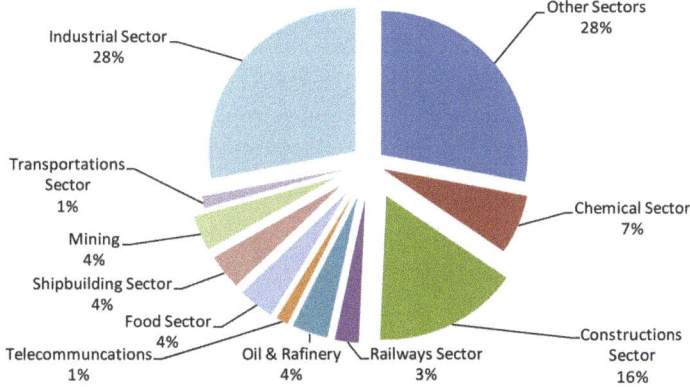

Figure 3. The pie-chart displays the distribution of papers with OHSMS standards (published by the 10 journals during 2006–2017) in association with the various fields of application.

5. Discussion

Taking into account that nearly 2.3 million of people die per year due to work-related accidents and/or diseases, occupational injuries and/or illnesses are significantly adverse for the employers and also for the state's economy, resulting in losses from early retirements, the staff absences and the high insurance costs [45,46]. Thus, occupational health/safety aspects represent some of the most crucial subjects of the social policy in the EU and as a consequence the Lisbon strategy was adopted for growth and employment, which aims to create jobs with care in health and safety for the employees.

Over the past three decades the use of Occupational Health and Safety Management Systems (OHSMS) has expanded in the workplaces of developed economies and their growing usage could be attributed to many reasons and factors. In particular, OHSMSs embody the principles of "continuous improvement" or "quality management" which have been used and applied by enterprises for improved business competitiveness. Taking into account that the principles of OHSMS are similar to methods such as Total Quality Management, it is presumable that experience with TQM has formed a basis for new applications of removing occupational hazards and improving safety awareness [47].

OHSMS differ as far as the miscellaneous techniques of implementation are concerned. According to Frick and Wren [40] there are three types of implementation, that is, voluntary, mandatory and hybrid. The first one exists when companies affiliate OHSMS on their own volition. On the other hand, mandatory systems have been enforced in many European countries (for example in Denmark and Norway) where the state's legislation requires establishment of a RAA system. Quasimandatory methodologies may also exist in case of external commercial requirements which take the form of legislative demands. Thus, many organizations settle OHSMS to comply with the demands of customers and suppliers, principal contractors and other commercial bodies. Hybrid methods require an admixture of voluntary motives and legislative requirements [47]. It is worth noting that, only a fraction of all employers and/or organisations have introduced what it is called as OHS management systems (such as OHSAS 18001) [2]. Yet, within the EU, California (USA) and in many other countries, employers are required by regulation mandated to organise a systematic OHS management (in the EU according to the Framework Directive of 89/391/EEC).

Compliance with these regulations will result in an effective prevention of OHS risks, even without introducing OHSAS or any other voluntary OHS management system. Some other distinctions that should be underlined are the following:

(i) The term "systematic" in mandatory systematic OHSM is not at all the same as the term "system" in voluntary OHSM systems. This means that the UK's HSG 65 is quite different from the other described "systems." HSG 65 constitutes guidelines (best practice) for how to comply with the mandatory regulation of systematic OHSM (transposing of EU directive 89/391/EEC into UK legislation) while all the other listed "systems" are voluntary and more or less commercial products (which still can be very useful).

(ii) Regulated mandatory systematic OHSM is not at all identical to voluntary OHSM standards. This article aims to cover both of them. In any case both mandatory OHSM (such as in all of the EU, since 1989, through 89/391/EEC but also in many other states, for example, in California since 1991) and the much more complex commercial products of voluntary OHSM systems (such as OHSAS 18001) are critically discussed and distinguished.

(iii) The term "Standard" is also a term with importantly different meanings. In Anglo-saxon countries this is often the same as a regulation but a voluntary standard (such as ISO) is (by definition) not a regulation.

The implementation and application of a range of efficient OHS management actions systematically, can contribute to optimal results for all interested parties. Organizations of all types establish a systematic process to manage OHS and develop OHSMS systems within the context of (i) the general growth of concern from all interested parties about OHS, (ii) changes to legislation and (iii) other measures to foster sustained OHS improvement. There are many reasons for organizations to develop OHSMSs (for example legal imperatives, ethical concerns, industrial considerations, financial performance improvement, etc.). Implementation of an efficient OHSMS will create a reduction of workplace illnesses and injuries, minimizing the expenses associated with accidents [28]. Consequently, the application of OHSMS standards provides a unique instrument for the development of a sustainable safety culture within enterprises. Besides, human factors (including the culture, politics, etc.) within organizations can make or break the effectiveness of any management system and needs to be considered very carefully when implementing a management system [22]. An OHSMS standard specifies requirements in order to help any enterprise to formulate a strategy taking into account legislation and information about risks and/or harm. It applies to those hazards over which the organization may exert control and over which it can be expected to have an influence [28].

Other standards (like ISO 31000, IEC31010) constitute an introduction to selected techniques and compare their possible applications, benefits and limitations. They also provide references to sources of more detailed information.

With the globalization of the economy and the success of Quality Management Systems (QMS) and of Environmental Management Systems (EMS), organizations require a simple, integrated and also a global management system. For this reason, and due to the lack of a model for an OHS management system that has been accepted worldwide, systems or models, guides or norms of the OHS management, have been propagated throughout the world.

Previous investigations have recognized a research gap, as far as the studies of the reliability and validity of OHSMS standards, are concerned [48]. In this study, the most considerable OHSMS standards created at the international level, are presented and analysed at their updated version. In other words, our article introduces different standardized occupational and safety management systems (guidelines), which are commonly used internationally. In particular, the purpose of this work is to collect important information regarding the use of OHSMS standards by achieving a scientific literature investigation of vicarious journals (published by Elsevier B.V., Taylor & Francis and the American Society of Safety Engineers) covering the time period of 2006–2017. This is an interesting topic which combines academic papers and practically used standards. Subsequently, the main aims of this exploratory study are: (i) the depiction of commercial and industrial trends, as far as the OHSMS standards are concerned, (ii) the implementation of these standards and (iii) the reinforcement of their application in the worksites of organizations.

It is worth mentioning that, few other systematic literature reviews (like the one of Robson et al. [3]) or narrative reviews (e.g., [40–43]) exist, concentrating on the topic of OHSMS standards and concerning the years before 2006. So this is the reason for conducting our review throughout the years 2006–2017. In addition, the literature overview of Li and Guldenmund [49] has a different aim and describes safety management systems (SMSs) as far as the SMSs models are concerned, that is, accident-related models and organizational models.

The methodological steps of the survey included: (i) research of the literature; (ii) screening the journals with the most important studies on S_OHSMS standards; (iii) selection of relevant studies; (iv) appraisal of the quality of the research evidence in the studies.

The review unveils the following:

- There are only a few publications including OHSMS standards (during 2006–2017), which are referred to a variety of occupational fields (like construction, industry, engineering, transportation, chemistry, oil and refinery, et cetera).
- These papers address techniques that have been emerged in areas such as design/development, quality-control and maintenance, as far as occupational risk assessment is concerned.
- "Safety Science" (SS) and "Journal of Loss Prevention in the Process Industries" (JLPPI) are the scientific journals which published (during 2006–2017) the most scientific papers concerning OHSMS.
- The sector of "Industry" (with 28%) and also the "Construction" (with 16%), accounted for the highest percentage, as far as the usage of OHSMS standards is concerned, presumably because their work conditions are more unstable and dangerous in comparison with other occupational sectors (e.g., telecomunications). In addition, these sectors are very hazardous worldwide, owing to their unique dynamic nature, poor conditions and tough environment [44,50].
- The comparison between the previous presented OHSMS standards (in Table 4), depicts an overview of their features, comparatively with various developed evaluation-criteria.
- The OHSAS 18001 standard, presents the higher relative occurrence-frequency (48%) in comparison with the other OHSMS standards, due to the fact that this international standard was produced by the co-operation of 13 different international organizations which represent 80% of the certification bodies.

- The percentage distribution of the papers (with OHSMS standards) reveals that "Empirical/Qualititative" is the most frequent type compared with other types of data.
- Our selected graphs, show (for the distribution of the publications with OHSMS standards) the existence of a long-term trend factor with positive inclination (during 2007–2012), with negative inclination (during 2013–2015) and also with a positive inclination (throughout the 2016–2017).
- In general, there is an increasing scientific interest for OHSMS standards, especially in period 2009–2012.
- The entire objective of the usage of OHSMS standards is to support and promote efficient OHS practices, in balance with socio-economic requirements [2].
- The OHSMS standards are not legally binding and have not got the intention to replace national laws or regulations and/or accepted standards.
- OHSMS have been proposed as a way to reduce injuries and illnesses for businesses of all types and sizes [51,52].
- The various OHSMS standards have common basic requirements and features but they are implemented in different processes. Besides, the different OHSMS standards follow, in general, similar paths between "start" and "finish."
- Many organizations apply an "integrated" management system which combines the requirements of two or more management system standards simultaneously (for example OHSMS with Environmental, or OHSMS with Quality management systems standards).

6. Conclusions

Governments, employers and workers recognize, day after day, the posistive impact of introducing S_OSHMS standards at the organization level, both on the reduction of hazards/risks and also on productivity. In particular the benefits of effective S_OSHMS standards to any commercial body or enterprise include: (i) more effective usage of resources, (ii) improved financial performance, (iii) improved risk management and (iv) increased capability to deliver consistent and improved services and products. OHSMS standards are not intended to replace national laws, regulations or accepted standards but on the other side, their main aim is to support and promote efficient OHS practices, in balance with socio-economic requirements.

Using the previous analysis and its main outcomes, the next dominant conclusions can be extracted:

- The application of OHSMS standards is not significantly extended in organizations and the knowledge about them has not been fully shared and expanded among the miscellaneous scientific fields, so we have the opinion, that the scientific community, have to transfer the similarities from one field to another.
- The implementation of a "integrated" management systems which combine the requirements of two or more management system standards simultaneously (for example OHSMS with Environmental, or OHSMS with Quality management systems standards) would enable organizations to achieve efficient results on the reduction of risks and also on productivity.
- OHSMS standards can be developed and implemented by organizations of any type and any size (large or small).
- As this study is not an empirical research but a review, the value of the article rests on clear distinctions, definitions and analysis. Moreover this exploratory study, provides a rich description for the usage of OHSMS standards in workplaces and lays the background for further research into the reliability and reinforcement of their application in any organization.

As a general conclusion, our literature survey shows that: (i) only a small number of published articles focusing on OHSMS standards (and concerning miscellaneous occupational fields) are available for the period of years 2006–2017 and (ii) the scientific community expanded its interest for the usage of OHSMS standards, during the years of 2009–2012.

Author Contributions: Conceptualization, P.M., D.K., I.N.; Data curation, P.M., S.T.; Formal analysis, P.M., D.K.; Funding acquisition, I.N.; Investigation, P.M., D.K.; Methodology, P.M., D.K., I.N.; Writing—original draft, P.M.; Writing—review & editing, P.M., D.K., I.N.

Conflicts of Interest: We declare no conflict of interest.

Abbreviations and Acronyms

ANSI	American National Standarization Institute
JAE	Applied Ergonomics
ASR	Architectural Science Review
AENOR	Asociación Española de Normalización/Certificación (Spanish Assoc. for Standards)
AS/NZS	Australian/New Zealand Standard
BSI	British Standards Institution
BSC	British Safety Council
CPQRA	Chemical Process Quantitative Risk Analysis
CCPS	left for Chemical Process Safety
UNI	Ente Nazionale Italiano di Unificazione (Italian National Unification)
EMS	Environmental Management Systems
HSE	Health and Safety Executive
IEC	International Electrotechnical Commission
IJIE	International Journal of Industrial Ergonomics
ILO	International Labour Organization
IMS	International Management Systems
ISO	International Organization for Standardization
JCP	Journal of Cleaner Production
JLPPI	Journal of Loss Prevention in the Process Industries
JOM	Journal of Operations Management
JSR	Journal of Safety Research
OHSAS	Occupational Health and Safety Assessment Series
OHSMS	Occupational health and safety management system
OHS	Occupational health-safety
JPS	Professional Safety
QMS	Quality Management Systems
RAA	Risk analysis and assessment
JSS	Safety Science
SSC	Singapore Standards Council
S_OHSMS	Sustainability and Occupational Health and Safety Management Systems
TQM	Total Quality Management

Appendix A

Table A1 presents the classification results of 75 papers [3,51,53–125] which are associated with OHSMS standards, covering the period of 2006–2017.

Table A1. The classification results of the 75 papers.

A/A	Paper Citation	Authors	Year of Publication	OHSMS Standard	Type of Paper Data or Material	Field of Application	Source
(A)	(B)	(C)	(D)	(E)	(F)	(G)	(H)
1	[53]	Jorgensen T.H. Arne Remmen M. Dolores Mellado	2006	ISO 14001; OHSAS 18001	Empirical/Qualitative	All Sectors	JCP
2	[54]	J.R. Santos-Reyes and & Beard, A. N.	2006	BS8800	Empirical/Qualitative	All Sectors	JPS
3	[55]	Adele Abrahms	2006	ANZI Z10-2005	Empirical/Qualitative	All Sectors	JPS
4	[56]	Manuele, Fred	2006	ANZI Z10-2005	Empirical/Qualitative	All Sectors	JPS
5	[57]	Wrona	2006	ANSI Z16.2	Empirical/Qualitative	All Sectors	JSR
6	[58]	Nielsen et al.	2006	OHSAS 18001	Empirical/Qualitative	Industrial sector	JSR
7	[59]	Cadieux et al.	2006	OHSAS 18001	Empirical/Qualitative	Industrial sector	JSR
8	[60]	Paivinen	2006	BS 8800	Empirical/Qualitative	Telecommunications	IJIE
9	[61]	S.X. Zeng, Jonathan J., Shi G.X. Lou	2007	OHSAS 18001	Questionnaires	All Sectors	JCP
10	[62]	Fernandez-Muniz et al.	2007	BS 8800, OHSAS 18001	Empirical/Qualitative	Industrial sector	JLPPI
11	[63]	Coleman and Kerkering	2007	ANSI Z16.1-1967	Accidents Data	Mining	JSR
12	[3]	Robson et al.	2007	ANSI Z10, OHSAS 18001, BS8800, HSG 65	Quantitative	All Sectors	JSS
13	[64]	Bevilacqua et al.	2008	OHSAS 18001	Quantitative	Oil & Refinery	AAP
14	[65]	Brooks	2008	ANSI Z10	Quantitative	All Sectors	JSS
15	[66]	Duijm et al.	2008	BS 8800, OHSAS 18000, ILO-OSH 2001, HSG 65	Questionnaires	Industrial sector	JSS
16	[67]	Lind, Kivisto-Rahnasto	2008	BS 8800, ILO-OSH2001, OHSAS 18001	Case Study	Transportations Sector	JSS
17	[68]	Nielsen et al.	2008	OHSAS 18001	Quantitative	Industrial sector	JSS
18	[69]	Schrover	2008	ISO-9001, ISO-14001, OHSAS-18001	Theoretical Foundations	Chemical Sector	JSS
19	[70]	Zeng et al.	2008	OHSAS 18001, ISO 9001	Accidents Data	Construction Sector	JSS
20	[71]	Komljenovic et al.	2008	AS/NZS; 1999, CAN/CSA Q850-97: 2002. ISO, 1999, 2000, ANSI: 2000, ASTME2205-02, 2002	Accidents Data	Mining	JSS
21	[72]	Caroly et al.	2009	OHSAS 18001	Empirical/Qualitative	Chemical Sector	JAE
22	[73]	Chang and Liang,	2009	OHSAS 18001, ILO-OSH-2001, BS 8800, HSG 65	Empirical/Qualitative	Chemical Sector	JLPPI
23	[74]	Santos-Reyes, Beard	2009	BS 8800, ILO-OSH: 2001, HSG 65	Empirical/Qualitative	Oil & Refinery	JLPPI
24	[75]	Crippa et al.	2009	OHSAS 18001	Quantitative	Oil & Refinery	JLPPI
25	[76]	Knegtering, Pasman,	2009	OHSAS 18001	Empirical/Qualitative	Industrial sector	JLPPI
26	[77]	Pasman et al.	2009	OHSAS 18001	Empirical/Qualitative	Industrial sector	JLPPI
27	[78]	Moriyama, Ohtani	2009	ILO-OSH 2001, BS 8800	Questionnaires	Food Sector	JSS
28	[79]	Fernandez-Muniz et al.	2009	BS 8800,OHSAS 18001	Quantitative	All Sectors	JSS
29	[80]	Celik M.	2009	OHSAS 18001:2007	Empirical/Qualitative	Shipbuilding Sector	JSS
30	[81]	Reniers et al.	2009	ILO-OSH 18000	Empirical/Qualitative	All Sectors	JSS
31	[82]	Baram	2009	ILO-OSH 2001	Empirical/Qualitative	Industrial sector	JSS
32	[83]	Shengli Niu,	2010	ILO-OSH 2001	Empirical/Qualitative	All Sectors	JAE
33	[84]	Gangolells et al.	2010	OHSAS 18001	Quantitative	Construction Sector	JSR
34	[85]	Cheng et al.	2010	ANSI Z10	Quantitative	Construction Sector	JSS
35	[86]	Jacinto, Silva	2010	BS 8800	Case Study	Shipbuilding Sector	JSS
36	[87]	Sgourou et al.	2010	BS 8800	Quantitative	All Sectors	JSS

Table A1. Cont.

A/A	Paper Citation	Authors	Year of Publication	OHSMS Standard	Type of Paper Data or Material	Field of Application	Source
(A)	(B)	(C)	(D)	(E)	(F)	(G)	(H)
37	[88]	Fera, Macchiaroli	2010	OsHAS 18001	Case Study	All Sectors	JSS
38	[89]	Lindhout et al.	2010	OHSAS 18001	Empirical/Qualitative	Chemical Sector	JSS
39	[90]	Celik M.	2010	ISO 9001:2000, ISO 14001:2004, OHSAS 18001:2007	Case Study	Shipbuilding sector	JSS
40	[91]	Dawson et al.	2011	AS/NZ 4360	Empirical/Qualitative	All Sectors	AAP
41	[92]	Low Sui Phen & Goh Kim Kwang	2011	ISO 14001; OHSAS 18001	Questionnaires	Construction Sector	ASR
42	[93]	Jallon et al.	2011	OHSAS 18001	Quantitative	Industrial sector	JSR
43	[94]	Leka et al.	2011	ILO-OSH 2001, BS OHSAS and ANSI Z-10	Empirical/Qualitative	Construction Sector	JSS
44	[95]	Blewett, O'Keeffe	2011	ANSI Z10, OHSAS 18001	Quantitative	Food Sector	JSS
45	[96]	Vinodkumar, Bhasi	2011	OHSAS 18001, BS 8800	Questionnaires	Chemical Sector	JSS
46	[97]	Hohnen, Hasle	2011	OHSAS 18001	Quantitative	All Sectors	JSS
47	[98]	Lu, Li	2011	OHSAS 18001	Case Study	Mining	JSS
48	[99]	Zwetsloot et al.	2011	OHSAS 18001	Case Study	Industrial sector	JSS
49	[100]	Frick	2011	ANSI Z-10, (2005), Australia and New Zealand (AS/NZS 4804), The Netherlands (NPR 5001), Spain (UNE 81900), UK (BS 8800), ISO 9000, ISO 14000, OHSAS:18001, ILO-OSH 2001	Theoretical Foundations	Industrial sector	JSS
50	[101]	Granerud, Rocha	2011	OHSAS 18001, ISO 14000, ISO 9000	Case Study	Industrial sector	JSS
51	[102]	Kristensen	2011	ISO 14001, OHSAS 18001	Case Study	Industrial sector	JSS
52	[103]	Zwetsloot et al.	2011	ISO 9000, ISO 14000, OHSAS 18000 series	Case Study	Industrial sector	JSS
53	[104]	Hasle Peter, Gerard Zwetsloot	2011	OHSAS 18001	Quantitative, Empirical/Qualitative	Industrial Sector	JSS
54	[105]	Beatriz Fernández-Muñiz, José Manuel Montes-Peón, Camilo José Vázquez-Ordás	2012	OHSAS 18001	Empirical/Qualitative	All Sectors	AAP
55	[106]	Goh, Y.M. et al.	2012	AS/NZS 4801	Empirical/Qualitative	Construction Sector	AAP
56	[107]	Cheng et al.	2012	ANSI Z.16.2-1995, ILO Guidelines	Quantitative	Construction Sector	AAP
57	[108]	Pinto et al.	2012	BS 8800:2004	Quantitative	Construction Sector	AAP
58	[109]	Badri et al.	2012	OHSAS 18000	Case Study	Industrial sector	AAP
59	[110]	Luria Gil, Ido Morag	2012	OHSAS 18001	Empirical/Qualitative	All Sectors	AAP
60	[111]	Beatriz Fernández Muñiz José Manuel Montes-Peón Camilo José Vázquez-Ordás	2012	OHSAS 18001	Questionnaires	All Sectors	JCP
61	[112]	Jeremy Mawhood & Claire Dickinson	2012	HSG 65	Empirical/Qualitative	Railways Sector	JPS
62	[113]	Lee et al.	2012	BS 8800, OHSAS 18001, ILO-OSH 2001	Accidents data	All Sectors	JLPPI
63	[114]	Gnoni, Lettera	2012	OHSAS 18001	Empirical/Qualitative	Industrial sector	JLPPI
64	[115]	Badri et al.	2012	ANSI Z10	Empirical/Qualitative	Industrial sector	JSS
65	[116]	Hamidi et al.	2012	OHSAS-18001, ANSI z10	Case Study	Construction Sector	JSS
66	[117]	Hsu et al.	2012	OHSAS: 18001, BS 8800,ILOOSH 2001	Quantitative	Food Sector	JSS
67	[118]	Wang, Liu	2012	OHSAS 18001, ILO-OSH 2001, HSG 65	Questionnaires	Railways Sector	JSS
68	[119]	Ismail et al.	2012	OHSAS 18001	Empirical/Qualitative	Construction Sector	JSS
69	[120]	Jesús Abad, Esteban Lafuente, Jordi Vilajosana	2013	OHSAS 18001	Accidents data	All Sectors	JSS
70	[121]	Santos G. et al.	2013	OHSAS 18001	Questionnaires	Industrial Sector	JSS

Table A1. *Cont.*

A/A	Paper Citation	Authors	Year of Publication	OHSMS Standard	Type of Paper Data or Material	Field of Application	Source
(A)	(B)	(C)	(D)	(E)	(F)	(G)	(H)
71	[122]	Chris K.Y. Lo, Mark Pagell, Di Fan, Frank Wiengarten, Andy C.L. Yeung	2014	OHSAS 18001	Case Study	Construction Sector	JOM
72	[123]	Yazdani A. et al.	2015	OHSAS18001, BS 8800:2004	Empirical/Qualitative	Industrial Sector	JAE
73	[51]	Autenrieth D.A. et al.	2016	OHSAS 18001, ANZI 10	Case Study	Industrial Sector	JSS
74	[124]	Miskeen Ali Gopang, et al.	2017	ILO-OSH 2001	Empirical/Qualitative	Other Sectors	JSS
75	[125]	Segarra Cañamares M. et al.	2017	OHSAS 18001	Case Study	Construction Sector	JSS

Appendix B

Table A2 presents statistical results of ten (10) journals investigation, concerning articles with OHSMS standards (during 2006–2017).

Table A2. Statistical results of ten (10) journals investigation, concerning articles with OHSMS.

Nr	Journal	Acronym	Number of Investigated Papers (Absolute Frequency N_i)	Relative Frequency ($F_i = N_i/N$) (%)	Number of Papers Concerning OHS Which Include or Use OHSMS Standards ($n_{ST(i)}$)	OHSMS Standard	Relative Occurrence Frequency of Papers (Referred to N) Which Include OHSMS-Standards ($f_{ST(i)} = n_{ST(i)}/N$) (%)	Normalized (Per Journal) Occurrence Frequency of Papers Which Include OHSMS Standards ($f_i{}^* = n_{ST(i)}/N_i$) (%)	Relative Occurrence Frequency of Papers (Referred to S) Which Include OHSMS Standards ($f^*_{ST(i)} = n_{ST(i)}/S$) (%)
(A)		(B)	(C)	(D) = (C)/N	(E)		(F) = (E)/N	(G) = (E)/(C)	(H) = (E)/S
1	Applied Ergonomics	JAE	886	9.02%	3		0.03%	0.34%	4.00%
2	Accident Analysis and Prevention	AAP	1522	15.50%	8		0.08%	0.53%	10.67%
3	Journal of Cleaner Production	JCP	1005	10.23%	3		0.03%	0.30%	4.00%
4	Journal of Operations Management	JOM	995	10.13%	1		0.01%	0.10%	1.33%
5	Architectural Science Review	ASR	1007	10.25%	1		0.01%	0.10%	1.33%
6	Safety Science	JSS	945	9.62%	40		0.41%	4.23%	53.33%
7	Professional Safety	JPS	998	10.16%	4		0.04%	0.40%	5.33%
8	Journal of Loss Prevention in the Process Industries	JLPPI	881	8.97%	8		0.08%	0.91%	10.67%
9	International Journal of Industrial Ergonomics	IJIE	955	9.72%	1		0.01%	0.10%	1.33%
10	Journal of Safety Research	JSR	628	6.39%	6		0.06%	0.96%	8.00%
	Totals:		9822	100.00%	75		0.76%	7.96%	100.00%

Annotations: The entire (total) absolute frequency (i.e., the total number of investigated articles) is: N = 9822; The number of articles that concern OHSS and include (or use or refer to) OHSMS standards is: S = 75.

References

1. Loke, Y.; Tan Wee, J.; Pegy, H. *Economic Cost of Work-Related Injuries and Health in Singapore*; WSH Institute Report; WSH: Singapore, 2013; pp. 1–29. Available online: https://www.wsh-institute.sg/~/media/wshi/past%20publications/2013/economic%20cost%20of%20work-related%20injuries%20and%20ill-health%20in%20singapore.pdf?la=en (accessed on 8 October 2018).
2. Occupational Health and Safety Assessment Series (OHSAS) Project Group. *OHSAS 18001:2007—Occupational Health and Safety Management Systems—Requirements*; OHSAS: Sydney, Australia, 2007; ISBN 978 0 580 50802 8.
3. Robson, L.S.; Clarke, J.A.; Cullen, K.L.; Amber, B.; Colette, S.; Bigelow, P.L.; Irvin, E.; Culyer, A.; Quenby, M. The effectiveness of occupational health and safety management system interventions: A systematic review. *Saf. Sci.* **2007**, *45*, 329–353. [CrossRef]
4. Sinelnikov, S.; Inouye, J.; Kerper, S. Using leading indicators to measure occupational health and safety performance. *Saf. Sci.* **2015**, *72*, 240–248. [CrossRef]
5. Høj, N.P.; Kröger, W. Risk analyses of transportation on road and railway from a European Perspective. *Saf. Sci.* **2002**, *40*, 337–357.
6. Woodruff, J.M. Consequence and likelihood in risk estimation: A matter of balance in UK health and safety risk assessment practice. *Saf. Sci.* **2005**, *43*, 345–353. [CrossRef]
7. Reniers, G.L.L.; Dullaert, W.; Ale, B.J.M.; Soudan, K. Developing an external domino accident prevention framework: Hazwim. *J. Loss Prev. Process Ind.* **2005**, *18*, 127–138. [CrossRef]
8. Haimes, Y.Y. *Risk Modeling, Assessment, and Management*, 3rd ed.; John Wiley & Sons Inc.: New York, NY, USA, 2009; pp. 154–196.
9. Marhavilas, P.K. Risk Assessment Techniques in the Worksites of Occupational Health-Safety Systems with Emphasis on Industries and Constructions. Ph.D. Thesis, Department of Production and Management Engineering, Democritus University of Thrace, Xanthi, Greece, March 2015.
10. Cliff, D. *The Management of Occupational Health and Safety in the Australian Mining Industry*; International Mining for Development Centre, Mining for Development, Guide to Australian Practice: Melbourne, Australia, 2012; pp. 100–200.
11. International Labour Organization (ILO). *Occupational Safety and Health: Synergies between Security and Productivity*; ILO's Governing Body Paper GB.295/ESP/3, GB; ILO: Geneva, Switzerland, 2006.
12. ISO/IEC. *Guide 51: Safety Aspects—Guidelines for Their Inclusion in Standards*, 2nd ed.; ISO/IEC: Geneva, Switzerland, 1999.
13. Lee, M. How Does Climate Change Affect the Assessment of Landslide Risk? 2006. Available online: http://cliffs.lboro.ac.uk/downloads/ML2006.pdf (accessed on 15 July 2017).
14. Olsson, F. *Tolerable Fire Risk Criteria for Hospitals*; Report 3101; Department of Fire Safety Engineering, Lund University: Lund, Sweden, 1999; ISSN 1402-3504.
15. IEC (International Electrotechnical Commission). *Risk Analysis of Technological Systems*; International Standard 60300-3-9, Dependability Management—Part 3: Application Guide—Section 9; IEC: Geneva, Switzerland, 1995.
16. Marhavilas, P.K.; Koulouriotis, D.E.; Gemeni, V. Risk Analysis and Assessment Methodologies in the Work Sites: On a Review, Classification and Comparative Study of the Scientific Literature of the Period 2000–2009. *J. Loss Prev. Process Ind.* **2011**, *24*, 477–523. [CrossRef]
17. Jonkman, S.N.; van Gelder, P.H.A.J.M.; Vrijling, J.K. An overview of quantitative risk measures for loss of life and economic damage. *J. Hazard. Mater.* **2003**, *99*, 1–30.
18. Vrouwenvelder, A.C.W.M.; Lovegrove, R.; Holicky, M.; Tanner, P.; Canisius, G. *Risk Assessment and Risk Communication in Civil Engineering*, 1st ed.; CIB General Secretariat: Rotterdam, The Netherlands, 2001; pp. 1–62. ISBN 90-6363-026-3.
19. CCPS. *Guidelines for Chemical Process Quantitative Risk Analysis*, 2nd ed.; Center for Chemical Process Safety (CCPS) of American Institute of Chemical Engineers: New York, NY, USA, 1989; pp. 1–748. ISBN 978-0-8169-0720-5.
20. International Organization for Standardization (ISO). Standards in Action. 2017. Available online: https://www.iso.org/standards-in-action.html (accessed on 15 July 2017).

21. Gallagher, C. Occupational Health and Safety Management Systems: System Types and Effectiveness. Ph.D. Thesis, Deakin University, Melbourne, Australia, 2000.
22. British Standards Institution (BSI). *BS 8800:1996—Guide to Occupational Health and Safety Management Systems*; BSI: London, UK, 1996; pp. 1–70. ISBN 0-580-25859-9.
23. British Standards Institution (BSI). *BS 18004:2008—Guide to Achieving Effective Occupational Health and Safety Performance*; BSI: London, UK, 2008; pp. 1–78. ISBN 978 0 580 52910 8.
24. British Standards Institution (BSI). *BS 8800:2004—Occupational Health and Safety Management Systems-Guide*; BSI: London, UK, 2004; pp. 1–87. ISBN 0 580 43987 9.
25. Health and Safety Executive (HSE). *Successful Health and Safety Management*; HSE: London, UK, 1997; pp. 1–80. ISBN 978-0-7176-1276-5.
26. Health and Safety Executive (HSE). *Managing for Health and Safety*; HSE: London, UK, 2013; pp. 1–66. ISBN 978-0-7176-6456-6.
27. International Labour Organization (ILO). *Guidelines on Occupational Safety and Health Management Systems—ILO-OSH 2001*; ILO: London, UK, 2009; ISBN 92-2-111634-4.
28. Australian/New Zealand Standard (AS/NZS). *AS/NZS 4801:2001-Occupational Health and Safety Management Systems-Specification with Guidance for Use*; AS/NZS: Melbourne, Australia, 2001; ISBN 0-7337-4092-8.
29. American Industrial Hygiene Association (AIHA). *ANSI/AIHA Z10-2005 Occupational Health and Safety Management Systems*; AIHA: New York, NY, USA, 2005; ISBN1 10:1931504644. ISBN2 13:978-1931504645.
30. SAI Global. ANSI/AIHA Z10-2005. Available online: https://www.saiglobal.com/assurance/ohs/Z10.htm (accessed on 15 July 2017).
31. Singapore Standards Council (SSC). *SS 506—Occupational Safety and Health (OSH) Management Systems—Part 1: Requirements*; SSC: Singapore, 2009; ISBN 978-981-4278-15-7.
32. Singapore Standards Council (SSC). *SS 506—Occupational Safety and Health (OSH) Management Systems—Part 2: Guidelines for the Implementation of SS 506: Part 1*; SSC: Singapore, 2009; ISBN 978-981-4278-16-4.
33. Romero, J.C.R. *Security Management Systems and Health at Work—Certified Or UNS Certified? ILO GuIdelines OHSAS 18001 Standard*; Industrial Security of the E.T.S.I.I. Malaga University: Malaga, Spain, 2001; pp. 4–13.
34. Barone, D.; Milano, Italy. Le nuove norme UNI 10617-2012 e UNI 10616-2012 relative ai Sistemi di Gestione della Sicurezza negli impianti a rischio di incidente rilevante. Personal communication, 2012.
35. International Organization for Standardization (ISO). ISO Survey. Available online: https://www.iso.org/the-iso-survey.html (accessed on 19 September 2018).
36. International Organization for Standardization (ISO). ISO 45001 Occupational Health and Safety—Briefing Notes. 2015. Available online: https://www.iso.org/iso-45001-occupational-health-and-safety.html (accessed on 19 September 2018).
37. International Organization for Standardization (ISO). ISO 45001—Occupational Health and Safety. Available online: https://www.iso.org/iso-45001-occupational-health-and-safety.html (accessed on 19 September 2018).
38. Gallagher, C. *Health and Safety Management Systems: An Analysis of System Types and Effectiveness*; National Key Centre in Industrial Relations, Monash University: Melbourne, Australia, 1997.
39. European Commission. Communication from the Commission Concerning Corporate Social Responsibility: A Business Contribution to Sustainable Development. 2002. Available online: https://eur-lex.europa.eu/LexUriServ/LexUriServ.do?uri=COM:2002:0347:FIN:EN:PDF (accessed on 26 September 2018).
40. Frick, K.; Wren, J. Reviewing occupational health and safety management: Multiple roots, diverse perspectives and ambiguous outcomes. In *Systematic Occupational Health and Safety Management: Perspectives on an International Development*; Frick, K., Jensen, P.L., Quinlan, M., Wilthagen, T., Eds.; Pergamon: Amsterdam, The Netherlands, 2000; pp. 17–42. ISBN 9780080434131.
41. Walters, D. (Ed.) *Regulating Health and Safety Management in the European Union: A Study of the Dynamics of Change*; Presses Interuniversitaires Europeenes: Brussels, Belgium, 2002; ISBN 90-5201-998-3. [CrossRef]
42. Gallagher, C.; Underhill, E.; Rimmer, M. Occupational safety and health management systems in Australia: Barriers to success. *Policy Pract. Health Saf.* **2003**, *1*, 67–81. [CrossRef]
43. Saksvik, P.O.; Quinlan, M. Regulating systematic occupational health and safety management: Comparing the Norwegian and Australian experience. *Relat. Ind./Ind. Relat.* **2003**, *58*, 33–59. [CrossRef]

44. Marhavilas, P.K.; Koulouriotis, D.E.; Spartalis, S.H. Harmonic Analysis of Occupational-Accident Time-Series as a Part of the Quantified Risk Evaluation in Worksites: Application on Electric Power Industry and Construction Sector. *Reliab. Eng. Syst. Saf.* **2013**, *112*, 8–25. [CrossRef]
45. International Labour Organization (ILO). *Emerging Risks and New Patterns of Prevention in Changing World of Work*, 1st ed.; ILO: Geneva, Switzerland, 2010; pp. 1–22, ISBN 978-92-2-123342-8 (print), ISBN 978-92-2-123343-5 (web pdf).
46. Jordan, J.R.; Letti, G.; Pinto, T.L. A proposal for the use of serious games in occupational safety. In *Occupational Safety and Hygiene II*; Arezes, P., Baptista, J.S., Barroso, M.P., Carneiro, P., Cordeiro, P., Costa, N., Melo, R.B., Miguel, S.A., Perestrelo, G., Eds.; Taylor and Francis Group: London, UK, 2014; ISBN 978-1-138-00144-2.
47. Gallagher, C.; Underhill, E.; Rimmer, M. *Occupational Health and Safety Management Systems: A Review of their Effectiveness in Securing Healthy and Safe Workplaces*; National Occupational Health and Safety Commission: Sydney, Australia, 2001; ISBN 0 642 70981 5.
48. Robson, L.S.; Macdonald, S.; Gray, G.C.; Van Eerd, D.L.; Bigelow, P.L. A descriptive study of the OHS management auditing methods used by public sector organizations conducting audits of workplaces: Implications for audit reliability and validity. *Saf. Sci.* **2012**, *50*, 181–189. [CrossRef]
49. Li, Y.; Guldenmund, F.W. Safety management systems: A broad overview of the literature. *Saf. Sci.* **2018**, *103*, 94–123. [CrossRef]
50. Aneziris, O.N.; Topali, E.; Papazoglou, I.A. Occupational risk of building construction. *Reliab. Eng. Syst. Saf.* **2012**, *105*, 36–46. [CrossRef]
51. Autenrieth, D.A.; Brazile, W.J.; Sandfort, D.R.; Douphrate, D.I.; Román-Muñiz, I.N.; Reynolds, S.J. The associations between occupational health and safety management system programming level and prior injury and illness rates in the U.S. dairy industry. *Saf. Sci.* **2016**, *84*, 108–116. [CrossRef]
52. Tsalis, T.A.; Stylianou, M.S.; Nikolaou, I.E. Evaluating the quality of corporate social responsibility reports: The case of occupational health and safety disclosures. *Saf. Sci.* **2018**, *109*, 313–323. [CrossRef]
53. Jørgensen, T.H.; Remmen, A.; Mellado, M.D. Integrated management systems—Three different levels of integration. *J. Clean. Prod.* **2006**, *14*, 713–722. [CrossRef]
54. Santos-Reyes, J.; Beard, A.N. Viability of a systemic safety management system. In Proceedings of the Safety and Reliability Conference, ESREL-2006, Estoril, Portugal, 18–22 September 2006.
55. Adele, A.L. Legal Perpspectives of ANZI Z10-2005. *Prof. Saf.* **2006**, 41–43. Available online: www.xprolegal.com/expertarticles/359-1312558749_ASSE%20Z10%20Article.pdf (accessed on 26 September 2018).
56. Fred, M. ANSI/AIHA Z10-2005: The new benchmark for safety management systems. *Prof. Saf.* **2006**, *51*, 25–33. Available online: https://aeassincludes.assp.org/professionalsafety/pastissues/051/02/020206as.pdf (accessed on 26 September 2018).
57. Wrona, R.M. The use of state workers' compensation administrative data to identify injury scenarios and quantify costs of work-related traumatic brain injuries. *J. Saf. Res.* **2006**, *37*, 75–81. [CrossRef] [PubMed]
58. Nielsen, K.J.; Carstensen, O.; Rasmussen, K. The prevention of occupational injuries in two industrial plants using an incident reporting scheme. *J. Saf. Res.* **2006**, *37*, 479–486. [CrossRef] [PubMed]
59. Cadieux, J.; Roy, M.; Desmarais, L. A preliminary validation of a new measure of occupational health and safety. *J. Saf. Res.* **2006**, *37*, 413–419. [CrossRef] [PubMed]
60. Päivinen, M. Electricians' perception of work-related risks in cold climate when working on high places. *Int. J. Ind. Ergon.* **2006**, *36*, 661–670. [CrossRef]
61. Zeng, S.X.; Shi, J.J.; Lou, G.X. A synergetic model for implementing an integrated management system: An empirical study in China. *J. Clean. Prod.* **2007**, *15*, 1760–1767. [CrossRef]
62. Fernández-Muñiz, B.; Montes-Peón, J.M.; Vázquez-Ordás, C.J. Safety management system: Development and validation of a multidimensional scale. *J. Loss Prev. Process Ind.* **2007**, *20*, 52–68. [CrossRef]
63. Coleman, P.J.; Kerkering, J.C. Measuring mining safety with injury statistics: Lost workdays as indicators of risk. *J. Saf. Res.* **2007**, *38*, 523–533. [CrossRef] [PubMed]
64. Bevilacqua, M.; Ciarapica, F.E.; Giacchetta, G. Industrial and occupational ergonomics in the petrochemical process industry: A regression trees approach. *Accid. Anal. Prev.* **2008**, *40*, 1468–1479. [CrossRef]
65. Brooks, B. Shifting the focus of strategic occupational injury prevention: Mining free-text, workers compensation claims data. *Saf. Sci.* **2008**, *46*, 1–21. [CrossRef]
66. Duijm, N.J.; Fiévez, C.; Gerbec, M.; Hauptmanns, U.; Konstandinidou, M. Management of health, safety and environment in process industry. *Saf. Sci.* **2008**, *46*, 908–920. [CrossRef]

67. Salla, L.; Kivistö-Rahnasto, J. Utilization of external accident information in companies' safety promotion—Case: Finnish metal and transportation industry. *Saf. Sci.* **2008**, *46*, 802–814. [CrossRef]
68. Nielsen, K.J.; Rasmussen, K.; Glasscock, D.; Spangenberg, S. Changes in safety climate and accidents at two identical manufacturing plants. *Saf. Sci.* **2008**, *46*, 440–449. [CrossRef]
69. Schrover, A.J.M. Ten years SHE-improvements on a chemical and nuclear research-site—Learning drivers. *Saf. Sci.* **2008**, *46*, 551–563. [CrossRef]
70. Zeng, S.X.; Vivian, W.Y.; Tam, C.M. Towards occupational health and safety systems in the construction industry of China. *Saf. Sci.* **2008**, *46*, 1155–1168. [CrossRef]
71. Komljenovic, D.; Groves, W.A.; Kecojevic, V.J. Injuries in U.S. mining operations—A preliminary risk analysis. *Saf. Sci.* **2008**, *46*, 792–801. [CrossRef]
72. Caroly, S.; Coutarel, F.; Landry, A.; Mary-Cheray, I. Sustainable MSD prevention: Management for continuous improvement between prevention and production-Ergonomic intervention in two assembly line companies. *Appl. Ergon.* **2009**, *41*, 591–599. [CrossRef] [PubMed]
73. Chang, J.I.; Liang, C.-L. Performance evaluation of process safety management systems of paint manufacturing facilities. *J. Loss Prev. Process Ind.* **2009**, *22*, 398–402. [CrossRef]
74. Santos-Reyes, J.; Beard, A.N. A SSMS model with application to the oil and gas industry. *J. Loss Prev. Process Ind.* **2009**, *22*, 958–970. [CrossRef]
75. Crippa, C.; Fiorentini, L.; Rossini, V.; Stefanelli, R.; Tafaro, S.; Marchi, M. Fire risk management system for safe operation of large atmospheric storage tanks. *J. Loss Prev. Process Ind.* **2009**, *22*, 574–581. [CrossRef]
76. Knegtering, B.; Pasman, H.J. Safety of the process industries in the 21st century: A changing need of process safety management for a changing industry. *J. Loss Prev. Process Ind.* **2009**, *22*, 162–168. [CrossRef]
77. Pasman, H.J.; Jung, S.; Prem, K.; Rogers, W.J.; Yang, X. Is risk analysis a useful tool for improving process safety? *J. Loss Prev. Process Ind.* **2009**, *22*, 769–777. [CrossRef]
78. Moriyama, T.; Ohtani, H. Risk assessment tools incorporating human error probabilities in the Japanese small-sized establishment. *Saf. Sci.* **2009**, *47*, 1379–1397. [CrossRef]
79. Fernández-Muñiz, B.; Montes-Peón, J.M.; Vázquez-Ordás, C.J. Relation between occupational safety management and firm performance. *Saf. Sci.* **2009**, *47*, 980–991. [CrossRef]
80. Celik, M. Designing of integrated quality and safety management system (IQSMS) for shipping operations. *Saf. Sci.* **2009**, *47*, 569–577. [CrossRef]
81. Reniers, G.L.L.; Ale, B.J.M.; Dullaert, W.; Soudan, K. Designing continuous safety improvement within chemical industrial areas. *Saf. Sci.* **2009**, *47*, 578–590. [CrossRef]
82. Baram, M. Globalization and workplace hazards in developing nations. *Saf. Sci.* **2009**, *47*, 756–766. [CrossRef]
83. Shengli, N. Ergonomics and occupational safety and health: An ILO perspective. *Appl. Ergon.* **2010**, *41*, 744–753. [CrossRef]
84. Gangolells, M.; Casals, M.; Forcada, N.; Roca, X.; Fuertes, A. Mitigating construction safety risks using prevention through design. *J. Saf. Res.* **2010**, *41*, 107–122. [CrossRef] [PubMed]
85. Cheng, C.-W.; Leu, S.-S.; Lin, C.-C.; Fan, C. Characteristic analysis of occupational accidents at small construction enterprises. *Saf. Sci.* **2010**, *48*, 698–707. [CrossRef]
86. Celeste, J.; Silva, C. A semi-quantitative assessment of occupational risks using bow-tie representation. *Saf. Sci.* **2010**, *48*, 973–979. [CrossRef]
87. Sgourou, E.; Katsakiori, P.; Goutsos, S.; Manatakis, E. Assessment of selected safety performance evaluation methods in regards to their conceptual, methodological and practical characteristics. *Saf. Sci.* **2010**, *48*, 1019–1025. [CrossRef]
88. Fera, M.; Macchiaroli, R. Appraisal of a new risk assessment model for SME. *Saf. Sci.* **2010**, *48*, 1361–1368. [CrossRef]
89. Lindhout, P.; Kingston-Howlett, J.C.; Ale, B.J.M. Controlled readability of Seveso II company safety documents, the design of a new KPI. *Saf. Sci.* **2010**, *48*, 734–746. [CrossRef]
90. Celik, M. Enhancement of occupational health and safety requirements in chemical tanker operations: The case of cargo explosion. *Saf. Sci.* **2010**, *48*, 195–203. [CrossRef]
91. Dawson, D.; Ian Noy, Y.; Härmä, M.; Åkerstedt, T.; Belenky, G. Modelling fatigue and the use of fatigue models in work settings. *Accid. Anal. Prev.* **2011**, *43*, 549–564. [CrossRef]
92. Pheng, L.S.; Kwang, G.K. ISO 9001, ISO 14001 and OHSAS 18001 Management Systems: Integration, Costs and Benefits for Construction Companies. *Arch. Sci. Rev.* **2005**, *48*, 145–152. [CrossRef]

93. Romain, J.; Imbeau, D.; de Marcellis-Warin, N. Development of an indirect-cost calculation model suitable for workplace use. *J. Saf. Res.* **2011**, *42*, 149–164. [CrossRef]
94. Leka, S.; Aditya, J.; Widerszal-Bazyl, M.; Żołnierczyk-Zreda, D.; Zwetsloot, G. Developing a standard for psychosocial risk management: PAS 1010. *Saf. Sci.* **2011**, *49*, 1047–1057. [CrossRef]
95. Verna, B.; O'Keeffe, V. Weighing the pig never made it heavier: Auditing OHS, social auditing as verification of process in Australia. *Saf. Sci.* **2011**, *49*, 1014–1021. [CrossRef]
96. Vinodkumar, M.N.; Bhasi, M. A study on the impact of management system certification on safety management. *Saf. Sci.* **2011**, *49*, 498–507. [CrossRef]
97. Pernille, H.; Hasle, P. Making work environment auditable—A 'critical case' study of certified occupational health and safety management systems in Denmark. *Saf. Sci.* **2011**, *49*, 1022–1029. [CrossRef]
98. Ying, L.; Xingdong, L. A study on a new hazard detecting and controlling method: The case of coal mining companies in China. *Saf. Sci.* **2011**, *49*, 279–285. [CrossRef]
99. Zwetsloot, G.I.J.M.; Zwanikken, S.; Hale, A. Policy expectations and the use of market mechanisms for regulatory OSH certification and testing regimes. *Saf. Sci.* **2011**, *49*, 1007–1013. [CrossRef]
100. Frick, K. Worker influence on voluntary OHS management systems—A review of its ends and means. *Saf. Sci.* **2011**, *49*, 974–987. [CrossRef]
101. Granerud, R.L.; Sø Rocha, R. Organisational learning and continuous improvement of health and safety in certified manufacturers. *Saf. Sci.* **2011**, *49*, 1030–1039. [CrossRef]
102. Kristensen, P.H. Managing OHS: A route to a new negotiating order in high-performance work organizations? *Saf. Sci.* **2011**, *49*, 964–973. [CrossRef]
103. Zwetsloot, G.I.J.M.; Hale, A.; Zwanikken, S. Regulatory risk control through mandatory occupational safety and health (OSH) certification and testing regimes (CTRs). *Saf. Sci.* **2011**, *49*, 995–1006. [CrossRef]
104. Hasle, P.; Zwetsloot, G. Editorial: Occupational Health and Safety Management Systems: Issues and challenges. *Saf. Sci.* **2011**, *49*, 961–963. [CrossRef]
105. Fernández-Muñiz, B.; Montes-Peón, J.M.; Vázquez-Ordás, C.J. Safety climate in OHSAS 18001-certified organisations: Antecedents and consequences of safety behavior. *Accid. Anal. Prev.* **2012**, *45*, 745–758. [CrossRef] [PubMed]
106. Goh, Y.M.; Love, P.E.D.; Stagbouer, G.; Annesley, C. Dynamics of safety performance and culture: A group model building approach. *Accid. Anal. Prev.* **2012**, *48*, 118–125. [CrossRef] [PubMed]
107. Cheng, C.-W.; Leu, S.-S.; Cheng, Y.M.; Wu, T.C.; Lin, C.C. Applying data mining techniques to explore factors contributing to occupational injuries in Taiwan's construction industry. *Accid. Anal. Prev.* **2012**, *48*, 214–222. [CrossRef]
108. Pinto, A.; Ribeiro, R.A.; Nunes, I.L. Fuzzy approach for reducing subjectivity in estimating occupational accident severity. *Accid. Anal. Prev.* **2012**, *45*, 281–290. [CrossRef] [PubMed]
109. Badri, A.; Nadeau, S.; Gbodossou, A. Proposal of a risk-factor-based analytical approach for integrating occupational health and safety into project risk evaluation. *Accid. Anal. Prev.* **2012**, *48*, 223–234. [CrossRef]
110. Luria, G.; Morag, I. Safety management by walking around (SMBWA): A safety intervention program based on both peer and manager participation. *Accid. Anal. Prev.* **2012**, *45*, 248–257. [CrossRef]
111. Fernández-Muñiz, B.; Montes-Peón, J.M.; Vázquez-Ordás, C.J. Occupational risk management under the OHSAS 18001 standard: Analysis of perceptions and attitudes of certified firms. *J. Clean. Prod.* **2012**, *24*, 36–47. [CrossRef]
112. Mawhood, J.; Dickinson, C. *Rail Staff Fatigue—The GB Regulators Perspective on Managing the Risks*; Dadashi, N., Scott, A., Wilson, J.R., Mills, A., Eds.; Taylor & Francis: New York, NY, USA, 2012; pp. 337–346, Print ISBN 978-1-138-00037-7, eBook ISBN 978-0-203-75972-1.
113. Lee, S.-W.; Kim, K.-H.; Kim, T.-G. Current situation of certification system and future improvements of the occupational health and safety management system for loss prevention in Korea—Focused on KOSHA 18001. *J. Loss Prev. Process Ind.* **2012**, *25*, 1085–1089. [CrossRef]
114. Gnoni, M.G.; Lettera, G. Near-miss management systems: A methodological comparison. *J. Loss Prev. Process Ind.* **2012**, *25*, 609–616. [CrossRef]
115. Badri, A.; Gbodossou, A.; Nadeau, S. Occupational health and safety risks: Towards the integration into project management. *Saf. Sci.* **2012**, *50*, 190–198. [CrossRef]
116. Hamidi, N.; Omidvari, M.; Meftahi, M. The effect of integrated management system on safety and productivity indices: Case study; Iranian cement industries. *Saf. Sci.* **2012**, *50*, 1180–1189. [CrossRef]

117. Hsu, I.-Y.; Su, T.-S.; Kao, C.-S.; Shu, Y.-L.; Lin, P.-R.; Tseng, J.-M. Analysis of business safety performance by structural equation models. *Saf. Sci.* **2012**, *50*, 1–11. [CrossRef]
118. Wang, C.-H.; Liu, Y.-J. Omnidirectional safety culture analysis and discussion for railway industry. *Saf. Sci.* **2012**, *50*, 1196–1204. [CrossRef]
119. Zubaidah, I.; Doostdar, S.; Harun, Z. Factors influencing the implementation of a safety management system for construction sites. *Saf. Sci.* **2012**, *50*, 418–423. [CrossRef]
120. Abad, J.; Esteban, L.; Jordi, V. An assessment of the OHSAS 18001 certification process: Objective drivers and consequences on safety performance and labour productivity. *Saf. Sci.* **2013**, *60*, 47–56. [CrossRef]
121. Santos, G.; Barros, S.; Mendes, F.; Lopes, N. The main benefits associated with health and safety management systems certification in Portuguese small and medium enterprises post quality management system certification. *Saf. Sci.* **2013**, *51*, 29–36. [CrossRef]
122. Lo, C.K.Y.; Pagell, M.; Fan, D.; Wiengarten, F.; Yeung, A.C.L. OHSAS 18001 certification and operating performance: The role of complexity and coupling. *J. Oper. Manag.* **2014**, *32*, 268–280. [CrossRef]
123. Yazdani, A.; Neumann, W.P.; Imbeau, D.; Bigelow, P.; Pagell, M.; Wells, R. Prevention of musculoskeletal disorders within management systems: A scoping review of practices, approaches, and techniques. *Appl. Ergon.* **2015**, *51*, 255–262. [CrossRef] [PubMed]
124. Gopang, M.A.; Nebhwani, M.; Khatri, A.; Marri, H.B. An assessment of occupational health and safety measures and performance of SMEs: An empirical investigation. *Saf. Sci.* **2017**, *93*, 127–133. [CrossRef]
125. Cañamares, S.M.; Escribano, B.M.V.; González-García, M.N.; Barriuso, A.R.; Rodríguez-Sáiz, A. Occupational risk-prevention diagnosis: A study of construction SMEs in Spain. *Saf. Sci.* **2017**, *92*, 104–115. [CrossRef]

© 2018 by the authors. Licensee MDPI, Basel, Switzerland. This article is an open access article distributed under the terms and conditions of the Creative Commons Attribution (CC BY) license (http://creativecommons.org/licenses/by/4.0/).

Article

The Sustainability of Romanian SMEs and Their Involvement in the Circular Economy

Ionica Oncioiu [1,*], Sorinel Căpușneanu [2], Mirela Cătălina Türkeș [2], Dan Ioan Topor [3], Dana-Maria Oprea Constantin [4], Andreea Marin-Pantelescu [5] and Mihaela Ștefan Hint [3]

1. Faculty of Finance-Banking, Accounting and Business Administration, Titu Maiorescu University, Bucharest 040051, Romania
2. Faculty of Finance, Banking and Accountancy, Dimitrie Cantemir Christian University, Bucharest 040051, Romania; sorinelcapusneanu@ucdc.ro (S.C.); mirela.turkes@ucdc.ro (M.C.T.)
3. Faculty of Economic Sciences, 1 Decembrie 1918 University, Alba-Iulia 510009, Romania; dan.topor@uab.ro (D.I.T.); mihacont73@gmail.com (M.S.H.)
4. Faculty of Geography, University of Bucharest, Bucharest 050107, Romania; danamartines@yahoo.com
5. Faculty of Business and Tourism, The Bucharest University of Economic Studies, Bucharest 010374, Romania; marin.andreea@com.ase.ro
* Correspondence: nelly_oncioiu@yahoo.com; Tel.: +40-744-322-911

Received: 22 June 2018; Accepted: 1 August 2018; Published: 4 August 2018

Abstract: Sustainability involves extending the relational framework of SMEs outside the sphere of economic activity by justifying and legitimizing actions with a social impact on the environment. Links with the circular economy are achieved through the economic and environmental dimensions and through corporate social responsibility as a component of sustainable development. The main purpose of the paper was to determine the level of involvement of Romanian SMEs in activities related to the circular economy. The sample survey conducted among SME managers offered the advantage of collecting a large amount of direct information on the activities undertaken, the size of the investments and the nature of the funding sources used over the last five years. In this descriptive research, the process of setting up a representative sample of 384 enterprises was carried out by random sampling. The major contributions of the research project are to outline the contribution of Romanian SMEs to the development of a sustainable economy through their involvement in specific activities, the size of the investments made, and the level of participation of representatives of the enterprises in courses in order to identify new sources of financing and positive solutions in order to implement the principles of the circular economy.

Keywords: sustainability; circular economy; investments; sources of funding; SME; sustainable economy

1. Introduction

The circular economy and sustainability are two concepts that outline an extended framework for sustainable development, through which the implementation of strategies capable of providing the enterprise with healthy development is also achieved by addressing the problems of environmental degradation and resource shortages [1–4]. Besides, the circular economy is a sustainable development strategy that tackles the problems of environmental degradation and natural resource shortages through three principles: reduction, reuse and recycling of materials [5–7]. These principles define a circular system where all the materials are recycled, and all the energy comes from renewable sources that support activities and rebuild the ecosystem as well as support human health, society and healthy resources that generate value [8–11].

Since small firms are an important engine of growth in the economy and sustainability is an essential input in the production process, identifying how firms respond to the circular economy is crucial to understanding growth in developing economies [12–15].

Through this study, the authors intended to highlight the extent and the degree of involvement of Romanian SMEs in the activities specific to the circular economy. To achieve this, quantitative research was carried out among SMEs in Romania using a survey with a representative sample drawn using the random numbering method. The degree of involvement of enterprises in the circular economy was first assessed with regard to issues such as the activities carried out, the level of investments made, the sources of financing attracted and the level of managers' interest in attending courses in order to implement new circular business models. In light of an apparent logical fault, the diversity of views expressed in the literature specifically referring to the circular economy helped us to determine the best approach to position our scientific research with regard to the new challenges of SMEs in the Romanian economy.

The results of the present study suggest that successful sustainability plays an important role in the survival and success of any organization in today's environment, which is extremely competitive and continually evolving. Finally, our findings are relevant for the transformation of Romanian SMEs by identifying the specific actions they take as part of their involvement in the circular economy.

The remainder of this study is organized as follows. Section 2 provides a brief literature review. Section 3 presents a description of the research methodology. The empirical results are presented and discussed in Section 4. Finally, conclusions and suggestions are presented in Section 5.

2. Literature Review

The circular economy concept has been debated in several schools of thought and theory that have challenged linear economic systems that suppose that resources are infinite [16–18]. According to their studies, some specialists consider the circular economy as a space economy that works by reproducing the limited initial input stock and recycling the waste produced [19–21]. Other specialists consider the circular economy to be an industrial economy that relies on the ability to restore natural resources [22,23] and aims to minimize (or eliminate) waste, use renewable energy sources and phase out the use of harmful substances [24].

The specialists considered that there was also the need to accept an economic model in which the materials and energy of waste products are reintroduced into the economic system [25]. Thus, a clear distinction was made between two different types of materials in a closed-loop economy: materials of biological origin and materials of non-biological origin. Materials of biological origin (forest products) can return to the biosphere as raw materials, but materials of non-biological origin (plastics or metals) cannot return to the biosphere and are not biodegradable [26,27]. This type of economy transcends the linear economy [28], seeking new transformations across the value chain to keep both types of material in the circular economy, preserving their value for as long as possible [29].

Different studies based on the design, investigation and creation of a general framework on the ecological side of the circular economy have been carried out by specialists around the world, including circular design [30], design for circular behavior [31,32], the incorporation of ecosystem services [33], evaluating the environmental dimension based on material efficiency strategy [34], and the analysis of consumer behavior related to the circular economy [35].

Implementing the concept of the circular economy requires a detailed analysis of the opportunities and benefits it can bring to a country's economy. According to specialists' studies, large enterprises have greater facility in adopting and realizing beneficial circular business models, such as creating new jobs [36,37], reducing costs in different sectors of the economy (cars, electric machines, machinery and equipment) [38], supply-side price mitigation on commodity markets, or supply risks [39]. Once large enterprises have adopted circular business models, SMEs become aware of the benefits of the circular economy and of improving their efficiency in using natural resources.

The European Commission report states that more than two-thirds of interviewed SMEs are satisfied with the return on investments made to improve resource efficiency, and have seen production cost reductions over the past two years [40]. Romanian SMEs are extremely different, so every branch of the national economy can benefit from the implementation of the principles of the circular economy in an adapted manner [41]. In these conditions, the extent to which SMEs are willing to adopt ecological measures and their attitude to green policies depends on the sector in which they operate [42].

Most studies undertaken by specialists have indicated that SMEs do not adopt and implement the principles of the cyclical economy due to the initial costs, the reimbursement period for investments, or the high costs of achieving resource efficiency [43–45]; the high cost of organic business models [46,47]; the impossibility of supporting profitable economic activity due to hidden costs, a lack of highly qualified employees, and sudden changes in the economic environment [48]; the lack of financial resources to establish and manage a recycling system [49]; a lack of information, including information on deviations from the ex-ante cost estimates of ecological procedures, which may induce uncertainty and harm the competitiveness of SMEs [38]; the production of a small amount of waste, so that the circular economy represents an economically unfavorable option [50]; the lack of internal competencies leading to a dependence on recommendations made by external actors [44]; and the limited influence of SMEs on suppliers' involvement in sustainable activities [51,52].

In addition, in the SME sector, there is only modest initiative from the government to support new investment, with no coherent legislative measures to encourage the circular economy convictions and principles [42]. In support of this, we can offer the example of the Ecological Management and Audit Scheme (EMAS), which has no clear delimitation between large businesses and the SME sector [53,54]. In this case, only the managers' commitment to sustainability contributes to the adaptation and implementation of the principles of the circular economy to the needs of the SMEs [55,56]. The need for a better regulatory agenda to design and implement environmental policies is highlighted by the first assessment of the EU Environmental Assistance Program for SMEs [57].

Research undertaken by the European Commission highlights the fact that some concepts and terms in the EU legislation are not clearly defined, namely provider responsibility, separate collection quality and the definitions of recycling, re-use and recovery [58]. In Romania between 2002 and 2008, the SME sector developed steadily, with the growth of more than 69% in the number of enterprises [41]. The number of active entities increased from 326,443 in 2002 to 557,189 in 2008. The impact of the financial and economic crisis was felt strongly among SMEs, with the loss of about 11.73% of the enterprises, with a total of 491,805 active entities registered at the end of 2010 [43]. Over the last four years, the SME sector has seen a slow growth of only 10%, with a distribution by representative sectors of trade (38.98%), industry (11.84%), construction (9.60%), transport (6.10%), hotel and restaurants (4.52%), agriculture, forestry and fishing (2.18%) and other services (26.77%) [59].

3. Research Methodology

To conduct this research, we considered the hypothesis that SMEs are the engine for the development of a circular economy. SMEs can make a major contribution to the development of a sustainable economy by gradually integrating the principles of the circular economy into their own business model. The determining role of SMEs remains: (1) producing beneficial effects for a country's economy by recycling waste and using it as a raw material in production processes; (2) developing products and services in symbiosis with other industries by reducing resource consumption; (3) creating customized, high quality and value-customized products; (4) job creation and staff qualification in the field of environmental protection; and (5) increasing competition in sustainable product markets. Taking into account this hypothesis, the purpose of the research was to determine the level of involvement of Romanian SMEs in the activities specific to a circular economy.

The main objectives of the research were as follows:

- Highlighting the activities related to the circular economy conducted by SME managers in Romania in the last five years;
- Identifying the size of current and future investments by allocating percentages of turnover both to businesses that have carried out circular economy activities over the past five years and those willing to develop circular business models in the years to come;
- Description of the funding sources used by SMEs in the last five years to ensure good functioning and to carry out activities related to the circular economy;
- Identifying the level of participation of Romanian managers in courses to acquire the knowledge and skills regarding the performance of some activities that promote resource efficiency, eco-innovation and the circular economy.

Thus, in order to achieve the objectives, quantitative marketing research was carried out among SME managers in Romania. The main considerations were: (1) the development of governmental and European programs that provide access to important sources of funding and create premises for coherent, systematic and coordinated actions aimed at fostering entrepreneurship and increasing the number of SMEs; (2) the interconnected functioning of productive SME chains, with a high potential for adding value at the national, regional and global levels; and (3) the massive contribution of the SME sector to the formation of national GDP, to the economic and sustainable growth of a country, generating social progress and social prosperity.

At the end of February 2017, the statistical metadata database of the National Institute of Statistics of Romania was consulted in order to obtain the information necessary for the realization of the quantitative research. According to the data provided by National Institute of Statistics [59], on 28 February 2017, a list was established in which 552,483 active enterprises with a minimum of five years of age and a number of employees ranging from 1–249 were identified in Romania (www.statistici.insse.ro). Micro-enterprises represent 89.12% of all SMEs in Romania (Figure 1).

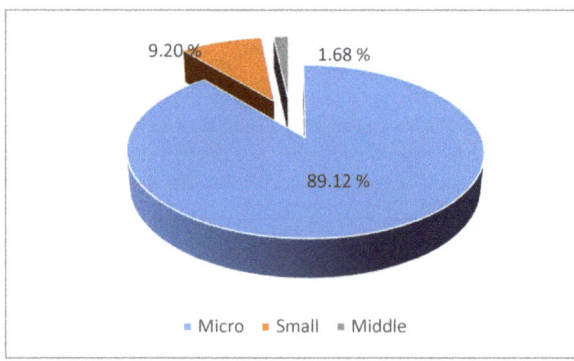

Figure 1. Distribution of economically active SMEs in Romania by their size on 31 January 2017 [59].

In Romania, micro-enterprises are defined as enterprises that have up to nine employees and achieve a net annual turnover of up to two million euro. Small enterprises are defined as enterprises that have up to 49 employees and achieve a net annual turnover of up to 10 million euro.

The research method used in the quantitative study was a survey by sampling, using a questionnaire as the data collection tool. The study was conducted between 12 March and 12 April 2017, with the support of eight interviewers with experience in the field who drafted the questionnaires for all eight development regions of Romania: North-East, South-East, South-Muntenia, South-West Oltenia, West, North-West, Center and Bucharest-Ilfov (Figure 2). Each interviewer held face-to-face interviews with business managers in one of Romania's development regions. For example, one interviewer covered the West region, another the South-West region.

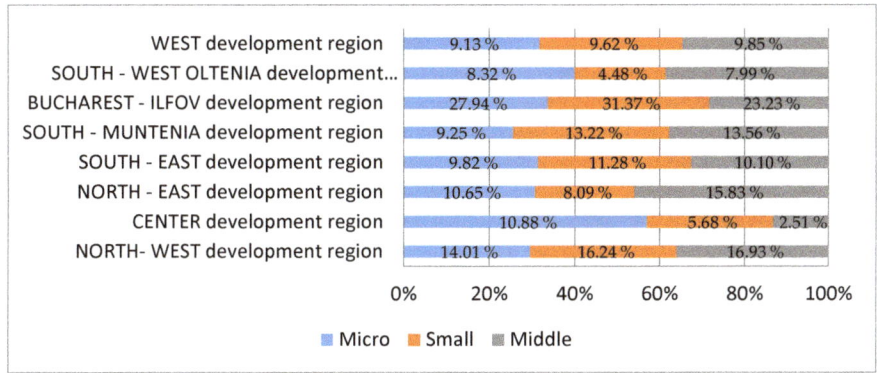

Figure 2. Distribution of economically active SMEs in the eight development regions of Romania on 31 January 2017 [59].

The sampling frame was made up of enterprises (SMEs), since these entities, both in terms of organization and functioning and through the activities carried out, were able to provide all the data and information necessary to achieve the intended purpose. The selection was based on simple random sampling, using a random number generator, from a list that included all Romanian SMEs with a minimum of five years of activity and a number of employees ranging from 1–249 people. The main criteria for structuring the enterprises were the number of employees, the development region; the field of activity and the year of establishment (minimum of five years) (Table 1 and Table A1).

It is necessary to clarify that the list was based on the statistical metadata from the National Institute of Statistics of Romania, which removed active SMEs with a working life of less than five years, inactive SMEs and those with a number of employees over 249 people. In this way, the structure of the research sample was a faithful reproduction of the structure of the reference population. The fundamental principle that was taken into account when using the sampling method was that the layers chosen were related to the dependent variable that was the object of the research. The proportion of subjects (SMEs) in each layer of the sample was proportional to that of the subjects at that layer level in the total population. To ensure a probability of guaranteeing 95% of the research results and obtaining an error margin of ±5% for a value $p = 0.50$, the sample size should be 384 observation units, thus the survey included 384 enterprises (SMEs). In the first phase, sub-samples were extracted from each layer, resulting in a high level of representativeness of the total sample, compared to simple random sampling, which can generate overrepresentation of some population groups and underrepresentation of others.

After identifying the enterprises, contact was established with their representatives to determine who should be surveyed, to obtain the survey participation agreement and to establish details of the meeting. The survey was the basis of the research; the process used to interview the SME managers was face-to-face interviews. The questionnaire was structured around four distinct objectives: (1) the activities undertaken related to the circular economy; (2) the size of the investments made; (3) the funding sources used for these types of activities; (4) participation in the acquisition of skills related to resource efficiency, eco-innovation and the circular economy. The information processing was based on the responses received from the SME managers and the information centralization was performed in relation to the consistency and convergence of the purpose of the research.

Table 1. Sample structure.

Number of Employees	Total SMEs Studied		Sample of SMEs Investigated	
	No.	%	No.	%
0–9 people (micro-enterprises)	465,621	89.12	343	89.32
10–49 people (small enterprises)	48,092	9.20	35	9.11
50–249 people (medium enterprises)	8770	1.68	6	1.56
Total	522,483	100	384	100.00
Development Region				
North-West development region	74,531	14.26	55	14.26
Center development region	53,596	10.26	39	10.26
North-East development region	54,846	10.50	40	10.50
South-East development region	52,057	9.96	38	9.96
South-Muntenia development region	50,624	9.69	37	9.69
Bucharest-Ilfov development region	147,210	28.18	108	28.18
South-West Oltenia development region	41,608	7.96	31	7.96
West development region	48,011	9.19	35	9.19
Total	522,483	100	384	100.00
Areas of Activity				
Agriculture, forestry and fishing	11,395	2.18	8	2.01
Industry	61,880	11.84	45	11.85
Construction	50,175	9.60	37	9.61
Trade	203,665	38.98	150	38.99
Hotels and restaurants	23,621	4.52	17	4.52
Transport	31,886	6.10	24	6.25
Other services	139,861	26.77	103	26.78
Total	522,483	100	384	100.00
Year of Establishment				
5–9 years	382,458	73.20	281	73.18
9–14 years	90,390	17.30	66	17.19
>15 years	49,636	9.50	37	9.64
Total	522,483	100	384	100

Source: Authors' calculation based on information extracted from the National Institute of Statistics of Romania [59].

4. Results and Discussion

The first objective was related to highlighting the main activities specific to the circular economy, carried out by SME managers in Romania (Figure 3).

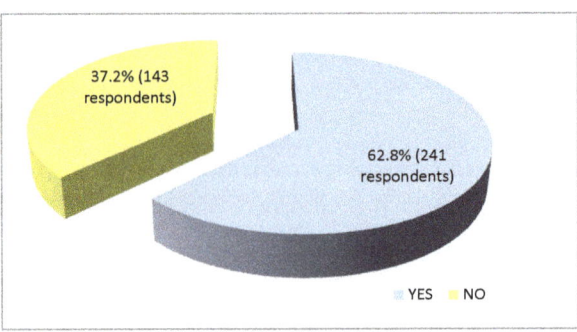

Figure 3. The share of SMEs that have carried out at least one activity related to the circular economy in the last five years.

In 241 (62.8%) of the 384 SMEs participating in the research, the managers said at least one activity related to the circular economy had taken place in the last five years. In total, 143 (37.2%) of 384 enterprises did not have at least one circular activity in the last five years. The management of the Romanian enterprises showed a positive attitude towards carrying out activities to support the circular economy at every stage of the value chain: production, consumption, repair and manufacture, waste management and secondary raw materials that are reintroduced into the economy.

In 62.8% of the SMEs surveyed in Romania, the managers said that they make real efforts to conduct activities related to the circular economy but face financial problems, with labor shortages and many legal barriers. However, the managers of these SMEs proposed the development of new strategies for the circular economy in the coming years and hoped to make a lot of progress in this regard.

The main activities related to the circular economy undertaken by Romanian SMEs in the last five years were (Figure 4) strengthening the guarantees offered to consumers who purchase goods online (14.10%), the use of renewable energy (12.78%), designing smart and green products and using energy labeling (12.33%), the use of advanced manufacturing facilities that generate cleaner production (10.13%), safe wastewater reuse (5.29%), the application of innovative techniques for the use of secondary raw materials/alternatives (3.08%) (2.64%), and the prevention of waste generation, the stimulation of recycling and the reduction of resource use (2.20%).

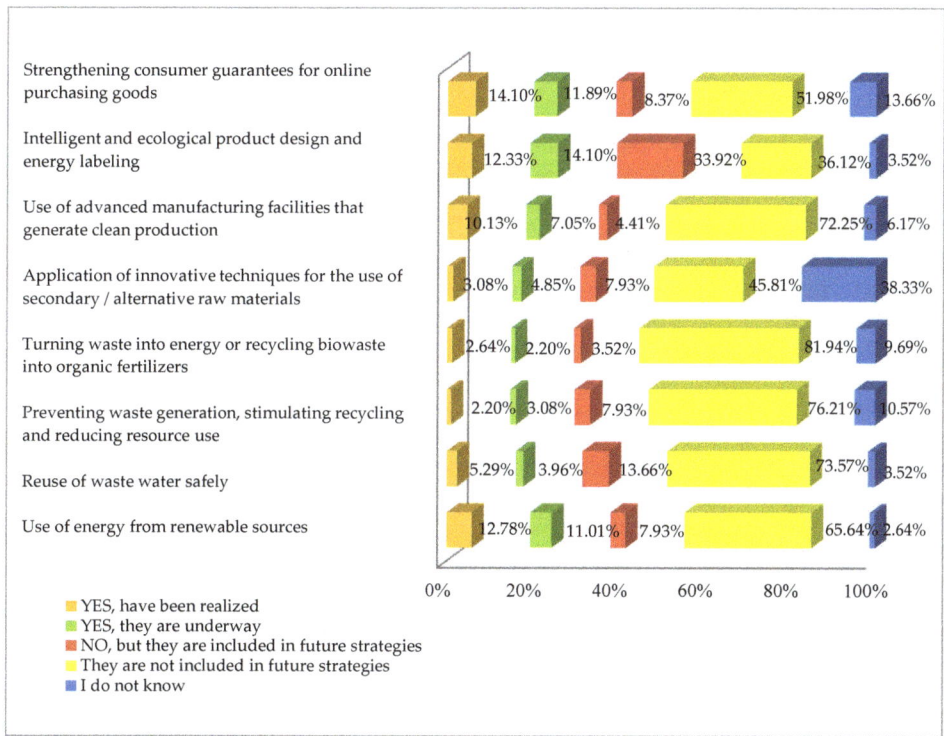

Figure 4. Description of activities related to the circular economy undertaken by Romanian SMEs in the last five years.

In order to identify the link between the three categories of enterprise and the activities related to the circular economy, a factorial analysis of correspondence (Figure 5) was used. The SMEs were grouped into three categories: micro-enterprises (0–9 employees), small enterprises (10–49 employees) and medium-sized enterprises (50–249 employees). Figure 5 shows the existence of certain associations between the three categories of enterprises and the activities related to the circular economy undertaken in the last five years.

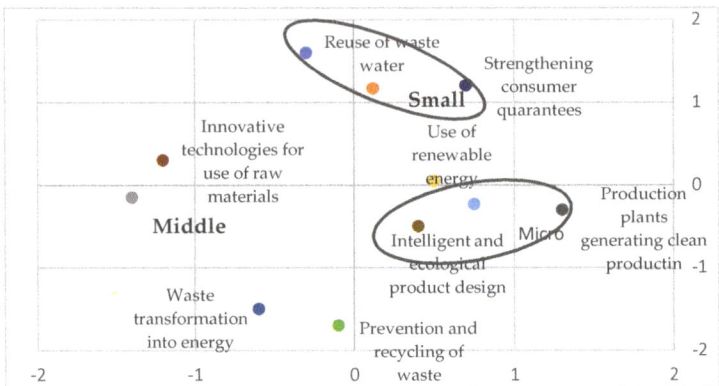

Figure 5. The correspondence between the three categories of enterprises and the activities related to the circular economy undertaken in the last five years.

A total of 219 of 343 micro-enterprise managers said that they already use advanced manufacturing facilities that generate clean production, carry out intelligent and environmentally-friendly product design, practice energy labeling, and use renewable energy constantly (see Table 1). In total, 5.29% of the 384 respondents claimed that they re-used waste water safely in the manufacturing process, the effects being reduced costs and reduced pressure on the resources used (see Figure 4). The two managers in the agricultural field claimed that water reuse contributes to the recycling of nutrients by replacing solid fertilizers. A further 22 small business managers claimed to have strengthened the guarantees offered to consumers who purchase goods online to provide better protection against defective products, thus contributing to sustainability and increased product repair potential. In this way, they claim they prevent the discarding of products and contribute significantly to the circular economy.

The managers of the SMEs surveyed said that they use innovative technologies that integrate into aspects relevant to the circular economy. At the level of their own businesses, the managers apply technologies to improve the use of secondary raw materials to increase energy efficiency and reduce wastewater generation, thereby helping to protect and reduce the use of available natural resources. Regarding waste reduction activities (recycling and reuse), most business managers have adopted sustainable and consistent waste management strategies. Some managers said they are trying to reduce the amount of waste by different methods: waste recycling, selling waste to certain specialized companies or re-using waste in the manufacturing process. Most respondents acknowledged that they do not carry out circular economy activities, but have planned future strategies based on concrete and measurable objectives. The second objective was to invest some percentage of the company's turnover in order to carry out activities related to the circular economy. Of the nearly 241 businesses that have developed circular economy activities over the past five years, most have invested an average of 1–5% of their turnover per year. Figure 6 shows that most of the investments were made by SMEs in the Bucharest-Ilfov and North-West regions. Almost 57.75% of the 241 enterprises that have carried out at least one activity related to the circular economy over the last five years have made investments of over 1%, while 31.69% of the SMEs have made no investments (Figure 6).

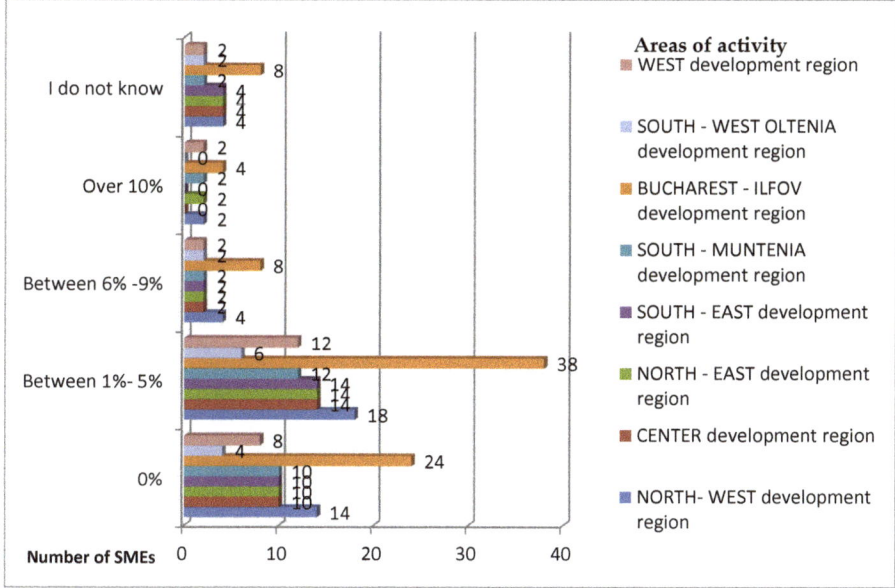

Figure 6. The distribution of SMEs that have carried out activities related to the circular economy in the eight development regions of the country, according to the share of investments made during the last five years.

In 10.56% (30 enterprises) of the SMEs participating in the research, managers did not keep a clear record of what percentage of their turnover they invested in circular economy activities over the past five years. Of the 158 enterprises that had not developed circular economy activities over the past five years, 30.38% would be willing to invest more than 1% of the turnover on average per year, 58.23% do not want to invest, and 11.39% did not know. Almost 34 enterprises invested between 1% and 5% of their turnover per year (21.52%), while 10 enterprises (6.33%) invested between 6% and 10% and 2.53% invested, on average, over 10% or more (Figure 6).

Figure 7 shows that 16 business managers in the Bucharest-Ilfov region who have not developed circular economy activities over the past five years, would be willing to invest more than 1% of their own turnover, on average, per year. Another 20 SME managers (North-West, Center, North-East, South-West Oltenia and West) said that they would be willing to invest part of the enterprise's turnover into circular economy activities over the next few years. Most activities undertaken by microenterprises that are related to the circular economy will be included in their future strategy; microenterprises represent 84.58% of all SMEs in Romania (Figure 7).

The third objective was related to the funding sources used by SMEs over the last five years to finance the activities related to the circular economy. Most of the SMEs surveyed funded their activities related to the circular economy from their own funds, i.e., turnover. Approximately 42.75% of the enterprises that have carried out at least one activity related to the circular economy over the last five years financed these types of activities from their own funds or from loans from close persons. Only 13.04% of enterprises used bank loans, while 10.51% benefited from government grants. Another 9.42% of the SMEs used various non-reimbursable funds from the EU, the EBRD, and the IMF, or had access to alternative sources of funding. Of the SMEs included in the survey, no enterprise had used a certain type of green technology investment (0%) for the circular economy activities undertaken over the last five years (Figure 8). Figure 8 shows that 125 enterprises of the 196 that used finances from their own funds come from two important sectors of the national economy: trade and services.

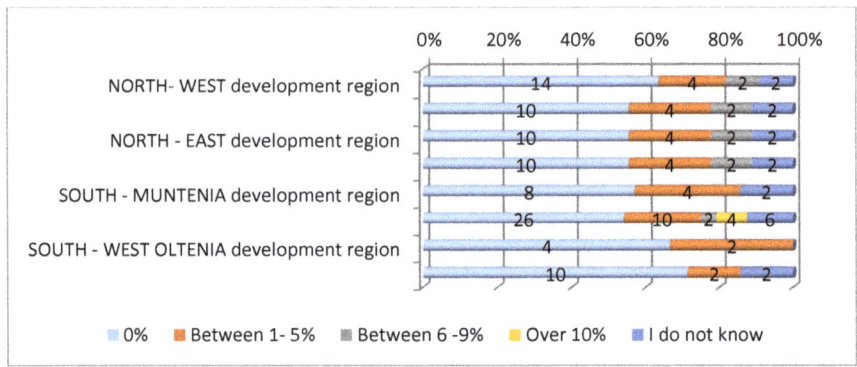

Figure 7. Distribution of SMEs not involved in circular economy activities in the eight development regions of the country, according to the share of future investments achievable.

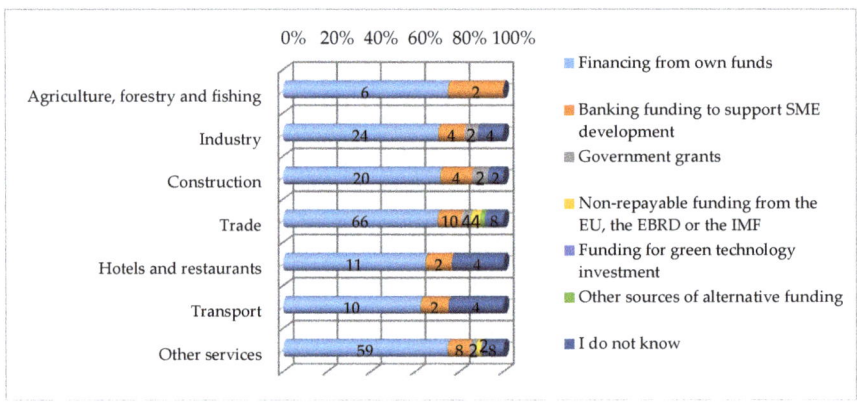

Figure 8. Distribution of SMEs according to the main fields of activity of the economy and by the way in which they finance activities related to the circular economy.

Of the 32 enterprises financed by bank loans, only eight come from industry and construction and six from agriculture, hotels and restaurants and transport. Non-reimbursable government funding benefited 10 enterprises from industry, construction, commerce and other services. Only six of the businesses analyzed, from trade and other services, used non-reimbursable grants from the EU, the EBRD or the IMF. Managers from 276 enterprises of the 384 surveyed who have been involved in the circular economy over the last five years specified the funding sources for these types of activity. For 42 of the 276 enterprises analyzed, managers did not want to indicate the sources of finance for the activities related to the circular economy undertaken over the last five years.

The last objective of the research was to identify the level of participation of SME representatives in courses to acquire new knowledge and skills regarding the implementation of resource efficiency, eco-innovation and circular economy activities and the determination of subjects of high interest for them (for example, the following courses are organized by the Chamber of Commerce and Industry, the National Center for Production and Sustainable Consumption Denkstatt Romania, the Ministry of Regional Development and Public Administration, the Ministry of European Funds and other public institutions and approved NGOs: Creative START, EU Ecolabel, START-UP Nation, GO Circular, capital markets and derivative financial instruments, etc.). Of the 384 SMEs surveyed, only 24.7% attended courses such as Geometric, 4th CSA, WaterWorks 2015, Synamera, Innovoucher, Columbus,

CoBioTech and others. Almost 64.8% of the SME managers did not attend courses, despite being aware of the running of governmental and European programs where free lectures are organized with the support of major institutions such as KPMG, Ecofys, CSR Netherlands and Circle Economy. Figure 9 shows that the managers who attended courses show an increased interest in topics such as EU-funded financial instruments to finance circular solutions (26.79%), participation in green public procurement (21.43%) and government programs to support SMEs related to circular actions (17.86%) (Figure 9).

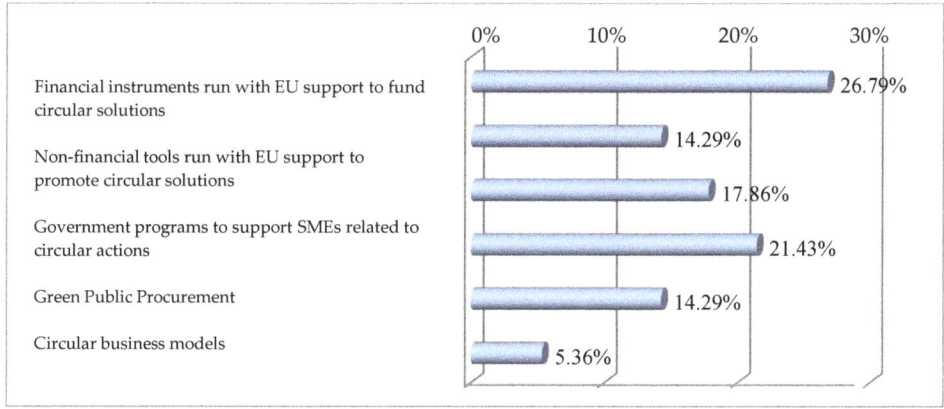

Figure 9. Assessing the interest of managers in the subjects treated in the specialization courses and in actions specific to the circular economy.

5. Conclusions

The results of this research highlighted the major contribution of SMEs to the development of a sustainable economy through their engagement in specific activities and through increasing the size of their investments. In the last five years, almost six out of ten Romanian enterprises (62.8%) engaged in activities specific to the circular economy [41]. The most frequent activities were the consolidation of guarantees for consumers who purchase goods online (14.10%), use of renewable energy (12.78%), smart and environmentally friendly product design and energy labeling (12.33%) and the use of advanced manufacturing facilities to achieve clean production (10.13%).

The research revealed that although more than half of the Romanian SMEs have undertaken at least one activity specific to the circular economy over the last five years, their level of involvement will remain moderate in the future. The main barriers to the development of a sustainable economy remain: (1) the low volume of future investments made by small- and medium-sized enterprises and micro-enterprises due to their small turnover; (2) the reduced rate of participation of business managers in non-reimbursable grant programs for circular actions and distinct SME programs that include courses necessary for the specialization and development of new circular business models.

Based on the review of the literature on the strategies, advantages and difficulties encountered in adopting the circular economy, a quantitative research study was carried out, including a wide range of areas of activity captured by SMEs, providing knowledge that can support successful actions for the implementation of the circular economy [60–62]. In Romania, according to the answers provided by the managers of the interviewed SMEs, the circular economy is seen as a significant strategic issue. As noted, circular economy activities in Romania's SMEs are still modest, and we believe that creating a fiscal, legal or organizational framework coupled with additional governmental actions to promote the principles of the circular economy would contribute to the successful implementation of the circular economy. Enhancing collaboration between micro, small- and medium-sized enterprises and providing support from large enterprises can also help to successfully implement the circular economy in Romania.

The size of the sample and the nature of the data used in this study did not allow for detailed research into the public or private sector, but this information is nonetheless very important and useful to policy-makers, professionals and economic agents in the business environment and academia. Research into the circular economy should be deepened at the national level.

Our suggestions for future research can be summarized as follows: (1) analyze how SMEs can access the human resources and technology needed to successfully adopt the circular economy; (2) determine the potential for internal and external optimization of the consumption of raw materials, water and energy; (3) analyze how SMEs can meet the needs of consumers, taking into account the principles of the circular economy; (4) analyze the effectiveness of the strategies aligned with the policies of the circular economy at the level of SMEs; (5) carry out a comparative analysis of the efficiency and the degree of successful implementation of the circular economy between different countries of the European Union.

Author Contributions: Conceptualization, I.O. and S.C.; Methodology, M.C.T. and D.I.T.; Validation, D.-M.O.C. and A.M.-P.; Formal Analysis, I.O. and M.S.H.; Writing—Original Draft Preparation, S.C. and M.C.T.; Writing—Review and Editing, I.O.

Funding: This research received no external funding.

Conflicts of Interest: The authors declare no conflict of interest.

Appendix A

Table A1. Sample structure.

Development Region	Number of Companies	Areas of Activity	Number of Companies	Number of Employees	Number of Companies	Weight in Total (%)	Sample Size
North-West development region	74,531	Agriculture, forestry and fishing	1625	0–9 people	1449	0.28	1
				10–49 people	150	0.03	0
				50–249 people	27	0.01	0
		Industry	8827	0–9 people	7866	1.51	6
				10–49 people	812	0.16	1
				50–249 people	148	0.03	0
		Construction	7157	0–9 people	6378	1.22	5
				10–49 people	659	0.13	0
				50–249 people	120	0.02	0
		Trade	29,052	0–9 people	25,891	4.96	19
				10–49 people	2674	0.51	2
				50–249 people	488	0.09	0
		Hotels and restaurants	3369	0–9 people	3003	0.57	2
				10–49 people	310	0.06	0
				50–249 people	57	0.01	0
		Transport	4548	0–9 people	4053	0.78	3
				10–49 people	419	0.08	0
				50–249 people	76	0.01	0
		Other services	19,951	0–9 people	17,780	3.40	13
				10–49 people	1836	0.35	1
				50–249 people	335	0.06	0

Table A1. *Cont.*

Development Region	Number of Companies	Areas of Activity	Number of Companies	Number of Employees	Number of Companies	Weight in Total (%)	Sample Size
Center development region	53,596	Agriculture, forestry and fishing	1169	0–9 people	1042	0.20	1
				10–49 people	108	0.02	0
				50–249 people	20	0.00	0
		Industry	6348	0–9 people	5657	1.08	4
				10–49 people	584	0.11	0
				50–249 people	107	0.02	0
		Construction	5147	0–9 people	4587	0.88	3
				10–49 people	474	0.09	0
				50–249 people	86	0.02	0
		Trade	20,892	0–9 people	18,618	3.56	14
				10–49 people	1923	0.37	1
				50–249 people	351	0.07	0
		Hotels and restaurants	2423	0–9 people	2159	0.41	2
				10–49 people	223	0.04	0
				50–249 people	41	0.01	0
		Transport	3271	0–9 people	2915	0.56	2
				10–49 people	301	0.06	0
				50–249 people	55	0.01	0
		Other services	14,347	0–9 people	12,785	2.45	9
				10–49 people	1321	0.25	1
				50–249 people	241	0.05	0
North-East development region	54,846	Agriculture, forestry and fishing	1196	0–9 people	1066	0.20	1
				10–49 people	110	0.02	0
				50–249 people	20	0.00	0
		Industry	6496	0–9 people	5789	1.11	4
				10–49 people	598	0.11	0
				50–249 people	109	0.02	0
		Construction	5267	0–9 people	4694	0.90	3
				10–49 people	485	0.09	0
				50–249 people	88	0.02	0
		Trade	21,379	0–9 people	19,052	3.65	14
				10–49 people	1968	0.38	1
				50–249 people	359	0.07	0
		Hotels and restaurants	2480	0–9 people	2210	0.42	2
				10–49 people	228	0.04	0
				50–249 people	42	0.01	0
		Transport	3347	0–9 people	2983	0.57	2
				10–49 people	308	0.06	0
				50–249 people	56	0.01	0
		Other services	14,681	0–9 people	13,084	2.50	10
				10–49 people	1351	0.26	1
				50–249 people	246	0.05	0

Table A1. Cont.

Development Region	Number of Companies	Areas of Activity	Number of Companies	Number of Employees	Number of Companies	Weight in Total (%)	Sample Size
South-East development region	52,057	Agriculture, forestry and fishing	1135	0–9 people	1012	0.19	1
				10–49 people	105	0.02	0
				50–249 people	19	0.00	0
		Industry	6165	0–9 people	5494	1.05	4
				10–49 people	567	0.11	0
				50–249 people	103	0.02	0
		Construction	4999	0–9 people	4455	0.85	3
				10–49 people	460	0.09	0
				50–249 people	84	0.02	0
		Trade	20,292	0–9 people	18,084	3.46	13
				10–49 people	1868	0.36	1
				50–249 people	341	0.07	0
		Hotels and restaurants	2353	0–9 people	2097	0.40	2
				10–49 people	217	0.04	0
				50–249 people	40	0.01	0
		Transport	3177	0–9 people	2831	0.54	2
				10–49 people	292	0.06	0
				50–249 people	53	0.01	0
		Other services	13,935	0–9 people	12,418	2.38	9
				10–49 people	1283	0.25	1
				50–249 people	234	0.04	0
South-Muntenia development region	50,624	Agriculture, forestry and fishing	1104	0–9 people	984	0.19	1
				10–49 people	102	0.02	0
				50–249 people	19	0.00	0
		Industry	5996	0–9 people	5343	1.02	4
				10–49 people	552	0.11	0
				50–249 people	101	0.02	0
		Construction	4862	0–9 people	4332	0.83	3
				10–49 people	447	0.09	0
				50–249 people	82	0.02	0
		Trade	19,733	0–9 people	17,586	3.37	13
				10–49 people	1816	0.35	1
				50–249 people	331	0.06	0
		Hotels and restaurants	2289	0–9 people	2040	0.39	1
				10–49 people	211	0.04	0
				50–249 people	38	0.01	0
		Transport	3089	0–9 people	2753	0.53	2
				10–49 people	284	0.05	0
				50–249 people	52	0.01	0
		Other services	13,551	0–9 people	12,077	2.31	9
				10–49 people	1247	0.24	1
				50–249 people	227	0.04	0

Table A1. *Cont.*

Development Region	Number of Companies	Areas of Activity	Number of Companies	Number of Employees	Number of Companies	Weight in Total (%)	Sample Size
Bucharest-Ilfov development region	147,210	Agriculture, forestry and fishing	3211	0–9 people	2861	0.55	2
				10–49 people	296	0.06	0
				50–249 people	54	0.01	0
		Industry	17,435	0–9 people	15,537	2.97	11
				10–49 people	1605	0.31	1
				50–249 people	293	0.06	0
		Construction	14,137	0–9 people	12,598	2.41	9
				10–49 people	1301	0.25	1
				50–249 people	237	0.05	0
		Trade	57,383	0–9 people	51,138	9.79	38
				10–49 people	5282	1.01	4
				50–249 people	963	0.18	1
		Hotels and restaurants	6655	0–9 people	5931	1.14	4
				10–49 people	613	0.12	0
				50–249 people	112	0.02	0
		Transport	8984	0–9 people	8006	1.53	6
				10–49 people	827	0.16	1
				50–249 people	151	0.03	0
		Other services	39,406	0–9 people	35,117	6.72	26
				10–49 people	3627	0.69	3
				50–249 people	661	0.13	0
South-West Oltenia development region	41,608	Agriculture, forestry and fishing	907	0–9 people	809	0.15	1
				10–49 people	84	0.02	0
				50–249 people	15	0.00	0
		Industry	4928	0–9 people	4392	0.84	3
				10–49 people	454	0.09	0
				50–249 people	83	0.02	0
		Construction	3996	0–9 people	3561	0.68	3
				10–49 people	368	0.07	0
				50–249 people	67	0.01	0
		Trade	16,219	0–9 people	14,454	2.77	11
				10–49 people	1493	0.29	1
				50–249 people	272	0.05	0
		Hotels and restaurants	1881	0–9 people	1676	0.32	1
				10–49 people	173	0.03	0
				50–249 people	32	0.01	0
		Transport	2539	0-9 people	2263	0.43	2
				10–49 people	234	0.04	0
				50–249 people	43	0.01	0
		Other services	11,138	0–9 people	9926	1.90	7
				10–49 people	1025	0.20	1
				50–249 people	187	0.04	0

Table A1. *Cont.*

Development Region	Number of Companies	Areas of Activity	Number of Companies	Number of Employees	Number of Companies	Weight in Total (%)	Sample Size
West development region	48,011	Agriculture, forestry and fishing	1047	0–9 people	933	0.18	1
				10–49 people	96	0.02	0
				50–249 people	18	0.00	0
		Industry	5686	0–9 people	5067	0.97	4
				10–49 people	523	0.10	0
				50–249 people	95	0.02	0
		Construction	4611	0–9 people	4109	0.79	3
				10–49 people	424	0.08	0
				50–249 people	77	0.01	0
		Trade	18,715	0–9 people	16,678	3.19	12
				10–49 people	1723	0.33	1
				50–249 people	314	0.06	0
		Hotels and restaurants	2171	0–9 people	1934	0.37	1
				10–49 people	200	0.04	0
				50–249 people	36	0.01	0
		Transport	2930	0–9 people	2611	0.50	2
				10–49 people	270	0.05	0
				50–249 people	49	0.01	0
		Other services	12,852	0–9 people	11,453	2.19	8
				10–49 people	1183	0.23	1
				50–249 people	216	0.04	0
Total	522,483		522,483		522,483	100.00	384

Source: Authors' calculation based on information extracted from the National Institute of Statistics of Romania [59].

References

1. Allwood, J.M. Squaring the circular economy: The role of recycling within a hierarchy of material management strategies. In *Handbook of Recycling State-of-the-Art for Practitioners, Analysts, and Scientists*; Worrel, E., Reuter, M.A., Eds.; Elsevier: Waltham, MA, USA, 2014; ISBN 978-0-12-396459-5.
2. Núñez-Cacho, P.; Molina-Moreno, V.; Corpas-Iglesias, F.A.; Cortés-García, F.J. Family businesses transitioning to a circular economy model: The case of "Mercadona". *Sustainability* **2018**, *10*, 538. [CrossRef]
3. Yuan, Z.; Bi, J.; Yuichi, M. The circular economy: A new development strategy in China. *J. Ind. Ecol.* **2014**, *10*, 4–8. [CrossRef]
4. Chamberlin, L.; Boks, C. Marketing approaches for a circular economy: Using design frameworks to interpret online communications. *Sustainability* **2018**, *10*, 2070. [CrossRef]
5. Velis, C.A.; Vrancken, K.C. Circular economy and global secondary material supply chains. *Waste Manag. Res.* **2015**, *33*, 389–391. [PubMed]
6. Scheepens, A.E.; Vogtländer, J.G.; Brezet, J.C. Two life cycle assessment (LCA) based methods to analyse and design complex (regional) circular economy systems. Case: Making water tourism more sustainable. *J. Clean. Prod.* **2015**, *114*, 257–268.
7. Molina-Moreno, V.; Leyva-Díaz, J.C.; Sánchez-Molina, J. Pelletasa technological nutrient within the circular economy model: Comparative analysis of combustion efficiency and CO and NOx emissions for pellets from olive and almond trees. *Energies* **2016**, *9*, 777. [CrossRef]
8. Biesbroek, G.R.; Klostermann, J.E.M.; Termeer, C.J.A.M.; Kabat, P. On the nature of barriers to climate change adaptation. *Reg. Environ. Chang.* **2013**, *13*, 1119–1129. [CrossRef]
9. Semenza, J.C. Climate Change and Human Health. *Int. J. Environ. Res. Public Health* **2014**, *11*, 7347–7353. [CrossRef] [PubMed]

10. Ziraba, A.K.; Haregu, T.N.; Mberu, B. A review and framework for understanding the potential impact of poor solid waste management on health in developing countries. *Arch. Public Health* **2016**, *74*, 55. [CrossRef] [PubMed]
11. Hashemi, H.; Pourzamani, H.; Rahmani Samani, B. Comprehensive planning for classification and disposal of solid waste at the industrial parks regarding health and environmental impacts. *J. Environ. Public Health* **2014**, *230*, 163. [CrossRef] [PubMed]
12. Li, L.; Li, Z.; Wu, G.; Li, X. Critical success factors for project planning and control in prefabrication housing production: A China study. *Sustainability* **2018**, *10*, 836. [CrossRef]
13. Olson, P.D.; Zuiker, V.; Danes, S.M.; Stafford, K.; Heck, R.; Duncan, K.A. The impact of the family and the business on family business sustainability. *J. Bus. Ventur.* **2003**, *18*, 639–666. [CrossRef]
14. Hadin, Å.; Hillman, K.; Eriksson, O. Prospects for increased energy recovery from horse manure—A case study of management practices, environmental impact and costs. *Energies* **2017**, *10*, 1935. [CrossRef]
15. Kovach, J.J.; Hora, M.; Manikas, A.; Patel, P.C. Firm performance in dynamic environments: The role of operational slack and operational scope. *J. Oper. Manag.* **2015**, *37*, 1–12. [CrossRef]
16. Meinel, U.; Schüle, R. The difficulty of climate change adaptation in manufacturing firms: Developing an action-theoretical perspective on the causality of adaptive inaction. *Sustainability* **2018**, *10*, 569. [CrossRef]
17. Liu, L. Evaluation of the sustainability of the industrial chain based on circular economy. *J. Shandong Jianzhu Univ.* **2012**, *27*, 88–91.
18. Gifford, R.; Kormos, C.; McIntyre, A. Behavioral dimensions of climate change: Drivers, responses, barriers, and interventions. *Wiley Interdiscip. Rev. Chang.* **2011**, *2*, 801–827. [CrossRef]
19. Lewandowski, M. Designing the business models for circular economy—Towards the conceptual framework. *Sustainability* **2016**, *8*, 43. [CrossRef]
20. Arushanyan, Y.; Björklund, A.; Eriksson, O.; Finnveden, G.; Söderman, M.L.; Sundqvist, J.-O.; Stenmarck, Å. Environmental assessment of possible future waste management scenarios. *Energies* **2017**, *10*, 247. [CrossRef]
21. Lazarevic, D.; Buclet, N.; Brandt, N. The application of life cycle thinking in the context of European waste policy. *J. Clean. Prod.* **2012**, *29*, 199–207. [CrossRef]
22. Montoya, F.; Peña-García, A.; Juaidi, A.; Manzano-Agugliaro, F. Indoor lighting techniques: An overview of evolution and new trends for energy saving. *Energy Build.* **2017**, *140*, 50–60. [CrossRef]
23. Jo, S.-H.; Lee, E.-B.; Pyo, K.-Y. Integrating a Procurement Management Process into Critical Chain Project Management (CCPM): A case-study on oil and gas projects, the piping process. *Sustainability* **2018**, *10*, 1817. [CrossRef]
24. Hoornweg, D.; Bhada, P. *What a Waste: A Global Review of Solid Waste Management*; Urban Development Series; Knowledge Papers No. 15; World Bank: Washington, DC, USA, 2012.
25. Bautista-Lazo, S.; Short, T. Introducing the all seeing eye of business: A model for understanding the nature, impact and potential uses of waste. *J. Clean. Prod.* **2013**, *40*, 141–150. [CrossRef]
26. Bicket, M.; Guilcher, S.; Hestin, M.; Hudson, C.; Razzini, P.; Tan, A.; Ten Brink, P.; van Dijl, E.; Vanner, R.; Watkins, E.; et al. *Scoping Study to Identify Potential Circular Economy Actions, Priority Sectors, Material Flows & Value Chains*; Study Prepared for the EU Commission, DG Environment; Publications Office of the European Union: Luxembourg, Luxembourg, 2014.
27. Ellen MacArthur Foundation. *Towards the Circular Economy. Opportunities for the Consumer Goods Sector*; Ellen MacArthur Foundation: Cowes, UK, 2013.
28. Heal, G.; Millner, A. Uncertainty and decision making in climate change economics. *Rev. Environ. Econ. Policy* **2014**, *8*, 120–137. [CrossRef]
29. Bocken, N.M.P.; Short, S.W.; Rana, P.; Evans, S. A literature and practice review to develop sustainable business model archetypes. *J. Clean. Prod.* **2014**, *65*, 42–56. [CrossRef]
30. Moreno, M.; De los Rios, C.; Rowe, Z.; Charnley, F. A Conceptual Framework for Circular Design. *Sustainability* **2016**, *8*, 937. [CrossRef]
31. Roos, G. Business model innovation to create and capture resource value in future circular material chains. *Resources* **2014**, *3*, 248–274. [CrossRef]
32. Wastling, T.; Charnley, F.; Moreno, M. Design for Circular Behaviour: Considering Users in a Circular Economy. *Sustainability* **2018**, *10*, 1743. [CrossRef]
33. Boons, F.; Lüdeke-Freund, F. Business models for sustainable innovation: State-of-the-art and steps towards a research agenda. *J. Clean. Prod.* **2013**, *45*, 9–19. [CrossRef]

34. Walker, S.; Coleman, N.; Hodgson, P.; Collins, N.; Brimacombe, L. Evaluating the Environmental Dimension of Material Efficiency Strategies Relating to the Circular Economy. *Sustainability* **2018**, *10*, 666. [CrossRef]
35. Govindan, K.; Soleimani, H.; Kannan, D. Reverse logistics and closed-loop supply chain: A comprehensive review to explore the future. *Eur. J. Oper. Res.* **2014**, *240*, 603–626. [CrossRef]
36. Oghazi, P.; Mostaghel, R. Circular business model challenges and lessons learned—An industrial perspective. *Sustainability* **2018**, *10*, 739. [CrossRef]
37. Talonen, T.; Hakkarainen, K. Elements of sustainable business models. *Int. J. Innov. Sci.* **2014**, *6*, 43–54. [CrossRef]
38. Jasch, C. How to perform an environmental management cost assessment in one day. *J. Clean. Prod.* **2006**, *14*, 1194–1213. [CrossRef]
39. Beuren, F.H.; Gomes Ferreira, M.G.; Cauchick Miguel, P.A. Product-service systems: A literature review on integrated products and services. *J. Clean. Prod.* **2013**, *47*, 222–231. [CrossRef]
40. European Commission. *Flash Eurobarometer 381-SMEs, Resource Efficiency and Green Markets*; European Commission: Brussels, Belgium, 2013. Available online: https://dbk.gesis.org/dbksearch/sdesc2.asp?no=5874&db=e (accessed on 22 March 2018).
41. Lakatos, E.S.; Cioca, L.-I.; Dan, V.; Ciomos, A.O.; Crisan, O.A.; Barsan, G. Studies and investigation about the attitude towards sustainable production, consumption and waste generation in line with circular economy in Romania. *Sustainability* **2018**, *10*, 865. [CrossRef]
42. Zamfir, A.-M.; Mocanu, C.; Grigorescu, A. Circular economy and decision models among European SMEs. *Sustainability* **2017**, *9*, 1507. [CrossRef]
43. Lakatos, E.S.; Dan, V.; Cioca, L.I.; Bacali, L.; Ciobanu, A.M. How supportive are romanian consumers of the circular economy concept: A survey. *Sustainability* **2016**, *8*, 789. [CrossRef]
44. Trianni, A.; Cango, E. Dealing with barriers to energy efficiency and SMEs: Some empirical evidences. *Energy* **2012**, *37*, 494–504. [CrossRef]
45. Grimmer, M.; Woolley, M. Green marketing messages and consumers' purchase intentions: Promoting personal versus environmental benefits. *J. Mark. Commun.* **2012**, *20*, 231–250. [CrossRef]
46. Lawrence, S.R.; Collins, E.; Pavlovich, K.; Arunachalam, M. Sustainability Practices of SMEs: The case of NZ. *Bus. Strategy Environ.* **2006**, *15*, 242–257. [CrossRef]
47. Stoeckl, N. The private costs and benefits of environmental self-regulation: Which firms have most to gain? *Bus. Strategy Environ.* **2004**, *13*, 135–155. [CrossRef]
48. Yacob, P.; Aziz, N.S.B.; bin Mohamad Makmur, M.F.; bin Mohd Zin, A.W. The policies and green practices of Malaysian SMEs. *Glob. Bus. Econ. Res. J.* **2013**, *2*, 52–74.
49. Rizos, V.; Behrens, A.; van der Gaast, W.; Hofman, E.; Ioannou, A.; Kafyeke, T.; Flamos, A.; Rinaldi, R.; Papadelis, S.; Hirschnitz-Garbers, M.; et al. Implementation of circular economy business models by small and medium-sized enterprises (SMEs): Barriers and enablers. *Sustainability* **2016**, *8*, 1212. [CrossRef]
50. Denafas, G.; Ruzgas, T.; Martuzevičius, D.; Shmarin, S.; Hoffmann, M.; Mykhaylenko, V.; Ogorodnik, S.; Romanov, M.; Neguliaeva, E.; Chusov, A.; et al. Seasonal variation of municipal solid waste generation and composition in four east European cities. *Resour. Conserv. Recycl.* **2014**, *89*, 22–30. [CrossRef]
51. Eltayeb, T.K.; Zailani, S. Going green through green supply initiatives towards environmental sustainability. *Oper. Supply Chain Manag.* **2009**, *2*, 93–110. [CrossRef]
52. Wooi, G.C.; Zailani, S. Green Supply Chain Initiatives: Investigation on the Barriers in the Context of SMEs in Malaysia. *Int. Bus. Manag.* **2010**, *4*, 20–27.
53. Studer, S.; Welford, R.; Hills, P. Engaging Hong Kong businesses in environmental change: Drivers and barriers. *Bus. Strategy Environ.* **2006**, *15*, 416–443. [CrossRef]
54. Velenturf, A.P.M.; Purnell, P.; Tregent, M.; Ferguson, J.; Holmes, A. Co-producing a vision and approach for the transition towards a circular economy: perspectives from government partners. *Sustainability* **2018**, *10*, 1401. [CrossRef]
55. Parker, C.M.; Redmond, J.; Simpson, M. A review of interventions to encourage SMEs to make environmental improvements. *Environ. Plan.* **2009**, *27*, 279–301. [CrossRef]
56. Liu, C.; Côté, R. A framework for integrating ecosystem services into China's Circular Economy: The case of eco-industrial parks. *Sustainability* **2017**, *9*, 1510. [CrossRef]

57. European Commission. *European Business Awards for the Environment Rewarding Eco-Innovation for Jobs and Growth*; European Commission: Brussels, Belgium, 2015. Available online: http://docplayer.net/16713447-European-commission-european-business-awards-for-the-environment-2014-2015-rewarding-eco-innovation-for-jobs-and-growth.html (accessed on 20 March 2018).
58. Study on Coherence of Waste Legislation, Study Prepared for the European Commission. Available online: http://ec.europa.eu/environment/waste/studies/index.htm (accessed on 20 March 2018).
59. National Institute of Statistics. Available online: http://statistici.insse.ro/shop/index.jsp?page=tempo3&lang=ro&ind=INT104B (accessed on 10 March 2018).
60. Ormazabal, M.; Prieto-Sandoval, V.; Puga-Leal, R.; Jaca, C. Circular Economy in Spanish SMEs: Challenges and opportunities. *J. Clean. Prod.* **2018**, *185*, 157–167. [CrossRef]
61. Ruggieri, A.; Braccini, A.M.; Poponi, S.; Mosconi, E.M. A meta-model of inter-organisational cooperation for the transition to a circular economy. *Sustainability* **2016**, *8*, 1153. [CrossRef]
62. Chen, J. Development of Chinese small and medium-sized enterprises. *J. Small Bus. Enterp. Dev.* **2006**, *13*, 140–147. [CrossRef]

© 2018 by the authors. Licensee MDPI, Basel, Switzerland. This article is an open access article distributed under the terms and conditions of the Creative Commons Attribution (CC BY) license (http://creativecommons.org/licenses/by/4.0/).

Article

Diffusion of Corporate Philanthropy in Social and Political Network Environments: Evidence from China

Wenqing Wu [1], Kexin Yu [1], Chien-Chi Chu [2,3,*], Jie Zhou [4,*], Hong Xu [4] and Sang-Bing Tsai [5,6,7,*]

1. College of Management and Economics, Tianjin University, Tianjin 300072, China; wenqingw@tju.edu.cn (W.W.); kionayu@163.com (K.Y.)
2. Department of Finance, Business School, Shantou University, Shantou 515063, China
3. Research Institute for Guangdong-Taiwan Business Cooperation, Shantou University, Shantou 515063, China
4. College of Tourism and Service Management, Nankai University, Tianjin 300071, China; xuhonhg@126.com
5. Zhongshan Institute, University of Electronic Science and Technology of China, Guangzhou 528400, China
6. China Academy of Corporate Governance, Nankai University, Tianjing 300071, China
7. College of Business Administration, Capital University of Economics and Business, Beijing 100070, China
* Correspondence: jqzhu@stu.edu.cn (C.-C.C.); zhoujie_1980@126.com (J.Z.); sangbing@hotmail.com (S.-B.T.)

Received: 5 April 2018; Accepted: 28 May 2018; Published: 6 June 2018

Abstract: Based on the strong influence social networks have on managerial decision-making, as an important aspect of the strategic decision of the company, it is necessary to study how corporate social responsibility (CSR) actions could be affected by social networks. An analysis of 1725 Chinese listed firms and 40,484 executives from 2010 to 2014 showed that corporate philanthropy behavior will diffuse in social networks; more concretely, the higher the degree of social network centrality, the higher the enterprise's donation level. Furthermore, the results also show that the role of social network centrality on corporate donation levels can be moderated by political connections. This study offers empirical evidence for developing a theoretical framework of CSR interaction and communication relevant to social networks, and offers insights into corporate philanthropy behavior based on social networks.

Keywords: corporate philanthropy; information transfer; political connections; social capital; social network; social network centrality

1. Introduction

Over the past several decades, an increasing amount of literature has focused on the influence of networks on corporate operations. Social networks have been used to explain a wide range of outcomes, including performance [1–4], firm policies [5–7], executive compensation [8], knowledge sharing [9,10], innovation [11], and ethical behavior [8,12]. Many scholars have studied the role of corporate philanthropy, an important aspect of a company's strategic decisions, on the enterprise itself and on society. From the perspective of the political relations, some scholars believe that corporate donations enhance a company's political connection with the government, thereby enhancing that company's competitiveness. For example, Hadani and Coombes [13] suggested that corporate donations may allow some firms to stand out from others when faced with political uncertainty. Porter and Kramer [14] found that a company's competitiveness can be improved by using corporate philanthropy as a tool. In addition, firms that are not politically well-connected were shown to benefit more from corporate donations, as gaining political resources is more critical for such firms [15]. On the other hand, some scholars doubt the positive relationship between nonmarket strategy such as corporate political activity and corporate performance. According to Mellahi, et al. [16], over one-third of the studies in their sample do not find positive performance effects of nonmarket strategies. In short, within the political and social network environment, corporate philanthropy is an important issue that needs further study. However, few of them observe the influence on corporate philanthropy from a network perspective, especially in developing countries.

Charitable donations are a strategic and salient component of corporate social performance [14], as well as an important aspect of corporate decision-making. We can analyze the relationship between social networks and corporate donations from two perspectives: information transfer and social capital theory [17]. On the one hand, the ways in which charitable donation behavior is affected by external information has commonly received attention from social psychology [18], economics [19], and other research fields. On the other hand, according to social capital theory [20], studies on the markets of developed countries reveal that social pressure and networks can increase a company's commercial philanthropy [21], or compel them to display "a greater sense of social responsibility" [22]. Therefore, it can be seen that the position of corporate executives in a social network will affect a company's donation level. In particular, the literature has confirmed that donations are affected by social networks from a personal perspective [23–25]. Thus, it is necessary to examine the important, yet understudied aspect of corporate social responsibility (CSR), that is, the relationship between executive social relations and corporate donation behavior.

The social network theory posits that individuals tend to change their preferences and decisions because of other people's actions [26]. In that way, the presence of social networks among senior executives can affect firms' corporate donation decisions [7]. As there is no literature on Chinese market data to study the relationship between the two, we investigate how social network links between pairs of firms influence their corporate donation decisions. In particular, we want to be the first to test whether executives who are socially connected make more similar decisions, which we call the "diffusion effect." To address the first component of the research issue, this study builds upon previous research on corporate philanthropy from an information transmission perspective. We use a matrix to represent the relationships between the listed firms, and then establish a two-stage firm pair model.

After confirming the diffusion effect of social networks [14,17], we investigate the overall effects of social network connections on corporate donations. In addition, we note that the firms' donation behavior is subject to pressure not only from social networks, but also from political networks, especially in developing countries [27]. Firms often build political connections to gain government support and favorable treatment that may effectively mitigate political uncertainties [15]. There are already studies proving that political connections can independently influence the level of corporate donations [27,28]. In this paper, we have chosen the executives' employment network to represent the social relationships between firms. Social networks can provide information and resources for executives, and ultimately affect executives' decision-making. From this perspective, we can see that both political connections and social networks can provide access to resources. Political connections also have social attributes: as Scott [29] mentioned in 2000, political associations and connections are also a kind of social network. Thus, political connections should be taken into consideration in our study, especially in China, where firms set a high value on them. Therefore, we believe that political connections have a moderate role in the impact of network connections on donation behavior.

To address the second component of the research question, an ordinary firm-level model is established to examine whether the position of firms in the social network influences donation behaviors. According to the analysis, we also added the interaction items of social network centrality and political connections to the framework to more thoroughly explore the impact mechanism of social networks.

In a nutshell, the goal of this paper is to analyze how social networks affect donation behavior. The core premise of this theoretical framework is that the research issue is a combination of two related problems: whether corporate charitable behavior diffuses through a social network; and if so, the means by which this is done. The results of an analysis of data on a sample of 1725 Chinese listed companies and 40,484 executives gathered between 2010 and 2014 suggest that the donation behavior of enterprises has a diffusion effect on social networks. Based on this, our results indicate that the position of a company in the social network is positively related to its donation level. We have also discovered that some political connections can reduce the effects of social network centrality on corporate donations.

The rest of the paper is organized as follows. The second section provides a review and discussion of the relevant literature, developing our theoretical arguments and hypotheses. The third section presents the methodology, including the sample data, variables, and models. The fourth section reveals the results of the analyses. Finally, the paper concludes with a discussion of the study's contributions, limitations, and implications.

2. Literature Review and Hypotheses

2.1. The Diffusion Effect of Social Networks on Corporate Donations

In the past, several studies have highlighted the role of social networks in the transfer of privileged information. For example, Cohen, Frazzini and Malloy [1] used social networks to identify information transfers in the stock market, and the results showed that social networks may be an important mechanism for how information flow affects asset prices. Cao, Dhaliwal, Li and Yang [3] found that independent directors with more social connections earn higher returns in their stock transitions than do others. Indeed, having additional internal and external information is beneficial to corporations in order to obtain key resources [30], reduce market transaction costs [31], and create competitive advantages [32]. Based on this, it can be concluded that external information has an impact on enterprise decision-making. It is also meaningful to understand how executives learn and imitate such information, because we are living through an unprecedented explosion of information today, and executives are vulnerable to being overwhelmed by too much complex information [26]. From the perspective of corporate donations, firms close to one another whose executives are affected by information from "neighboring firms" make similar decisions [33,34].

Recent studies also reveal that social information affects personal donations [24]; however, few studies have been conducted on this topic in the corporate world. It is important to explore the relationship between the two on the corporate level, as donation behavior is a typical decision-making behavior found within corporations. It is natural for corporations to receive donation behavior information from a "neighboring firm" through their social network, and this information, in turn, controls the company's donation levels by affecting its executives' behavior.

At the same time, scholars have highlighted that CSR's influence on companies' management and strategy is growing [35,36]. Thus, it is important to continue to explore the motivation of corporate donations. While there is extensive literature related to driving factors, such as firm value [37], moral motivations [38], and political motivations [27], an examination of the ways in which social interactions affect people's charitable behavior has not been fully developed [39]. Such studies mainly focus on psychological and cultural backgrounds, emphasizing the interaction and connection between people. Gautier and Pache [40] reviewed about 30 years' worth of academic research on corporate philanthropy, and found that there were only five articles that investigated the impact of executives' social networks on firm donations. So far, the existing literature has proved that social networks have a direct impact on CSR [17,41], but most of these studies are rather limited at the firm level, as well as in terms of case studies.

Based on these arguments, we address this gap by considering the diffusion effect of the social network on donation behavior within firm pairs. To compare differences between two companies' donation levels caused by a social network, we implemented the two-stage model used by Fracassi [26] in the first part of our research. If the expected impact exists, the firm pairs with social connections are bound to have similar donation amounts. On these grounds, our first hypothesis is as follows:

Hypothesis 1. *The donation behavior of corporations has a diffusion effect in their social networks; two corporations with social connections are more similar in terms of their donation levels.*

2.2. The Effect of Social Networks Centrality on Corporate Donations

As we know, social networks have always been fiercely debated upon in the literature. Additionally, concepts of social capital have frequently been considered in the literature [42–44]. Such studies mainly focus on the relationships between different kinds of social networks and the social capital embedded in them. Several scholars have also combined donation studies with related issues [45].

Social capital refers to the resources embedded in personal social networks, such as power, wealth, prestige, etc.; these resources exist in the relationships between people, and one must be connected with others to obtain them [20]. Witt [46] pointed out that network conditions can help enterprises integrate into the market and the local network, and adapt to local political, cultural, and economic conditions. Huggins [47] proposed that firms can gain access to knowledge through network capital to enhance expected economic returns. As a result, social contact can become an integral part of a business that cannot be separated from a firm's overall strategy. Therefore, as an entrepreneur's management decision, the influence of social relations should be taken into consideration when analyzing the level of corporate donation. As firms usually aspire to achieve a positive reputation and prominent status in their networks [17,48], they often aim to meet the expectations of the other members in the network by donating more.

In order to measure the degree of social connections of the corporations, we introduced a social network centrality variable. Social network centrality is a collection of measures that describe an individual's position in a social network [4], representing their ability to influence economic decision-making. Several common measures of centrality have been constructed in previous literature [49]. In this paper, we use the number of direct ties a corporation has with other firms in the network to measure the network centrality. In general, high centrality means higher status and power. This definition is similar to that provided by Lu, Shailer and Wilson [17], whose study also examined some kinds of donations. They used the number of directors' relationships with other corporations, finding that these networks influence corporate political donations. The above analysis indicates that enterprises in important positions need to meet the expectations of other members to a higher degree. Thus, our second hypothesis is as follows:

Hypothesis 2. *The higher the degree of a social network's centrality, the higher the enterprise's donation level.*

2.3. The Moderating Effect of Political Connections

Political connections are crosslinks between business and government; they are an important asset or political strategy [28] for the economy as a whole, as well as for individual businesses within a given country or emerging economy [50,51]. Some scholars believe that political connections provide access to information and other licenses for corporations [52], which can reduce the increasing donation behavior of executives aiming to obtain added resources and prestige. However, some scholars have come to the opposite conclusion [53], which can reduce the increasing donation behavior of executives aiming to obtain added resources and prestige. However, some scholars have come to the opposite conclusion [53]: they believe that these political connections will lead to an increase in corporate donations. The existence of these contradictory ideas leads us to consider the classification of political connections. Zhang, Marquis and Qiao [27] distinguished two types of managerial political connections—achieved political connections and ascribed bureaucratic connections—and found that the two react differently to government pressure to donate. Marquis and Qian [53] summed up that there are generally two types of political connections. Thus, we divided them into two categories according to the type of connection: Connections with People's Congress/Chinese People's Political Consultative Conference (NPC/CPPCC) members and connections with government officials [53]. The literature suggests that the former may have greater symbolic meaning and the latter greater substantive meaning.

First, for the NPC/CPPCC member connections, executives are generally chosen by the government as corporate elites. In other words, these types of political connections are generally obtained after company executives have attained certain achievements and reached a particular social status during their tenure. Therefore, executives with this background may pay more attention to their social reputation and public image, and tend to increase their donation level to achieve a positive public image [27]. This connection's influence on corporate donation is in accordance with social network connections. When companies obtain the resources they require through political connections, their motivation for obtaining resources through social connections will be reduced. In addition, it is thought that when the corporate donation level is increased due to influence from one of the two sides, the influence of the other side will be greatly reduced. Thus, we propose the following hypothesis:

Hypothesis 3A. *The higher the degree of the NPC/CPPCC member connections, the weaker the impact of social network centrality on the level of corporate donations.*

Second, contrary to the NPC/CPPCC member connections, executives with connections of government officials usually have access to the government prior to becoming involved with the enterprise, and thus, there is no need to maintain ties with the government [27]; even if executives have left the government, the listed companies can still be supported by the government, and the company does not have to rely on a commitment to social responsibility in order to receive "favors" from the government. Theoretically, previous government work experience has made such individuals very aware of government officials and activities [54], and stable ties secured by connections with government officials will reduce their eagerness to improve a corporation's value through donations. Compared to the former type of political connections, relationships with government officials reveal that the enterprise attaches great importance to the establishment of a political background from the very beginning, and will ensure the introduction of managers with political experience. Therefore, when they are influenced by donation information from other companies, they will be more active in social responsibility. Thus, we propose the following hypothesis:

Hypothesis 3B. *The higher the degree of connections with government officials, the stronger the impact of social network centrality on the level of corporate donations.*

3. Research Method

3.1. Sample and Data Collection

All companies listed in China's A share market before 2010 were included, with a sample interval of 2010 to 2014. The listed companies' Directors Council members and senior management personnel were used as the basis to build the company employment network; senior management personnel include the chairman, CEO, general manager, deputy general manager, financial manager, board secretary, and other personnel stipulated in the company's articles of association. Our initial sample contained 2044 listed companies. The samples were screened, and the following were removed: (1) financial companies and (2) special treatment companies. The reporting structure of China's financial sector is different from others, and so it is generally not compared with other sectors. ST companies usually refer to listed companies with abnormal financial or other situations. They are more likely to have false information disclosure problems, and are not suitable for comparison with other listed companies. After screening, the samples included 1725 listed companies and 40,484 senior managers and directors (hereinafter referred to as executives). The data is mainly from the CSMAR database (China Stock Market and Accounting Research Database), which is widely used in the study of listed companies in China. In addition, data not found in the CSMAR database were obtained by analyzing the annual reports of the listed companies, retrieving Sina Financial websites, and searching web pages.

Information on political connections came from the executives' resume projects on CSMAR; the biographical items are text information; therefore, we were able to manually sort out the executives' political connections. According to the actual conditions, we set up two variables to measure the political connections: political connections with government officials and political connections with NPC/CPPCC members.

3.2. Dependent Variable

We obtained the company donation data from the financial statements database within the CSMAR database. The database includes two items: donation expenditure and public welfare donations. If the public welfare donation is a detailed account of the donation expenditure, the donation expenditure data were regarded as the amount of the donation. If the two subjects were parallel, both were regarded as the amount of the donation. Furthermore, the donation data were recorded as zero when the donation expenditure data were zero, and it was regarded as missing data if the donation expenditure data were missing; the corresponding companies were deleted in the subsequent study. The logarithm of the donation amount as the index was used to measure the level of corporate donation (DON_LE$_{i,t}$). Since the independent variables in our study lag for one year, the sample interval of donation level data is from 2011 to 2015.

3.3. Independent Variables

The CSMAR database provides annual resume data for executives. For Hypothesis 1, we used annual employment information to establish a dummy variable at the firm pair level to measure the social connections between the two firms. In order to ensure the quality of the data, we distinguished between executives with the same names by giving each executive set a specific name. We first established a 40,484-by-40,484 undirected (symmetric) binary adjacency matrix according to the society connections between all the firm pairs. If two firms appointed one executive at the same time, or two executives from two firms worked for a third firm at the same time, it is believed that the two firms are socially related.

As this section studies the relationship between social connections and donation levels at the firm pair level, a social network dummy variable (NET) was established between firms based on the executive network. If there is a social connection between employees of the two companies, the value of the dummy variable is one; otherwise, it is zero.

In order to test Hypotheses 2 and 3, we established an index to measure the degree of social network centrality between the firms. After generating an undirected employment network matrix diagram for all the firm samples, the centrality of each node (firm) was used as an index to measure the degree of social relevancy. For the undirected matrix with n nodes, the degree of node 1 was the ratio of the sum of the direct connections between node i and other $N - 1$ nodes, divided by the total number of nodes in the network, expressed in Model (1). Firms with strong centrality often connect with many other firms, and have more opportunities to obtain resources and learn behavior from other firms through the network.

$$Degree_i = \frac{\sum_{j=1}^{N} NET_{ij}(i \neq j)}{N - 1} \quad (1)$$

3.4. Moderator Variable

To the above analysis, researchers added the interaction items of DEGREE and political connections to examine Hypotheses 3A and 3B. Political connections are a common phenomenon in enterprises at home and abroad. Broadly speaking, they can be regarded as a kind of social connection. Political connections were divided into two categories: connections with NPC/CPPCC members and connections with government officials. According to the resume information of executives, we manually sorted the government job information for executives.

The NPC/CPPCC member connections refer to the connections executives build with the government by being members of political councils, such as the National People's Congress (NPC), the only legislative body in China, or the Chinese People's Political Consultative Conference (CPPCC), an advisory board for the Chinese government. When constructing variables, in order to ensure that only the political elites were considered, we selected only national or provincial-level political connections. This variable is represented by REPRESENT_PC, and was calculated by the number of executives divided by the number of firm executives.

Connections with government officials refer to the connections between individuals who have been or are being employed by government departments and the government. In China, the national administration can be roughly divided into five levels: state, department, bureau, division, and section. If an executive has served or is serving at the division level or above, we believe there is a connection with a government official. The proportion of executives who have political connections in the listed companies is used to measure political connections with government officials, and the variable is represented by BUREAU_PC.

4. Empirical Results

4.1. The Empirical Model of Network Diffusion Effect on Firm Pair Level

We set up a two-stage firm pair model to examine whether two socially connected firms had more similar donation levels than those that did not have social connections. In the first stage of the model, the firm's donation level ($DON\_LE_{i,t}$) is used as the dependent variable, and the variables that influence the level of corporate donation are used as control variables ($CON_{i,t-1}$) for regression, as shown in Model (2). The residual of Model (2) is the part of the donation level that cannot be explained by the selected control variables: that is, as a measure of the impact of the social network on the donation level. Next, we define the absolute difference of residuals obtained by the two firms in Model (3) ($DON\_DS_{i,j,t}$) as the donation level difference of the two.

$$Don\_LE_{i,t} = \alpha_0 + \alpha_1 CON_{i,t-1} + \epsilon_{i,t} \qquad (2)$$

$$Don\_DS_{i,j,t} = abs(\epsilon_{i,t} - \epsilon_{j,t}) \qquad (3)$$

Through the literature review, we determined that we must add the following control variables in Model (2); all of these variables can theoretically affect the level of corporate donation, and their impact has been proven by empirical evidence. We measured the company size (FIRM_SIZE) by the logarithm of the total assets of the company at the end of the year. The slack cash (SLACK_CASH) is expressed by the total cash flow of the company divided by the total assets [15]. Advertising intensity (ADVE_INTE) is expressed in terms of the logarithm of annual sales expenses. We use the date of the sample year minus the company's IPO date to represent the company's IPO age as (AGE_IPO), which can control the impact of the stock market on the company [27]. We controlled for the size of executive boards by the number of executives in the company, denoted as NO_EXEC. Whether the chairman and the general manager (CEO) positions are held by the same person determines the concentration of management power [55]. If the company chairman and general manager (CEO) is the same person, the dummy variable DUALITY value is one; conversely, if the positions are held by different people, the value is zero. We use the return on assets (ROA) to measure company performance FIRM_PERF, and measure solvency LEVERAGE with the asset liability ratio.

Finally, we add the region-year and industry-year dummy to control the effects of industry and region on regression results. For industry, we use the China Securities Regulatory Commission Industry Classification Standards to classify the samples. For region, we divide the firms into three categories according to registered places. The Chinese regional economy is formed according to the long-term evolution of the different levels of economic development and geographical location. Mainland China can be divided into three major economic regions: eastern, central, and western.

In the second-stage model, the $DON\_DS_{i,j,t}$ calculated is the dependent variable, which measures the similarity between the donation levels of the companies. The employment network variable NET is the independent variable of the second-stage model, and it is used to measure whether there is a social connection between two companies. The second-stage model is shown as Model (4):

$$Don\_DS_{i,j,t} = \beta_0 + \beta_1 NET_{i,j,t-1} + \beta_2 C_{i,j,t-1} + \eta_{i,t} \tag{4}$$

The control variables ($C_{i,j,t-1}$) are also added to linear regression Model (4), where ABS_ASEET represents the difference in total assets between the two firms; the variable is used to control the impact of the firm size on the regression results. ABS_AGE_EXEC represents the difference between the average ages of the executives in the two firms. ABS_NO_EXEC and ABS_WOMEN represent the difference in the number of executives and the proportion of female executives, respectively. We also added NO_EXEC and AGE_EXEC, which are the sum of the executives and the average age of the executives in the two firms, respectively.

In addition to the full sample regression, we also conducted regression analysis on samples from the same industries/regions as well as different industries/regions.

4.2. The Empirical Model of Network Effect on Firm Level

In the last section, we analyze whether the existence of social networks can affect corporate donation behavior. However, we cannot judge whether the impact of the network on corporate donation levels is positive or negative through the results. Therefore, this section will analyze the impact of social network centrality on corporate donation. Simultaneously, we will add political connection variables into the model to study their impact on corporate donations, and further explore the effect of the interaction terms between them and the social network centrality.

In this section, we establish linear regression Model (5) to test Hypothesis 2. Consistent with the first-stage model in the previous section, corporate donation level (the logarithm of donation amount) is used as a dependent variable, and explanatory variables and control variables are added to the regression to examine the impact of social network centrality on corporate donation behavior, as shown in Model (5):

$$Don\_LE_{it} = \beta_0 + \beta_1 DEGREE_{i,t-1} + \beta_2 C_{i,t-1} + \eta_{i,t} \tag{5}$$

In Models (6) and (7), we continue to add moderator variables based on Model (5) to test the impact of political connections on the relationship between ownership concentration and corporate donation.

$$Don\_LE_{it} = \beta_0 + \beta_1 DEGREE_{i,t-1} + \beta_2 REPRESENT\_PC_{i,t-1} + \beta_3 C_{i,t-1} + \eta_{i,t} \tag{6}$$

$$Don\_LE_{it} = \beta_0 + \beta_1 DEGREE_{i,t-1} + \beta_2 BUREAU\_PC_{i,t-1} + \beta_3 C_{i,t-1} + \eta_{i,t} \tag{7}$$

Models (8) and (9) are based on Models (6) and (7), with the addition of the interaction items of the moderator variables and the dependent variable, to test whether they affect the relationship between social network centrality and the donation level, and draw more conclusions about how social network centrality affect the corporate donation level.

$$\begin{aligned}Don\_LE_{it} = {} & \beta_0 + \beta_1 DEGREE_{i,t-1} + \beta_2 BUREAU\_PC_{i,t-1} \\ & + \beta_5 DEGREE_{i,t-1}*BUREAU_{i,t-1} + \beta_7 C_{i,t-1} + \eta_{i,t}\end{aligned} \tag{8}$$

$$\begin{aligned}Don\_LE_{it} = {} & \beta_0 + \beta_1 DEGREE_{i,t-1} + \beta_2 REPRESENT\_PC_{i,t-1} \\ & + \beta_6 DEGREE_{i,t-1}*REPRESENT\_PC_{i,t-1} + \beta_7 C_{i,t-1} + \eta_{i,t}\end{aligned} \tag{9}$$

Similar to the last section, the main control variables that affect the level of corporate donation are determined, which include the firm size, slack cash, advertising intensity, IPO age, duality, corporate performance, solvency, and the number of executives. We also added dummy variables to reduce the impact of year, region, and industry on regression results. In addition, according to the estimated results of the second-stage model in the last section, the proportion of female executives also affects the donation level; therefore, we added the proportion of female executives in the control variables (WOMEN).

The summary statistics of the variables are shown in Tables 1 and 2:

Table 1. Summary statistics of firm level variables.

Variables	Mean	Std. Dev	No. of. Obs
DON_LE	10.770	4.838	6581
FIRM_SIZE	22.106	1.302	6581
SLACK_CASH	0.040	0.083	6581
ADVE_INTE	17.868	3.038	6581
AGE_IPO	10.220	6.073	6581
FIRM_PERF	0.042	0.074	6581
NO_EXCE	7.762	1.287	6581
DUALITY	0.137	0.3435	6581
LEVERAGE	0.489	1.548	6581

Table 2. Summary statistics of firm pair level variables.

Variables	Mean	Std. Dev	No. of. Obs
NET	0.012	0.109	4,697,720
NET (within industry)	0.012	0.108	1,675,248
NET (cross industry)	0.012	0.110	3,022,472
NET (within region)	0.014	0.118	2,355,591
NET (across region)	0.010	0.099	2,342,129
Same industry	0.357	0.479	4,697,720
Same region	0.501	0.500	4,697,720
NO_EXEC	40.721	7.164	4,697,720
AGE_EXEC	48.663	2.254	4,697,720
ABS_ASSET	2.11×10^{10}	1.03×10^{11}	4,697,720
ABS_NO_EXEC	5.322	4.510	4,697,720
ABS_WOMEN	0.114	0.0882	4,697,720
ABS_AGE_EXEC	3.451	2.631	4,697,720

4.3. The Empirical Results of Network Diffusion Effect

The regression results of the first-stage model are shown in Table 3. The coefficients and significance results are basically consistent with previous scholars' conclusions. Table 4 lists the regression results of the second-stage models, where the dependent variable is the difference in the level of donation between companies ($DON\_DS_{i,j,t}$). Our main focus in this regression is the coefficient of NET. As reflected in Hypothesis 1, we believe that corporate donations can be spread through a social network existing between executives. When such a social network between two companies is small, the disparity in donation levels is much smaller. Therefore, the employment network variables (NET) that represent social networks should have a significant negative coefficient. If the results indicate otherwise, this means that our hypothesis is invalid. It is worth mentioning that this study cannot draw the direction of network impact on the donation level; the purpose of the analysis is to prove the impact of the network on corporate donations.

Table 3. Network diffusion effect results: first-stage regression.

	DON_LE
FIRM_SIZE	0.976 ***
	(10.84)
SLACK_CASH	2.453 ***
	(2.95)
ADVE_INTE	0.142 ***
	(4.44)
AGE_IPO	−0.064 ***
	(−4.66)
FIRM_PERF	5.433 ***
	(3.71)
NO_EXCE	0.248 ***
	(2.76)
DUALITY	0.208(1.05)
LEVERAGE	0.076 ***
	(2.71)
Industry-year fe	yes
Region-year fe	yes
R2-sq	0.143
No. of obs.	6581

The dependent variable is the companies' donation level. *** indicate significance at the 10, 5, and 1 percent levels, respectively. The OLS (We analyze our questions by ordinary least squares regression, hereinafter referred to as OLS.) coefficients are reported, with the t-statistics in parentheses. Standard errors are corrected for by clustering the error term at the firm level (Petersen 2009). The constant is omitted.

In Model (1) in Table 4, only the explanatory variable NET was added. In theory, the second-stage regression no longer needs control variables, because the first-stage regression already controls for industry, region, year, size, and other necessary factors. The coefficient is significant at the 1 percent level, and its value is −0.2005498, indicating that two enterprises with social connections are more similar on the donation level. Corporate donation information can be spread through the social network between executives, thereby confirming Hypothesis 1. In Model (2), some control variables were added based on Model (1). First of all, the size and average age of the executives for each firm pair (NO_EXEC and AGE_EXEC) was controlled. These two variables were negatively correlated with the dependent variable. Additionally, the relationship between the NO_EXEC and the dependent variable is not significant, while the AGE_EXEC coefficient is negative and significant, which proves that the aging management team is more stable in terms of donation behavior [26]. Furthermore, the results revealed that the proportion of female executives will affect the similarity of the two donation levels; ABS_WOMEN is the female ratio dissimilarity of the number of executives between firm pairs. The size control variables (asset and executive number) are not significant; however, this does not prove that the impact of size on the level of donation is not significant, as the first-stage model has controlled the firm size. A dummy year was also added to the models.

Furthermore, from the descriptive statistics of Table 2, social connections are more common in the same region. In the first stage of the model, we have controlled the industries and regions, but in order to control the possible heteroscedasticity, two dummy variables (Same industry and Same region) were added in the second stage of the model. Hence, we assume that when the two companies are in the same industry or region, the dummy variable value is 1; otherwise, the dummy variable value is 0. In Table 2, regarding Model (2) we simultaneously added Same industry and Same region dummy variables. Table 4 Model (2) shows that the Same industry and Same region coefficients were negative and correlated significantly with the dependent variables in the full sample. Models (3)–(6) regress the same industry/region models and cross industry/region models, except for Model (3). The conclusions, however, still significantly support Hypothesis 1.

In the second-stage regression, the R2-sq is low, but this is not necessarily caused by the insufficient explanatory power of independent variables. On the one hand, our sample size is very large, which may affect the goodness of fit of the model; on the other hand, this once again reminds us that there are many factors that influence the relationship between social relationships and behavior. It is a field that warrants further study.

Table 4. Network diffusion effect results: second-stage regression.

	(1)	(2)	(3) Same Industry	(4) Cross Industry	(5) Same Region	(6) Cross Region
NET	−0.201 ***	−0.142 ***	−0.042	−0.196 ***	−0.134 **	−0.161 **
	(−3.65)	(−2.70)	(−2.75)	(−3.50)	(−2.36)	(−2.35)
Same industry		−0.250 ***			−0.317 ***	−0.177 **
		(−3.48)			(−3.85)	(−2.22)
Same region		−0.182 ***	−0.285 **	−0.122 **		
		(−3.04)	(−3.42)	(−1.99)		
NO_EXEC		−0.007	−0.003	−0.008	0.006	−0.017 ***
		(−1.07)	(−0.38)	(−1.29)	(−0.83)	(−2.68)
AGE_EXEC		−0.062 ***	−0.037	−0.075 ***	−0.077 ***	−0.048 **
		(−3.07)	(−1.37)	(−3.55)	(−3.42)	(−2.19)
ABS_ASSET		-5.39×10^{-14}	-2.92×10^{-14}	-1.89×10^{-14}	-1.18×10^{-13}	-3.26×10^{-14}
		(−0.15)	(−0.03)	(−0.05)	(−0.31)	(−0.10)
ABS_NO_EXEC		$-0.6.93 \times 10^{-4}$	−0.011	0.004	0.005	−0.006
		(−0.10)	(−1.12)	(0.57)	(0.61)	(0.87)
ABS_WOMEN		0.519 *	0.547	0.515	0.493	0.553
		(1.74)	(1.32)	(1.62)	(1.41)	(1.63)
ABS_AGE_EXEC		0.010	0.018	0.006	0.010	0.011
		(0.91)	(1.19)	(0.49)	(0.79)	(0.95)
Year FE	no	Yes	Yes	Yes	Yes	Yes
R2-sq	0.000024	0.003	0.003	0.002	0.003	0.003
No. of obs	4697720	4697720	1675248	3022472	2355591	2342129

The dependent variable is the difference of the donation level between the companies. *, **, and *** indicate significance at the 10, 5, and 1 percent levels, respectively. The OLS coefficients are reported, with the t-statistics in parentheses. Standard errors are corrected for by clustering the error term at the firm level. The constant is omitted.

4.4. The Empirical Results of the Network Centrality Effect

The results of Table 5 verify Hypotheses 2 and 3A. For Hypothesis 2, the coefficient of network centrality DEGREE is significantly positive (for models (1) to (5): $\beta > 0$, $p < 5\%$). This indicates that the higher the centrality of social network, the higher the donation level of the enterprise. Overall, we find strong evidence that companies that are more centrally positioned in a network have higher donation levels, suggesting that social connections not only affect the similarity of firms' donation behavior, bu also influence their total donations. Model (2) confirms that there is a significant positive correlation between the NPC/CPPCC connections and the corporate donation levels. Model (4) contains the interaction item of NPC/CPPCC member connections and social network centrality, and the coefficient of the item is negative ($\beta = -0.7385246$, $p < 1\%$). It is concluded that the political connections of NPC/CPPCC members will buffer the network motivation of corporate donations. Thus, in China, a listed company with higher network centrality will provide more donations, but if the company has high NPC/CPPCC member connections at the same time, the positive impact of the political background on donations will be reduced.

For Hypothesis 3B, the results of Table 5 show that although connections with government officials are negatively correlated with the level of corporate donations, the coefficient is not significant. Hypothesis 3B is reversed; however, this does not prove that there is no relationship between the two. Zhang, Marquis and Qiao [27] only considered the political connections of the CEO; their research identified a significant negative correlation between connections with government officials and corporate donation levels in private listed companies in China. The difference indicates that there may be a more complex relationship between connections with government officials and donation levels. This remains to be further studied.

Table 5. Network effect on firm level.

	(1)	(2)	(3)	(4)	(5)
DEGREE	0.0193 ***	0.019 ***	0.019 ***	0.027 ***	0.015 **
	(3.29)	(3.20)	(3.27)	(4.09)	(2.02)
FIRM_SIZE	1.241 ***	1.229 ***	1.248 ***	1.228 ***	1.248 ***
	(16.42)	(16.29)	(16.47)	(16.27)	(16.47)
SLACK_CASH	2.984 ***	2.919 ***	2.985 ***	2.942 ***	2.979 ***
	(3.64)	(3.57)	(3.64)	(3.61)	(3.63)
ADVE_INTE	0.166 ***	0.162 ***	0.166 ***	0.163 ***	0.165 ***
	(5.12)	(5.01)	(5.12)	(5.06)	(5.11)
AGE_IPO	−0.063 ***	−0.060 ***	−0.063 ***	−0.059 ***	−0.062 ***
	(−4.58)	(−4.31)	(−4.53)	(−4.28)	(−4.53)
FIRM_PERF	5.020 ***	4.94 ***	5.000 ***	4.973 ***	4.988 ***
	(3.41)	(3.38)	(3.40)	(3.41)	(3.38)
NO_EXCE	-7.66×10^{-6} **	-8.06×10^{-6} **	-7.30×10^{-6} *	-8.15×10^{-6} **	-7.28×10^{-6} *
	(−2.01)	(−2.12)	(−1.92)	(−2.11)	(−1.92)
DUALITY	0.174	0.178	0.175	0.181	0.174
	(0.89)	(0.91)	(0.90)	(0.93)	(0.93)
LEVERAGE	0.113 ***	0.112 ***	0.115 ***	0.112 ***	0.115 ***
	(4.17)	(4.20)	(4.20)	(4.21)	(4.20)
WOMEN	2.26 ***	2.212 ***	2.230 ***	2.188 ***	2.236 ***
	(3.08)	(3.01)	(3.03)	(2.98)	(3.04)
REPRESENT_PC		6.573 ***		5.708 ***	
		(3.48)		(3.04)	
BUREAU_PC			−1.330		−1.234
			(−1.06)		(−0.98)
DEGREE * REPRESENT_PC				−0.739 ***	
				(−3.06)	
DEGREE * BUREAU_PC					0.0760
					(0.79)
Industry fe	Yes	Yes	Yes	Yes	Yes
Region fe	Yes	Yes	Yes	Yes	Yes
Year fe	Yes	Yes	Yes	Yes	Yes
R2-sq	0.139	0.141	0.139	0.142	0.139
No. of obs	6581	6581	6581	6581	6581

The dependent variable is the companies' donation level. *, **, and *** indicate significance at the 10, 5, and 1 percent levels, respectively. The OLS coefficients are reported, with the *t*-statistics in parentheses. Standard errors are corrected for by clustering the error term at the firm level. Constant is omitted.

5. Discussion and Conclusions

Hypothesis 1 in this paper argues that firm pairs with social connections are more similar in their donation level than those without social connections. Resume information was collected from the executives from listed firms in China to establish a matrix of connections between them and demonstrate that the linkages cause their donation behaviors to be more similar. The results from our analysis confirm this hypothesis. Since the Wenchuan earthquake in 2008, charitable donations have become a common method used by Chinese corporations to fulfill their social responsibilities. The level of corporate donations has been a significant concern among the public and stakeholders alike [56]. The externality of the donation decision-making in China deserves serious research. However, the related research mainly focuses on the plight of culture and system [36,57] and provides suggestions on countermeasures [58,59]. Thus, this paper further proves that donation decision-making is affected by social networks, making policy-making externalities seem sensible.

Moreover, Hypothesis 2 argues that there is a positive relationship between social network centrality and the corporate donation level. This hypothesis is confirmed as well. Previous research has concluded that the individual pursuit of prestige and status in society can encourage executives to make decisions in accordance with group ethics. Under these circumstances, executives will try to match the expectations of the network members to their own behavior [17,25]. Based on this, we also tested the relationship between corporate donations and the social network centrality, and proved that the higher the level of social network centrality, the higher the social status, and the higher the level of donation. This finding is in line with another study that examined the impact of director influences on political party donation activity from a network perspective [17]. Similar to our research, the researchers suggested a positive influence of managers' and directors' social network centrality on the corporate political donations in their corporations.

Next, Hypotheses 3A and 3B state that the relationships between social network centrality and donation level will be affected by political connections. In other words, we want to examine the impact of the combination of social networks and political networks on corporate donations. The related issues have not been discussed in previous literature. However, previous studies have shown that the government can shape corporate philanthropy through political connections [15,60]. All corporations need to adopt appropriate political strategies to cope with the external environment, and political connections are one of the most important representatives. Our findings partially confirm this hypothesis: we have only found a significant relationship between the NPC/CPPCC member connections and the effect of social network centrality. Therefore, the effects of connections with government officials still require further research.

5.1. Theoretical and Practical Implications

Our study contributes to the literature associated with corporate donation, social networks, and political connections in several ways. To start, we examine the diffusion effect of corporate donations through a social network's perspective. That verifies some of the views on the externality of decision-making, which are widespread in society [26]. This paper notes that a firm's donation decisions may be affected by their peers through the social network, providing evidence that decision externalities could also play an important role in corporate donation behavior. The conclusions not only broaden the existing research on charitable donation motivation, but also provide significant references for the government and related institutions to utilize in the formulation of social responsibility management systems. Furthermore, our results complement studies that suggest that corporate executives actively commit to proper social responsibility to enhance their reputation [61]. Companies in the center of a network will receive more attention from their peers, and therefore, may choose to donate in the belief that philanthropy can improve the company's reputation and status within their network. In particular, we examine the impact of social networks under different political connections; which highlights the importance of the institutional context of CSR studies. The institutional context is especially important in transition countries (e.g., China), as social economic activities can be affected. This leads to the final point: as far as we know, research on the relationship between social networks and corporate donation policies based on the latest data from the Chinese market has not been performed to date. This study is the first attempt in this field; thus, we are able to contribute to theory development while introducing academic audiences to this increasingly relevant domain. We believe that researchers cannot ignore the importance of social and political network in influencing corporate contributions and other corporate public service activities. We can see that social networks, political connections, and donation-related issues are attracting more and more attention in other countries' markets [26,57,62,63]. Prior literature has focused on the sharing of information between firms' management [3], as well as the influence of network resources on people's behavior. Thus, we have reason to believe that our results are interesting enough to motivate further inquiry and research on these issues.

To the best of our knowledge, no prior study has explored the diffusion effect of social networks on corporate donations. Our research may alert firm executives to the significant effect of social networks on donation behavior. In general, executives should realize that different firms may be in different network positions to take advantage of social networks. Such awareness may help to explain the unknown motivation factors of CSR. Then, as both political and corporate networks have significant roles in promoting donations, the government and related agencies should pay attention to the coordination and promotion of political and network conditions while deploying charitable activities. According to the empirical analysis conducted for this article, for the companies with the NPC/CPPCC connections, the government should pay attention to reducing the mutual influence between the social relations and the NPC/CPPCC connections in the process of promoting a company's philanthropy. In addition, highly interactive platforms for charity should be regarded as a current focus. According to our results, if the government can set up a more convenient donation strategy and a more transparent financial plan for corporations, the spread of philanthropy may be expanded. Social philanthropy would then experience a positive change, and corporations' influence would be maximized.

5.2. Limitations and Future Research

This study has several limitations. First, we have only explored one dimension of CSR: corporate donation amounts. We are not aware of whether or not a social network would have similar effects on other corporate socially responsible actions. Future studies could explore the effects of the network on other CSR activities.

Second, this study only confirms the impact of social networks on corporate donations. Nowadays, there are many studies on charitable motives. Although this provides a lot of reference for practice and theory in related fields, it also makes the knowledge framework too broad in scope. Future research can dig deeper into the association mechanisms between different motivations. Based on the certain accumulation of research, we can revise and perfect the existing theoretical model, establish new models that are integrated with other motivations, and promote the development of related research.

Finally, nowadays there is a growing interest in social relationships in academia [3,64]. This is because the relationship is a very important element in the governance of firms. With the rise of stakeholder management theory [65,66], it is necessary for companies to use a variety of network relationships to obtain resources needed for development. Thus, the influence of social networks on charity is related to the interaction between enterprises and their stakeholders. Therefore, we account for a measure of the external interaction taking the degree of political connections. In this case, another question that should be raised is whether other variables affect the relationship between network centrality and donations.

Author Contributions: W.W. and S.-B.T. conceived and designed the experiments; K.Y. performed the experiments; W.W. and K.Y. analyzed the data; W.W. and S.-B.T. contributed reagents/materials/analysis tools; W.W. and K.Y. wrote the paper. W.W., C.-C.C., J.Z., H.X. revised the paper. All authors read and approved the manuscript.

Acknowledgments: This paper is supported by the National Social Science Foundation of China (Grant No. 17BGL025), National Natural Science Foundation of China (No. 71672089), Zhongshan City Science and Technology Bureau Project (No. 2017B1015) ,Shantou University Innovative and Strong School Project (No. 2015WQNCX031), National Natural Science Foundation of China (Grant No. 71602105), Natural Science Foundation of Guangdong Province of China (Grant No. 2016A030313073)and the Institute of Guangdong and Taiwan of Shantou University for their financial support.

Conflicts of Interest: The authors declare no conflict of interest.

References

1. Cohen, L.; Frazzini, A.; Malloy, C. The small world of investing: Board connections and mutual fund returns. *J. Political Econ.* **2008**, *116*, 951–979. [CrossRef]
2. Cai, Y.; Sevilir, M. Board connections and m&a transactions. *J. Financ. Econ.* **2012**, *103*, 327–349. [CrossRef]
3. Cao, Y.; Dhaliwal, D.; Li, Z.Q.; Yang, Y.G. Are all independent directors equally informed? Evidence based on their trading returns and social networks. *Manag. Sci.* **2015**, *61*, 795–813. [CrossRef]
4. El-Khatib, R.; Fogel, K.; Jandik, T. Ceo network centrality and merger performance. *J. Financ. Econ.* **2015**, *116*, 349–382. [CrossRef]
5. Fracassi, C.; Tate, G. External networking and internal firm governance. *J. Financ.* **2012**, *67*, 153–194. [CrossRef]
6. Wong, L.H.H.; Gygax, A.F.; Wang, P. Board interlocking network and the design of executive compensation packages. *Soc. Netw.* **2015**, *41*, 85–100. [CrossRef]
7. Shue, K. Executive networks and firm policies: Evidence from the random assignment of mba peers. *Rev. Financ. Stud.* **2013**, *26*, 1401–1442. [CrossRef]

8. Butler, A.W.; Gurun, U.G. Educational networks, mutual fund voting patterns, and ceo compensation. *Rev. Financ. Stud.* **2012**, *25*, 2533–2562. [CrossRef]
9. Leonardi, P.M. Social media, knowledge sharing, and innovation: Toward a theory of communication visibility. *Inf. Syst. Res.* **2014**, *25*, 796–816. [CrossRef]
10. Tortoriello, M.; Reagans, R.; McEvily, B. Bridging the knowledge gap: The influence of strong ties, network cohesion, and network range on the transfer of knowledge between organizational units. *Org. Sci.* **2012**, *23*, 1024–1039. [CrossRef]
11. Rodan, S.; Galunic, C. More than network structure: How knowledge heterogeneity influences managerial performance and innovativeness. *Strateg. Manag. J.* **2004**, *25*, 541–562. [CrossRef]
12. Caskey, J.; Hughes, J.S.; Liu, J. Strategic informed trades, diversification, and expected returns. *Acc. Rev.* **2015**, *90*, 1811–1837. [CrossRef]
13. Hadani, M.; Coombes, S. Complementary relationships between corporate philanthropy and corporate political activity: An exploratory study of political marketplace contingencies. *Bus. Soc.* **2015**, *54*, 859–881. [CrossRef]
14. Brammer, S.; Millington, A. Does it pay to be different? An analysis of the relationship between corporate social and financial performance. *Strateg. Manag. J.* **2008**, *29*, 1325–1343. [CrossRef]
15. Wang, H.L.; Qian, C.L. Corporate philanthropy and corporate financial performance: The roles of stakeholder response and political access. *Acad. Manag. J.* **2011**, *54*, 1159–1181. [CrossRef]
16. Mellahi, K.; Frynas, J.G.; Sun, P.; Siegel, D.S. A review of the nonmarket strategy literature: Toward a multi-theoretical integration. *J. Manag.* **2016**, *42*, 143–173. [CrossRef]
17. Lu, Y.; Shailer, G.; Wilson, M. Corporate political donations: Influences from directors' networks. *J. Bus. Ethics* **2016**, *135*, 461–481. [CrossRef]
18. Yoon, Y.; Gurhan Canli, Z.; Schwarz, N. The effect of corporate social responsibility (csr) activities on companies with bad reputations. *J. Consum. Psychol.* **2006**, *16*, 377–390. [CrossRef]
19. Grau, S.L.; Folse, J.A.G. Cause-related marketing (crm)—The influence of donation proximity and message-framing cues on the less-involved consumer. *J. Advert.* **2007**, *36*, 19–33. [CrossRef]
20. Lin, N. *Social Resources and Instrumental Action*, 1st ed.; Sage Publications: Los Angeles, CA, USA, 1981.
21. Nielsen, A.E.; Thomsen, C. Sustainable development: The role of network communication. *Corp. Soc. Resp. Environ. Manag.* **2011**, *18*, 1–10. [CrossRef]
22. Guimaraes-Costa, N.; Cunha, M.P.E. The atrium effect of website openness on the communication of corporate social responsibility. *Corp. Soc. Resp. Environ. Manag.* **2008**, *15*, 43–51. [CrossRef]
23. Apinunmahakul, A.; Devlin, R.A. Social networks and private philanthropy. *J. Public Econ.* **2008**, *92*, 309–328. [CrossRef]
24. Martin, R.; Randal, J. How is donation behaviour affected by the donations of others? *J. Econ. Behav. Org.* **2008**, *67*, 228–238. [CrossRef]
25. Saxton, G.D.; Wang, L.L. The social network effect: The determinants of giving through social media. *Nonprof. Volunt. Sec. Q.* **2014**, *43*, 850–868. [CrossRef]
26. Fracassi, C. Corporate finance policies and social networks. *Manag. Sci.* **2016**, *63*, 2420–2438. [CrossRef]
27. Zhang, J.J.; Marquis, C.; Qiao, K.Y. Do political connections buffer firms from or bind firms to the government? A study of corporate charitable donations of chinese firms. *Org. Sci.* **2016**, *27*, 1307–1324. [CrossRef]
28. Francis, B.B.; Hasan, I.; Sun, X. Political connections and the process of going public: Evidence from china. *J. Int. Money Financ.* **2009**, *28*, 696–719. [CrossRef]
29. Scott, J. *Social Network Analysis: A Handbook*, 2nd ed.; Sage Publications: Newbury, Park, CA, USA, 1991.
30. Konrad, A.M.; Radcliffe, V.; Shin, D. Participation in helping networks as social capital mobilization: Impact on influence for domestic men, domestic women, and international mba students. *Acad. Manag. Learn. Educ.* **2016**, *15*, 60–78. [CrossRef]
31. Chen, W.J.; Kamal, F. The impact of information and communication technology adoption on multinational firm boundary decisions. *J. Int. Bus. Stud.* **2016**, *47*, 563–576. [CrossRef]
32. Lin, Y.N.; Wu, L.Y. Exploring the role of dynamic capabilities in firm performance under the resource-based view framework. *J. Bus. Res.* **2014**, *67*, 407–413. [CrossRef]
33. Haunschild, P.R. Interorganizational imitation: The impact of interlocks on corporate acquisition activity. *Admin. Sci. Q.* **1993**, *38*, 564–592. [CrossRef]

34. Westphal, J.D. Collaboration in the boardroom: Behavioral and performance consequences of ceo-board social ties. *Acad. Manag. J.* **1999**, *42*, 7–24. [CrossRef]
35. Carroll, A.B.; Shabana, K.M. The business case for corporate social responsibility: A review of concepts, research and practice. *Int. J. Manag. Rev.* **2010**, *12*, 85–105. [CrossRef]
36. Miras-Rodriguez, M.D.; Carrasco-Gallego, A.; Escobar-Perez, B. Are socially responsible behaviors paid off equally? A cross-cultural analysis. *Corp. Soc. Resp. Environ. Manag.* **2015**, *22*, 237–256. [CrossRef]
37. Godfrey, P.C.; Merrill, C.B.; Hansen, J.M. The relationship between corporate social responsibility and shareholder value: An empirical test of the risk management hypothesis. *Strateg. Manag. J.* **2009**, *30*, 425–445. [CrossRef]
38. Reynolds, S.J.; Ceranic, T.L. The effects of moral judgment and moral identity on moral behavior: An empirical examination of the moral individual. *J. Appl. Psychol.* **2007**, *92*, 1610–1624. [CrossRef] [PubMed]
39. Richman, S.B.; DeWall, C.N.; Wolff, M.N. Avoiding affection, avoiding altruism: Why is avoidant attachment related to less helping? *Pers. Indiv. Differ.* **2015**, *76*, 193–197. [CrossRef]
40. Gautier, A.; Pache, A.C. Research on corporate philanthropy: A review and assessment. *J. Bus. Ethics* **2015**, *126*, 343–369. [CrossRef]
41. Atkinson, L.; Galaskiewicz, J. Stock ownership and company contributions to charity. *Admin. Sci. Q.* **1988**, *33*, 82–100. [CrossRef]
42. Hampton, K.N.; Lee, C.J.; Her, E.J. How new media affords network diversity: Direct and mediated access to social capital through participation in local social settings. *New Media Soc.* **2011**, *13*, 1031–1049. [CrossRef]
43. Hochberg, Y.V.; Ljungqvist, A.; Yang, L.U. Whom you know matters: Venture capital networks and investment performance. *J. Financ.* **2007**, *62*, 251–301. [CrossRef]
44. Brass, D. Connecting to brokers: Strategies for acquiring social capital. In *Social Capital: Reaching out, Reaching in*, 1st ed.; Bartkus, V., Davis, J., Eds.; Edward Elgar Press: Cheltenham, UK, 2009; pp. 260–274.
45. Oto-Peralias, D.; Romero-Avila, D. The consequences of persistent inequality on social capital: A municipal-level analysis of blood donation data. *Econ. Lett.* **2017**, *151*, 53–57. [CrossRef]
46. Witt, P. Entrepreneurs' networks and the success of start-ups. *Entrep. Reg. Dev.* **2004**, *16*, 391–412. [CrossRef]
47. Huggins, R. Forms of network resource: Knowledge access and the role of inter-firm networks. *Int. J. Manag. Rev.* **2010**, *12*, 335–352. [CrossRef]
48. Lin, C.P.; Hung, W.T.; Chiu, C.K. Being good citizens: Understanding a mediating mechanism of organizational commitment and social network ties in ocbs. *J. Bus. Ethics* **2008**, *81*, 561–578. [CrossRef]
49. Costenbader, E.; Valente, T.W. The stability of centrality measures when networks are sampled (vol 25, pg 283, 2003). *Soc. Netw.* **2004**, *26*, 351. [CrossRef]
50. Faccio, M. Politically connected firms. *Am. Econ. Rev.* **2006**, *96*, 369–386. [CrossRef]
51. Gao, Y. Philanthropic disaster relief giving as a response to institutional pressure: Evidence from china. *J. Bus. Res.* **2011**, *64*, 1377–1382. [CrossRef]
52. Lester, R.H.; Hillman, A.; Zardkoohi, A.; Cannella, A.A. Former government officials as outside directors: The role of human and social capital. *Acad. Manag. J.* **2008**, *51*, 999–1013. [CrossRef]
53. Marquis, C.; Qian, C.L. Corporate social responsibility reporting in china: Symbol or substance? *Org. Sci.* **2014**, *25*, 127–148. [CrossRef]
54. Hillman, A.J. Politicians on the board of directors: Do connections affect the bottom line? *J. Manag.* **2005**, *31*, 464–481. [CrossRef]
55. Sauerwald, S.; Lin, Z.; Peng, M.W. Board social capital and excess ceo returns. *Strateg. Manag. J.* **2016**, *37*, 498–520. [CrossRef]
56. Jia, M.; Zhang, Z. Managerial ownership and corporate social performance: Evidence from privately owned chinese firms' response to the sichuan earthquake. *Corp. Soc. Responsib. Environ. Manag.* **2013**, *20*, 257–274. [CrossRef]
57. Jamali, D.; Neville, B. Convergence versus divergence of csr in developing countries: An embedded multi-layered institutional lens. *J. Bus. Ethics* **2011**, *102*, 599–621. [CrossRef]
58. Xu, S.K.; Yang, R.D. Indigenous characteristics of chinese corporate social responsibility conceptual paradigm. *J. Bus. Ethics* **2010**, *93*, 321–333. [CrossRef]

59. Kao, T.Y.; Chen, J.C.H.; Ben Wu, J.T.; Yang, M.H. Poverty reduction through empowerment for sustainable development: A proactive strategy of corporate social responsibility. *Corp. Soc. Resp. Environ. Manag.* **2016**, *23*, 140–149. [CrossRef]
60. Ma, D.L.; Parish, W.L. Tocquevillian moments: Charitable contributions by chinese private entrepreneurs. *Soc. Forces* **2006**, *85*, 943–964. [CrossRef]
61. Dean, D.H. Consumer perception of corporate donations—Effects of company reputation for social responsibility and type of donation. *J. Advert.* **2003**, *32*, 91–102. [CrossRef]
62. Collins, C.J.; Clark, K.D. Strategic human resource practices, top management team social networks, and firm performance: The role of human resource practices in creating organizational competitive advantage. *Acad. Manag. J.* **2003**, *46*, 740–751.
63. Kostovetsky, L. Political capital and moral hazard. *J. Financ. Econ.* **2015**, *116*, 144–159. [CrossRef]
64. Webb, D.A.; Sweeney, J.C. How functional, psychological, and social relationship benefits influence individual and firm commitment to the relationship. *J. Bus. Ind. Mark.* **2007**, *22*, 474–488. [CrossRef]
65. Hörisch, J.; Freeman, E.; Schaltegger, S. Applying stakeholder theory in sustainability management. Links, similarities, dissimilarities, and conceptual framework. *Org. Environ.* **2014**, *27*, 328–346. [CrossRef]
66. Mok, K.Y.; Shen, G.Q.; Yang, J. Stakeholder management studies in mega construction projects: A review and future directions. *Int. J. Proj. Manag.* **2015**, *33*, 446–457. [CrossRef]

© 2018 by the authors. Licensee MDPI, Basel, Switzerland. This article is an open access article distributed under the terms and conditions of the Creative Commons Attribution (CC BY) license (http://creativecommons.org/licenses/by/4.0/).

Article

The Difficulty of Climate Change Adaptation in Manufacturing Firms: Developing an Action-Theoretical Perspective on the Causality of Adaptive Inaction

Ulrike Meinel [1,2,3,*] and Ralf Schüle [3]

1. alpS—Center for Climate Change Adaptation, Grabenweg 68, 6020 Innsbruck, Austria
2. Institute of Geography, University of Innsbruck, Innrain 52f, 6020 Innsbruck, Austria
3. Wuppertal Institute for Climate, Environment and Energy, Döppersberg 19, 42103 Wuppertal, Germany; ralf.schuele@wupperinst.org
* Correspondence: ulrike.meinel@wupperinst.org; Tel.: +49-202-2492-184

Received: 5 January 2018; Accepted: 22 February 2018; Published: 24 February 2018

Abstract: Climate change induces various risks for supply chains of manufacturing firms. However, surveys have suggested that only a minority of firms conducts strategic adaptations, which we define as anticipatory and target-oriented action with the purpose of increasing resilience to climate change. While several barrier-centered studies have investigated the causality of non-adaptation in industry, the examined barriers are often not problem-specific. Furthermore, it has been shown that even in cases when managers perceive no barriers to adaptation at all, strategic adaptations may still not be conducted. On this background, the present analysis focuses on the logic of adaptive inaction, which we conceive, in particular, as inaction with regard to strategic adaptations. Adopting an action-theoretical perspective, the study examines (a) which aspects may shape the rationality of adaptive inaction among managers, (b) which more condensed challenges of conducting strategic adaptations emerge for managers, and (c) how the theoretical propositions can be tested. For this purpose, the study employs an exploratory approach. Thus, hypotheses on such aspects are explored, which may shape the rationality of adaptive inaction among managers. Subsequently, predictions are inferred from the theoretical propositions, which allow testing their empirical relevance. Methodologically, the hypotheses are explored by reexamining existing explanatory approaches from literature based on a set of pretheoretical assumptions, which include notions of bounded rationality. As a result, the study proposes 13 aspects which may constrain managers in conducting adaptations in such a way, which serves the economic utility of the firm. By condensing these aspects, 4 major challenges for managers are suggested: the challenges of (a) conducting long-term adaptations, of (b) conducting adaptations at an early point in time, of (c) conducting adaptations despite uncertain effects of the measures, and of (d) conducting adaptations despite cross-tier dependencies in supply chains. Finally, the study shows how the propositions can be tested and outlines a research agenda based on the developed theoretical suggestions.

Keywords: climate change; adaptation; manufacturing firms; strategic management; action theory

1. Introduction

Within research on climate adaptation in industry, adaptation has broadly been conceived as a question of economic self-interest of firms. Thus, it has been argued that climate change can induce various risks for firms, which arise from potential climatic impacts on the firm, on its supply chain network and on its political, economic and natural environment [1–5]. However, data from initial firm surveys on climate adaptation suggest that only a minority of managers is engaging their firms in strategic adaptations [6–8], thus in 'anticipatory and target-oriented action with the purpose of increasing resilience to climate change' [7]. Such findings of a frequently occurring lack of strategic adaptations in industry have mostly been explained by lists of barriers to adaptation [3,4,9]. At the same time, a recent study has provided empirical evidence that even if managers do not perceive any barriers at all, strategic adaptations may still not be conducted [7]. Insights into problem-specific rationales of managers, which may explain such findings, yet have hardly been developed [7,10].

Based on this background, the present analysis examines the problem-specific logic of adaptive inaction, which will be defined as inaction with regard to strategic adaptations. Therefore, the study develops an action-theoretical perspective on the causality of adaptive inaction, which presumes utility-maximizing targets of managers and adopts notions of bounded rationality [11]. More specifically, the study examines (a) which aspects may shape the rationality of adaptive inaction among managers, (b) which more condensed challenges of conducting strategic adaptations emerge for managers, and (c) how the theoretical propositions can be tested. For this purpose, the study employs an exploratory approach. Thus, hypotheses on such aspects are explored, which may shape, in particular, the utility-oriented rationales behind adaptive inaction among managers. Subsequently, predictions are inferred from the theoretical propositions, which allow testing their empirical relevance. Methodologically, the hypotheses are explored by reexamining explanatory approaches from literature based on a set of pretheoretical assumptions.

While in reality, other forms of rationales, such as rationales shaped by values, emotions or habits [12], may similarly influence decisions of managers, the present study exclusively focuses on utility-oriented, instrumental rationales. Thus, the study only examines a part of reality, though one which is known to play an important role in strategic decision-making processes in firms [13] including such decision processes, which relate to climate adaptations [7].

In conceptual terms, the study applies the term agency restraints for designating those aspects, which shape the rationality of adaptive inaction. In particular, such agency restraints are explored, which may be faced by managers of manufacturing firms, as the latter have been conceived as being at risk of various potential impacts of climate change [1].

As an initial step, arguments underlining the importance of strategic adaptations in manufacturing firms are outlined. Afterwards, the state of research on the causality of adaptive inaction is depicted. Then, the applied model of action is described and its implications for employing the term agency are discussed. Subsequently, the employed exploratory approach is depicted. Thereafter, agency restraints are proposed and testable predictions are inferred. In a final discussion, the proposed agency restraints are condensed to more abstract challenges of conducting strategic adaptations, and next steps for analyzing climate adaptations in firms are suggested.

2. The Relevance of Strategic Adaptations for Manufacturing Firms

Various studies have so far examined climate risks for firms, which are linked to potential climatic impacts on firms, on their supply chain network and on their political, economic and natural environment [1,2,4,5,8,14–16]. Thus, it has been argued that climate change may influence businesses in various ways, for example by affecting the reliability of transportation or of water and energy supply, by affecting work productivity, or by inducing changes in political and economic framework conditions [1]. Referring to the discussed risks, it has been proposed that non-strategic forms of adaptations, such as 'hidden adaptations' [17,18] respectively adaptations by co-benefits [2,19] may not suffice, particularly if current emission trends persist, such as projected in the business-as-usual scenario of the IPCC (RCP8.5) [20]. With regard to strategic adaptations, various measures have been discussed. For example, adaptations which are supposed to increase the robustness of firms against direct (biophysical) impacts of climate change have been explored, such as proofing the built infrastructure of the firm against weather extremes [1,15,21,22]. Furthermore, adaptation options have been analyzed which may allow business resilience to indirect impacts of climate change, such as to climatic impacts on economic and political framework conditions. For example, resilience-increasing product and management innovations have been discussed in this regard (for a broader overview of discussed adaptation measures, see [1,15,16,21,22]). However, management practices seem to predominantly contrast the suggested business relevance of climate risks as a minority of firms is currently implementing strategic adaptations, according to surveys [6,7].

3. The State of Research on the Causality of Adaptive Inaction

With regard to the causality of adaptive inaction, studies have predominantly examined barriers to adaptation, thus 'factors and conditions which hamper the process of developing and implementing climate change adaptations' [23]. A literature review, which summarizes the identified barriers to adaptation in industry, has been conducted elsewhere [7]. However, the barrier-centered perspective has recently been criticized as it would neglect actors, their strategies and motivation [7,10,23]. Furthermore, barrier-centered studies would tend to neglect the question *why* and *how* the barriers emerge [10,23]. Finally, some barriers, such as financial restrictions, would tend to be suggested regardless of the problem at hand [23]. The present study takes into account the depicted gaps by focusing on problem-specific aspects which may shape the rationales behind adaptive inaction.

Only in exceptional cases, studies have developed more theoretically condensed explanations of adaptive inaction. These exceptional studies have either employed sociological perspectives on communicative processes and discourses in municipal adaptation politics [10,24] or have examined the interplay of different psychological factors which can affect intention to adapt [7]. The mentioned studies especially allow understanding how perceptions relating to climate adaptations emerge as a consequence of psychological [7] or social [10,24] processes. However, these studies hardly provide general insights or predictions with regard to the behavior of actors within a particular population. The present study provides a first contribution to address this gap by developing propositions on utility-oriented rationales of managers and by inferring testable predictions on prerequisites of adaptations.

In literature, some scattered insights into utility-oriented rationales behind adaptive inaction can be detected. For example, rationales induced by the uncertainty of climatic developments have been analyzed [25–28] and possible strategies of risk reduction have been outlined [29]. Furthermore, some ideas on different time horizons of economically and climatically induced requirements of action have been suggested [27,30]. Finally, constraints for adaptations emerging from interdependencies between actors with diverging interests have been discussed [10,24,31].

The present study aims at systematically examining such utility-oriented aspects of choice with regard to climate adaptations. For this purpose, the study reviews explanatory approaches of adaptive inaction which have emerged in adaptation, sustainability and strategic management literature and reexamines the approaches based on simplifying assumptions about actors and their rationality.

4. Conceptual Foundations: The Employed Model of Action and Its Implications for Applying the Term Agency

In order to examine the logic of adaptive inaction among managers, the study employs a set of simplifying assumptions. In particular, it will be assumed that actors aim at utility-maximization, yet are restricted in this attempt due to incomplete information. The study thus adopts notions of bounded rationality [11], which have been widely applied in economic theory [13] and have also stimulated extensive experimental investigations [13].

Furthermore, it will be assumed that managers aim at maximizing the corporate utility of the firm. Though in reality, managers may also pursue personal ambitions [32], equating the interests of managers with the corporate interest of the firm has been considered as one way of approaching reality, at least when conceiving of managers as CEOs [33].

In addition, examining actor rationales implies a focus on a particular kind of non-adapting managers because rational action requires intention to act [12,34]. Therefore, the study conceives actors as such managers, who intend to increase their firm's resilience, yet abstain from climate adaptations due to considerations of utility maximization.

Taking together the outlined assumptions, actors will thus be conceived as CEOs of manufacturing firms

- who intend to increase the resilience of their firms to climate change,
- who aim at maximizing the economic utility of the firm, and
- who are disposing of limited information.

Applying these assumptions, the study may contribute to existing debates on the causality of adaptive inaction in various ways. Thus, *by presuming the existence of intention to adapt*, the study may complement insights obtained from studies which have focused on causes for lacking intention to adapt [7,10,25,35,36]. Thus, the study may show that even if managers consider climate risks as a relevant business factor, strategic adaptations may still not be conducted due to various rationales. Furthermore, *assuming a utility-maximizing logic of action* may extend insights from barrier-centered analyses of non-adaptation (for reviews, see [7,23]) as removing barriers may foremost support adaptations if actors expect the adaptation to allow higher utility than inaction.

When examining the logic of adaptive inaction, the study especially focuses on the problem-specific aspects which shape this logic and terms these aspects *agency restraints*. Out of the various notions of agency, which have evolved in the social sciences (for an overview, see [37]), the study conceptualizes 'agency' closely attached to such notions of the term, which have been developed in rational choice theory [32,38,39], as the latter suit analyses of instrumental rationales. In this sense, *agency* will be defined as an actor's capability to choose deliberately between alternatives in order to pursue a specific target, while aiming at economic utility maximization. More specifically, *adaptive agency* will be defined as the respective capability to deliberately take adaptive action in order to increase resilience towards climate change, again given the assumption of instrumental rationality. Vice versa, *agency restraints* will be conceived as those aspects, which shape the rationality of inaction with regard to a particular problem by suggesting a higher utility of inaction than of action.

5. Outlining the Explorative Research Design

Due to the infant state of research into utility-oriented rationales behind adaptive inaction, an exploratory research design is applied. In particular, two steps of argumentation are taken. As a *first* step, hypotheses on agency restraints are explored. For this purpose, existing explanatory approaches from sustainability, adaptation and strategic management literature are reexamined from the outlined, action-theoretical perspective (see Section 4). In order to activate the invention of propositions on agency restraints [40], the literature review aimed, in particular, at discovering discrepancies between requirements and opportunities of conducting adaptations, which may be faced by managers, who aim at maximizing the economic utility of the firm. In order to detect such potential requirement-opportunity discrepancies, a combination of keyword and snowball research was applied in the literature review. Based on a discussion of each of the emerging requirement-opportunity discrepancies, agency restraints are proposed.

As a *second* step of argumentation, predictions are deductively inferred from each of the propositions on agency restraints. The predictions allow testing the empirical effects which the proposed agency restraints have on adaptation-related choices of managers [40].

It may be noted that conducting respective empirical tests represents the third step of argumentation within the iterated process of developing (falsifiable) theoretical explanations [40], yet rests beyond the scope of the present study.

In order to categorize the emerging agency restraints, the latter were assigned to three dimensions, thus 'bare aspects' [41], of the scrutinized problem in line with the topics emerging in the reviewed literature: the dimensions of time, knowledge and system boundaries, which are conceived as boundaries between actors in supply chain networks for the purpose of the present study. In the following sections, potential agency restraints will be explored along these dimensions. An overview of the proposed agency restraints and of the inferred predictions will be provided in Section 7.

6. Emerging Agency Restraints along the Dimension of Time

Along the dimension of time, eight potential agency restraints were identified. The suggested agency restraints refer to two challenges of conducting adaptations, which may arise to managers: first, the challenge of conducting such adaptations, which may possibly yield benefits only in the long term (see Sections 6.1–6.4); and second, the challenge of conducting adaptations at an early point in time (see Section 6.5).

6.1. Managers' Adaptive Agency with Regard to Investment Horizons

Discussion of Requirements and Opportunities of Action

In order to become effective, some adaptations require long-term investment horizons. This may concern adaptations of the built infrastructure of the firm or such changes of the product portfolio, which require long innovation lead time, which is the time needed for research, development, distribution and product launch [42]. However, employing long-term investment horizons may often lack expectations of satisfying returns, particularly if high discount rates are assumed. In particular, discounting relates to the rationale that expectable future returns of investments decrease when losses are subtracted which occur over time. Such losses are typically conceived as opportunity costs, which are the missed returns from alternative investments, such as from investments at capital markets [43,44].

Proposition of Agency Restraints

In the described sense, the proposition (PRO) emerges that opportunity costs of lost returns from alternative investments constrain the agency of managers to engage their firms in long-term adaptations ($PRO_{Time\ (T)1}$).

Deduction of Predictions

The relevance which managers attribute to opportunity costs should decrease if the expected advantages of the investment increase. Thus, the prediction (PRE) follows that managers who ascribe a high business relevance to climate risks are more likely to engage their firms in long-term adaptations (PRE_{T1}).

6.2. Managers' Adaptive Agency with Regard to Long-Term Strategic Planning

Discussion of Requirements and Opportunities of Action

Applying long-term strategic planning horizons can be required for developing some adaptations, such as construction measures, some product innovations, or changes in long-term contracts. However, the utility of conducting long-term strategic planning can be questioned by several rationales.

One potential rationale refers to innovation lead time. In particular, short innovation lead time may question the utility of long-term strategic planning as flexible responses to changing demands are facilitated. In fact, only 10% of small and medium-sized companies (SMEs) have innovation lead times of more than 3 years and 28% have respective lead times of less than a year according to a survey among German SMEs [45], notwithstanding sectoral differences [42,46].

A second potential, adverse rationale concerns the lifespan of businesses. As only 50% of businesses survive the first five years of their existence [47], perceived risks of short-term mortality may question benefits of conducting long-term planning, in particular among managers of start-ups.

Proposition of Agency Restraints

Based on the outlined considerations, the proposition emerges that an uncertain value contribution of conducting long-term strategic planning restrains the agency of managers to engage their firms in long-term adaptations (PRO_{T2}).

Deduction of Predictions

The prediction follows that long-term adaptations become more likely if managers expect a higher value contribution of long-term strategic planning. In this sense, long-term adaptations should become more likely if managers

- consider long innovation lead time to prevail in their firm ($PRE_{T2.1}$) or
- have high confidence in experiencing a long lifespan of the firm ($PRE_{T2.2}$).

6.3. Managers' Adaptive Agency with Regard to Institutionalized Time Horizons

Discussion of Requirements and Opportunities of Action

Conducting long-term adaptations may also be disincentivized by institutions, which in line with a very short definition will be conceived as formal and informal rules [48]. Thus, mismatching institutionalized time horizons may decrease the utility of integrating long-term perspectives in corporate decision-making processes in various ways.

For example, *electoral cycles and related political variability* may question the utility of long-term adaptations due to uncertain future changes of framework conditions, such as potential future developments of emission trading, CO_2 taxes, or caps.

Furthermore, *annual or even quarterly reporting obligations* for companies at capital markets may incentivize the optimization of short-term instead of long-term business figures. Respective tendencies may moreover be supported by requirements regarding the *content of the reporting*. For example, companies at capital markets in the European Union (EU) are obliged to report on figures of the past business year but scarcely on future risks. Thus, in line with the EU directive on annual financial statements (Directive 2013/34/EU) or its national transposition laws, such as the German trade law (particularly §289 HGB), SMEs are not obliged to report on future risks. Even large listed companies are only obliged to disclose such environmental risks that represent relevant business information (§289 HGB). Notably, considerations of the 'relevance' of the respective information are subject to the discretion of the reporting company (§289 HGB). Moreover, reporting duties concern only such risks, which may become effective in a time horizon of 2–3 years in future in line with the interests of investors (International Financial Reporting Standards (IFRS) 7). The Corporate Social Responsibility Directive of the EU (Directive 2014/95/EU) similarly entails very limited duties of disclosing environmental risks as only such risks have to be reported, which are connected to high impacts *and* to high probability.

In addition, *time horizons imposed by informal institutions* might constrain long-term adaptations. This may include management customs, such as the focus of entrepreneurial crisis management on instant reactions to immediate crisis situations [2]. Similarly, routines of planners to extrapolate past data in forecasting models [42] may hardly provide the insights required for long-term adaptations to climate change as the latter is dynamic in nature [49].

Proposition of Agency Restraints

Based on these considerations, the following aspects are supposed to constrain the adaptive agency of managers:

- risks that adaptations induce sunk costs due to uncertain changes in political framework conditions (PRO_{T3}),
- uncertain value contribution of long-term adaptations due to the content of, and due to the time horizons imposed by, reporting legislations (PRO_{T4}), and

- enforcement costs of imposing a long-term orientation to informal institutions (PRO_{T5}).

Deduction of Predictions

Three predictions follow. Namely, that managers are more likely to engage their firms in long-term adaptations

- if they are aware of adaptation options whose effectiveness and efficiency is scarcely depending on political framework conditions (PRE_{T3}),
- if participating in voluntary reporting schemes, which incentivize the reporting of climate risks and adaptations, such as the Carbon Disclosure Project (PRE_{T4}), or
- if they are in charge of such companies, which already dispose of informal institutions oriented at long-term risk management (PRE_{T5}).

6.4. Managers' Adaptive Agency with Regard to Time Horizons Imposed by Significant Other Actors

Discussion of Requirements and Opportunities of Action

Furthermore, the expected utility of adapting to long-term risks can decrease if significant other actors impose short time horizons. For example, an average orientation of investment funds at profits at a horizon of less than 2 years has been suggested by figures of Morningstar. Thus, the average stock holding period by investment managers was 1.4 years across the 25 largest open-end mutual fund categories in 2013 [50]. Thus, allocating resources to the development of long-term resilience, and not to the maximization of short-term success, may induce risks of reduced attractiveness for some investors.

Proposition of Agency Restraints

Thus, the proposition emerges that risks of reduced attractiveness for some investors, such as investment funds, restrain the agency of managers to engage their firms in long-term adaptations (PRO_{T6}).

Deduction of predictions

The prediction follows that managers of family-owned firms or of firms financed by strategic investors are more likely to engage their firms in strategic adaptations than managers of firms financed by free-floating shares (PRE_{T6}).

6.5. Managers' Adaptive Agency with Regard to Early Mover Disadvantages

Discussion of Requirements and Opportunities of Action

As potential impacts of climate change may occur at an uncertain point of time in future, taking precaution may require early action. However, the expected utility of early action can be lower than that of waiting due to several rationales.

Such rationales may, for example, be shaped by the early state of political and economic developments in the field of climate adaptation [27,51]. For example, legal obligations for firms to disclose their climate risks have scarcely evolved so far (at least in the EU, see Section 6.3). Furthermore, such pressure on companies to adapt, which is emerging from financial markets, is only beginning to develop, notwithstanding first voluntary reporting initiatives referring to climate risks and adaptations, such as the Carbon Disclosure Project. In particular, legislative *obligations* for institutional investors to report on climate risks relating to their assets—which would induce subsequent pressure on companies to engage in adaptations—are lacking, notwithstanding one first approach in France [51]. As a consequence of these aspects, it can be assumed that early adaptations of firms may be linked to lower effects on the attractiveness for investors and business partners than adaptations under more developed framework conditions.

Furthermore, early action in a field of high uncertainty and learning can raise concerns to miss future technological and scientific developments, which possibly allow a higher effectiveness and efficiency of the adaptation in question [52]. For example, such beneficial developments might include future access to improved 'climate services', thus to tools, products and information which support practitioners in dealing with climate change [53]. Thus, managers engaging their firms in strategic adaptations at present, may face risks of adjustment costs, which have been defined as costs incurred while learning about new climate conditions [52].

Proposition of Agency Restraints

On this background, the proposition emerges that the agency of managers to conduct strategic adaptations at an early point in time is constrained

- by efficiency risks of conducting adaptations under immature framework conditions (PRO_{T7}) and
- by risks of facing adjustment costs (PRO_{T8}).

Deduction of Predictions

The prediction follows that such managers are more likely to engage their firms in strategic adaptations

- who already perceive relevant business impacts of framework conditions, which relate to climate adaptations (PRE_{T7}), or
- who consider the available practical knowledge required for adaptations as sufficiently developed (PRE_{T8}).

7. Emerging Agency Restraints along the Dimension of Knowledge

Based on literature research, 2 potential agency restraints were identified along the dimension of knowledge. Both emerging agency restraints refer to the challenge of conducting adaptations despite discrepancies between required and feasibly achievable levels of knowledge.

7.1. Managers' Adaptive Agency with Regard to Risks of Negative Externalities

Discussion of Requirements and Opportunities of Action

Conducting adaptations in a utility-maximizing way requires the avoidance of negative externalities [27,29], which can sometimes be complex. For example, some adaptations, such as the installation of cooling systems, may increase CO_2 emissions. Furthermore, externalities of adaptations may induce negative social, ecological or economic developments in the supply chain network and its societal environment. For example, abandoning suppliers from climate-sensitive regions, such as vulnerable suppliers from arid or flood-prone areas, may increase the resilience of the supply chain but may worsen the vulnerability of these suppliers. In the same sense, developing 'green-tech' innovations, which may also be interpreted as adaptations to climatically induced shifts in demand, can induce side-effects, such as rebound effects [54] or the consumption of raw materials that are extracted in a socially or ecologically detrimental way. While such negative externalities of adaptations may induce reputational risks, their avoidance may in some cases surpass the informatory capacity of managers.

Proposed Agency Restraints

Thus, the proposition emerges that reputational risks of inducing negative externalities restrain the adaptive agency of managers ($PRO_{Knowledge\ (K)1}$).

Deduction of Predictions

The prediction follows that managers are more likely to engage their firms in strategic adaptations if

- estimating to dispose of high informational capacities ($PRE_{K1.1}$) or
- if estimating that the measure in question is connected to a low complexity and to a low severity of potential ecological, social and economic consequences ($PRE_{K1.2}$).

7.2. Managers' Adaptive Agency with Regard to Risks of Inducing Vulnerability-Increasing Lock-In Effects

Discussion of Requirements and Opportunities of Action

Discrepancies between required and achievable levels of knowledge may moreover induce risks of maladaptive lock-in effects that potentially decrease future resilience. Lock-in effects have been defined as constraining effects of decisions, events or outcomes at one point in time on options available at a later point in time [55]. For example, the installation of cooling systems may be costly to reverse if climate change requires stronger systems at a later point in time. Similarly, if innovations with long lead times are aiming at incremental changes, a strategy turn may be difficult if transformational innovations are instead required in future [56,57].

Proposed Agency Restraints

Thus, the proposition emerges that risks of inducing lock-in effects constrain the adaptive agency of managers (PRO_{K2}).

Deduction of Predictions

The prediction follows that managers are less likely to engage their firms in such adaptations, which presumably induce lock-in effects (for an overview of critical aspects, see [29]). Thus, such adaptations should become less likely to occur,

- which are connected to long-lived investments ($PRE_{K2.1}$) and
- which are only reversible at high costs ($PRE_{K2.2}$).

8. Emerging Agency Restraints along the Dimension of System Boundaries

Along the dimension of system boundaries, which are here conceived as boundaries between actors in supply chain networks (see Section 5), 3 potential agency restraints were discovered. The identified potential agency restraints all refer to the challenge of conducting strategic adaptations despite dependencies in supply chain networks.

8.1. Managers' Adaptive Agency with Regard to Cross-Tier Dependencies of Resilience-Increasing Innovations

Discussion of Requirements and Opportunities of Action

Increasing resilience based on product or management innovations may often depend on cross-tier cooperation in supply chains. At the same time, the influence that managers can exert on business partners can strongly vary, for example, depending on the power position which a firm holds within the respective chain. For example, a small furniture manufacturer intending to change to a more resilient type of wood may be hampered in this attempt if the furniture retailer rejects the product innovation supposing that his customers won't buy the new furniture. Similarly, management innovations aiming at increased resilience to climate change [1,15,16,21,22] can be hampered. For example, employing slack time in production processes in order to increase resilience can be inhibited if powerful customers demand the uninterrupted possibility of just-in-time delivery [1,58].

Proposed Agency Restraints

Subsequently, the proposition emerges that a low power position in the supply chain network constrains the agency of managers to initiate and develop resilience-increasing innovations across tiers ($PRO_{System\ Boundaries\ (SB)1}$).

Deduction of Predictions

The prediction follows that managers are more likely to engage their firms in cross-tier, resilience-increasing innovations if considering the position of their company within the supply chain as powerful (PRE_{SB1}).

8.2. Managers' Adaptive Agency with Regard to Cross-Tier Dependencies of Climate-Sensitive Risk Management

Discussion of Requirements and Opportunities of Action

An effective management of climate risks, too, can depend on cross-tier cooperation. Thus, it has been repeatedly argued that the effects of hazardous impacts of climate change can ripple throughout supply chains [1,4,5,8]. In fact, various options of developing cross-tier risk management exist, such as conducting risk audits among suppliers or supporting vulnerable suppliers in increasing their resilience [59–61]. However, opportunities to conduct such actions can be limited. For example, the power position in the supply chain can decide whether risk audits or the adoption of resilience-increasing measures can be imposed on suppliers [62] and risks reported by suppliers may not be controllable [63,64].

Proposed Agency Restraints

Subsequently, three potential agency restraints emerge:

- limitations due to having a low power position in the supply chain network (=PRO_{SB1}),
- costs of developing, introducing and enforcing a climate-sensitive management of cross-tier risks (PRO_{SB2}), and
- risks of trusting the information disclosed in cross-tier risk management (PRO_{SB3}).

Deduction of Predictions

The predictions follow that managers are more likely to engage their firms in cross-tier risk management

- if considering the position of their company within the supply chain as powerful (=PRE_{SB1}),
- if considering the company to be disposing of large (e.g., financial and personnel) capacities ($PRE_{SB2.1}$),
- if considering climate risks as highly relevant for business success ($PRE_{SB2.2}$), and
- if trust exists in business partners along the supply chain (PRE_{SB3}).

9. Discussion

The study has explored agency restraints for managers who aim at increasing the resilience of their firms to climate change. Table 1 summarizes the proposed agency restraints and the inferred, testable predictions. Furthermore, Table 1 condenses the proposed agency restraints to more abstract, major challenges of adaptations, which may be faced by managers.

Table 1. Condensed challenges, proposed agency restraints, and inferred predictions.

Challenges	Proposed Agency Restraints		Predicted Prerequisites for Adaptations	
Conducting long-term adaptations (T)	PRO_{T1}	Opportunity costs of lost returns from investments with shorter time horizons	PRE_{T1}	Consideration of climate risks as highly relevant for business success
	PRO_{T2}	Uncertain value contribution of long-term strategic planning	$PRE_{T2.1}$ $PRE_{T2.2}$	Consideration that long innovation lead time prevails in the firm; High confidence in a long lifespan of the firm
	PRO_{T3}	Risks that long-term adaptations induce sunk costs if changes in political framework conditions occur	PRE_{T3}	Awareness of adaptation options whose effectiveness and efficiency is scarcely depending on changing political framework conditions
	PRO_{T4}	Uncertain value contribution of conducting long-term adaptations due the content of, and due to the time horizons imposed by, reporting legislations	PRE_{T4}	Situation of the firm to be participating in voluntary reporting schemes which incentivize the disclosure of climate risks
	PRO_{T5}	Enforcement costs of imposing a long-term orientation to informal institutions	PRE_{T5}	Situation of the firm to be already disposing of informal institutions oriented at long-term risk management
	PRO_{T6}	Risks of reduced attractiveness for some investors if allocating resources to optimizing long-term instead of short-term performance	PRE_{T6}	Situation of the firm to be family-owned or financed by strategic investors
Conducting adaptations at an early point in time (T)	PRO_{T7}	Efficiency risks of conducting adaptations under immature framework conditions	PRE_{T7}	Consideration of present framework conditions, which relate to adaptations, as already having a relevant impact on business success
	PRO_{T8}	Risks of facing future adjustment costs	PRE_{T8}	Consideration to dispose of sufficient practical knowledge on expectable climate impacts and on adaptation options
Conducting adaptations despite uncertain effects of the measure (K)	PRO_{K1}	Reputational risks of inducing negative externalities	$PRE_{K1.1}$ $PRE_{K1.2}$	Consideration of the firm to dispose of high informational capacities; Estimation that the respective measure is connected to a low complexity and low severity of ecological, social and economic consequences
	PRO_{K2}	Risks of inducing vulnerability-increasing lock-in effects	$PRE_{K2.1}$	Awareness of adaptation options which are not connected to long-lived investments
			$PRE_{K2.2}$	Awareness of adaptation options which are reversible at low costs
Conducting adaptations despite cross-tier dependencies in supply chains (SB)	PRO_{SB1}	Constraints to conducting resilience-increasing innovations and cross-tier risk management in case of adverse actor constellations	PRE_{SB1}	Consideration of the company's position within the supply chain as powerful
	PRO_{SB2}	Costs of developing a climate-sensitive, cross-tier risk management	$PRE_{SB2.1}$	Consideration that the firm disposes of large (e.g., financial and personnel) capacities
	PRO_{SB3}	Risks of trusting information disclosed in cross-tier risk management	$PRE_{SB2.2}$ PRE_{SB3}	Consideration of climate risks as highly relevant for business success; Perception of trust in business partners

(T)/(K)/(SB): Challenge occurs along the dimension of time/knowledge/system boundaries.

As Table 1 shows, 13 potential agency restraints were detected, which can be condensed to 4 major challenges.

First, 6 agency restraints (see PRO_{T1} to PRO_{T6} in Table 1) can be subsumed under the major challenge of *conducting long-term adaptations*. In line with the content of the subsumed agency restraints, the challenge refers to difficulties of conducting such adaptations, which improve resilience in the long term, yet may not necessarily yield short-term benefits. At the same time, the results depicted in Table 1 suggest that adaptations may nonetheless occur, depending on considerations of managers with regard to firm characteristics ($PRE_{T2.1/T2.2}$; PRE_{T4} to PRE_{T6}), with regard to recognized adaptation options (PRE_{T3}) and with regard to the extent of perceived climate risks (PRE_{T1}).

Second, *conducting adaptations at an early point in time* emerged as a further major challenge. The challenge relates to aspects, which imply a higher utility of waiting than of conducting adaptations (see PRO_{T7} and PRO_{T8}). Again, Table 1 also proposes that the occurrence and the effects of this challenge are variable and may, in particular, depend on the state of developed political and economic framework conditions (see PRE_{T7}) as well as on the quality of already accessible, practical information (see PRE_{T8}).

Third, the emerging major challenge of *conducting adaptations despite uncertain effects of the measures* refers to discrepancies between the level of knowledge required for conducting specific adaptations and the level of knowledge achievable for managers, given limited informatory capacities (see PRO_{K1} and PRO_{K2}). However, Table 1 suggests that this challenge may lose its adverse effects on the implementation of strategic adaptations depending on considerations of managers with regard to adaptation options ($PRE_{K1.2}$ to $PRE_{K2.2}$) and depending on firm characteristics (see $PRE_{K1.1}$).

As a fourth challenge, Table 1 suggests that *conducting adaptations despite cross-tier dependencies in supply chains* may impede strategic adaptations in manufacturing firms. Again, Table 1 suggests that the effects of this challenge are variable and may depend on actor constellations within the supply chain network (see PRE_{SB1}), on perceived trust in business partners (see PRE_{SB3}), on the expected impacts of climate change (see $PRE_{SB2.2}$), and on firm characteristics (see $PRE_{SB2.1}$).

However, as empirical tests are still lacking, the propositions on agency restraints as well as the condensed, major challenges still remain tentative. Therefore, future research may conduct statistical analyses which examine the empirical relevance respectively compare the effects of the proposed agency restraints on choices of managers relating to climate adaptations.

Depending on the results of such empirical tests, two basic directions emerge for further research. *First*, if tests falsify the propositions, new propositions may be invented, which allow explanations of adaptive inaction in firms. *Second*, if tests provide some evidence that the proposed agency restraints actually do affect strategic adaptations in firms, the following, subsequent questions emerge.

Thus, (a) empirically supported insights on agency restraints could be employed as given starting points for analyzing political and economic framework conditions in support of climate adaptations in firms. In this sense, it might be examined how framework conditions could be developed in a way, which is sensitive to the logic of action among managers and which could thus increase the effects of the respective policies on firm behavior.

In addition, (b) it could be analyzed how framework conditions can induce deeper changes in the logic of action by influencing the agency restraints themselves. For example, informatory framework conditions could be analyzed, which aim at an increased awareness of managers towards climate risks and at improved practical knowledge about adaptation options, as several agency restraints (as suggested in PRO_{T1}, PRO_{T3}, PRO_{T8}, PRO_{K2}, PRO_{SB2}; see Table 1) might be addressed in this way.

Limitations of the study especially refer to the employed pretheoretical assumptions. Thus, the rationales of managers may not only be shaped by the instrumental aim of maximizing the economic utility of the firm. Instead, rationales may also be shaped by value-oriented, affectual or traditional (i.e., habitual) rationality [12]. Furthermore, while intention to adapt is presumed in the present study, such intention can be questioned by various motivational factors as shown in existing, psychological adaptation studies [7,25]. In addition, differences in actor rationales may occur depending on the sector because climate risks can vary between different manufacturing sectors [1]. Finally, social interactions may shape perceptions of climate risks and of adaptations among managers, and may subsequently influence the rationales.

10. Conclusions

Employing an action-theoretical perspective, the study has shown that despite considerable climate risks for manufacturing firms [1], various problem-specific aspects might make managers attribute a higher utility to inaction than to conducting strategic adaptations. In this regard, the study has suggested challenges of adaptations as well as conditions, under which the utility of conducting strategic adaptations may increase. However, empirical evidence of the propositions is still lacking. Generating such evidence might not only improve insights into the causality of adaptive inaction but may also promote the further analysis of framework conditions in support of climate adaptations in industry.

Acknowledgments: This work was conducted within the project T2B-CCA, which is supported by the COMET funding program of the Austrian Ministry for Transport, Innovation and Technology, the Austrian Ministry of Science, Research and Economy, the state of Tyrol, and the state Vorarlberg; the COMET program is processed by the Austrian Research Promotion Agency (FFG). Moreover, the work was promoted by the University of Innsbruck within the scope of a doctoral scholarship. The authors would particularly like to thank Holger Berg, Wuppertal Institute for Climate, Environment and Energy, for helpful comments and discussions.

Author Contributions: Ulrike Meinel conceptualized and wrote the study. Ralf Schüle supported the study with regard to its methodological foundations and proofread the manuscript.

Conflicts of Interest: The authors declare no conflict of interest.

References

1. Meinel, U.; Abegg, B. A multi-level perspective on climate risks and drivers of entrepreneurial robustness—Findings from sectoral comparison in alpine Austria. *Glob. Environ. Chang.* **2017**, *44*, 68–82. [CrossRef]
2. Linnenluecke, M.K.; Griffiths, A.; Winn, M. Extreme weather events and the critical importance of anticipatory adaptation and organizational resilience in responding to impacts. *Bus. Strateg. Environ.* **2012**, *21*, 17–32. [CrossRef]
3. Ford, J.D.; Pearce, T.; Prno, J.; Duerden, F.; Ford, L.B.; Smith, T.R.; Beaumier, M. Canary in a coal mine: Perceptions of climate change risks and response options among Canadian mine operations. *Clim. Chang.* **2011**, *109*, 399–415. [CrossRef]
4. Fleming, A.; Hobday, A.J.; Farmery, A.; van Putten, E.I.; Pecl, G.T.; Green, B.S.; Lim-Camacho, L. Climate change risks and adaptation options across Australian seafood supply chains—A preliminary assessment. *Clim. Risk Manag.* **2014**, *1*, 39–50. [CrossRef]
5. Lim-Camacho, L.; Hobday, A.J.; Bustamante, R.H.; Farmery, A.; Fleming, A.; Frusher, S.; Green, B.S.; Norman-López, A.; Pecl, G.T.; Plagányi, É.E.; et al. Facing the wave of change: Stakeholder perspectives on climate adaptation for Australian seafood supply chains. *Reg. Environ. Chang.* **2015**, *15*, 595–606. [CrossRef]
6. Crawford, M.; Seidel, S. *Weathering the Storm: Building Business Resilience to Climate Change*; C2ES Center for Climate and Energy Solutions: Arlington, VA, USA, 2013; Available online: https://www.c2es.org/publications/weathering-storm-building-business-resilience-climate-change (accessed on 3 January 2018).
7. Meinel, U.; Höferl, K.-M. Non-adaptive behavior in the face of climate change: First insights from a behavioral perspective based on a case study among firm managers in alpine Austria. *Sustainability* **2017**, *9*, 1132. [CrossRef]
8. Loechel, B.; Hodgkinson, J.; Moffat, K. Climate change adaptation in Australian mining communities: Comparing mining company and local government views and activities. *Clim. Chang.* **2013**, *119*, 465–477. [CrossRef]
9. Fichter, K.; Schneider, T. *Wie Unternehmen den Folgen des Klimawandels Begegnen: Ergebnisse der Panelbefragung 2010 und 2012*; Nordwest2050-Werkstattbericht (No. 24); Universität Oldenburg: Oldenburg, Germany, 2013.

10. Biesbroek, G.R.; Termeer, C.J.A.M.; Klostermann, J.E.M.; Kabat, P. Rethinking barriers to adaptation: Mechanism-based explanation of impasses in the governance of an innovative adaptation measure. *Glob. Environ. Chang.* **2014**, *26*, 108–118. [CrossRef]
11. Simon, H.A. Bounded Rationality and Organizational Learning. *Organ. Sci.* **1991**, *2*, 125–134. [CrossRef]
12. Weber, M. *Wirtschaft und Gesellschaft*; J.C.B Mohr: Tübingen, Germany, 1922.
13. Heukelom, F. *Behavioral Economics: A History*; Cambridge University Press: New York, NY, USA, 2014.
14. Ridoutt, B.; Sanguansri, P.; Bonney, L.; Crimp, S.; Lewis, G.; Lim-Camacho, L. Climate Change Adaptation Strategy in the Food Industry—Insights from Product Carbon and Water Footprints. *Climate* **2016**, *4*, 26. [CrossRef]
15. Agrawala, S.; Carraro, M.; Kingsmill, N.; Lanzi, E.; Mullan, M.; Prudent-Richard, G. *Private Sector Engagement in Adaptation to Climate Change*; OECD Publishing: Paris, France, 2011.
16. UK Environment Agency. *Assessing and Managing Climate Change Risks in Supply Chains*; Environment Agency: Bristol, UK, 2013.
17. Grüneis, H.; Penker, M.; Höferl, K.-M. The full spectrum of climate change adaptation: Testing an analytical framework in Tyrolean mountain agriculture (Austria). *Springerplus* **2016**, *5*, 1848. [CrossRef] [PubMed]
18. Hopkins, D. The sustainability of climate change adaptation strategies in New Zealand's ski industry: A range of stakeholder perceptions. *J. Sustain. Tour.* **2014**, *22*, 107–126. [CrossRef]
19. Berkhout, F. Adaptation to climate change by organizations. *Wiley Interdiscip. Rev. Clim. Chang.* **2011**, *3*, 91–106. [CrossRef]
20. The Intergovernmental Panel on Climate Change (IPCC). *Climate Change 2013: The Physical Science Basis. Contribution of Working Group I to the Fifth Assessment Report of the Intergovernmental Panel on Climate Change*; Cambridge University Press: Cambridge, UK, 2013.
21. BSR. Adapting to Climate Change: A Guide for the Consumer Products Industry. 2011. Available online: http://www.bsr.org/reports/BSR_Climate_Adaptation_Issue_Brief_CP.pdf (accessed on 3 January 2018).
22. The Global Compact; UNEP. Business and Climate Change Adaptation: Toward Resilient Companies and Communities. 2012. Available online: http://caringforclimate.org/wp-content/uploads/Business_and_Climate_Change_Adaptation.pdf (accessed on 3 January 2018).
23. Biesbroek, G.R.; Klostermann, J.E.M.; Termeer, C.J.A.M.; Kabat, P. On the nature of barriers to climate change adaptation. *Reg. Environ. Chang.* **2013**, *13*, 1119–1129. [CrossRef]
24. Gottschick, M. How stakeholders handle uncertainty in a local climate adaptation governance network. *Clim. Chang.* **2015**, *132*, 445–457. [CrossRef]
25. Gifford, R. The dragons of inaction. *Am. Psychol.* **2011**, *66*, 290–302. [CrossRef] [PubMed]
26. Heal, G.; Millner, A. Uncertainty and decision making in climate change economics. *Rev. Environ. Econ. Policy* **2014**, *8*, 120–137. [CrossRef]
27. The Intergovernmental Panel on Climate Change (IPCC). Economics of adaptation. In *Climate Change 2014: Impacts, Adaptation, and Vulnerability. Part A: Global and Sectoral Aspects. Contribution of Working Group II to the Fifth Assessment Report of the Intergovernmental Panel of Climate Change*; IPCC, Ed.; Cambridge University Press: Cambridge, UK; New York, NY, USA, 2014; pp. 945–977.
28. Pahl, S.; Sheppard, S.; Boomsma, C.; Groves, C. Perceptions of time in relation to climate change. *Wiley Interdiscip. Rev. Clim. Chang.* **2014**, *5*, 375–388. [CrossRef]
29. Hallegatte, S. Strategies to adapt to an uncertain climate change. *Glob. Environ. Chang.* **2009**, *19*, 240–247. [CrossRef]
30. Baum, S.D. Description, prescription and the choice of discount rates. *Ecol. Econ.* **2009**, *69*, 197–205. [CrossRef]
31. Adger, W.N.; Arnell, N.W.; Tompkins, E.L. Successful adaptation to climate change across scales. *Glob. Environ. Chang.* **2005**, *15*, 77–86. [CrossRef]
32. Eisenhardt, K.M. Agency Theory: An Assessment and Review. *Acad. Manag. Rev.* **1989**, *14*, 57–74. [CrossRef]
33. Bertrand, M.; Schoar, A. Managing with Style: The Effect of Managers on Firm Policies. *Q. J. Econ.* **2003**, *118*, 1169–1208. [CrossRef]
34. Straub, J. Handlungstheorie. In *Handbuch Qualitative Forschung in der Psychologie*; Mey, G., Mruck, K., Eds.; VS Verlag für Sozialwissenschaften: Wiesbaden, Germany, 2010; pp. 107–123.
35. Gifford, R.; Kormos, C.; McIntyre, A. Behavioral dimensions of climate change: Drivers, responses, barriers, and interventions. *Wiley Interdiscip. Rev. Chang.* **2011**, *2*, 801–827. [CrossRef]

36. Feng, X.; Liu, M.; Huo, X.; Ma, W. What motivates farmers' adaptation to climate change? The case of apple farmers of Shaanxi in China. *Sustainability* **2017**, *9*, 519. [CrossRef]
37. Helfferich, C. *Agency. Die Analyse von Handlunsgfähigkeit und Handlungsmacht in Qualitativer Sozialforschung und Gesellschaftstheorie*; Bethmann, S., Helfferich, C., Hoffmann, H., Niermann, D., Eds.; Beltz Juventa Verlag: Weinheim, Germany; Basel, Switzerland, 2012.
38. Little, D. Social Agency and Rational Choice. 2009. Available online: https://understandingsociety.blogspot.de/2009/01/social-agency-and-rational-choice.html (accessed on 4 January 2018).
39. Paternoster, R.; Bachman, R.; Bushway, S.; Kerrison, E.; O'Connell, D. Human agency and explanations of criminal desistance: Arguments for a rational choice theory. *J. Dev. Life Course Criminol.* **2015**, *1*, 209–235. [CrossRef]
40. Popper, K.R. *The Logic of Scientific Discovery*; First Published by Julius Springer: Vienna, Austria, 1935.
41. Vervoort, J.M.; Rutting, L.; Kok, K.; Hermans, F.L.P.; Veldkamp, T.; Bregt, A.K.; van Lammeren, R. Exploring Dimensions, Scales, and Cross-scale Dynamics from the Perspectives of Change Agents in Social-Ecological Systems. *Ecol. Soc.* **2012**, *17*, 1708–3087. [CrossRef]
42. Hammoud, M.S.; Nash, D.P. What corporations do with foresight. *Eur. J. Futures Res.* **2014**, *2*, 1–20. [CrossRef]
43. Götze, U. *Investitionsrechnung. Modelle und Analysen zur Beurteilung von Investitionsvorhaben*; Springer: Berlin/Heidelberg, Germany, 2014.
44. Poggensee, K. *Investitionsrechnung: Grundlagen—Aufgaben—Lösungen. Investitionsrechnung*; Springer: Berlin/Heidelberg, Germany, 2011.
45. Jannek, K.; Burmeister, K. Corporate Foresight in Small and Medium-Sized Enterprises. 2007. Available online: http://www.foresight-platform.eu/wp-content/uploads/2011/04/EFMN-Brief-No.-101-Corporate-Foresight-SME.pdf (accessed on 4 January 2018).
46. Vecchiato, R. Environmental uncertainty, foresight and strategic decision making: An integrated study. *Technol. Forecast. Soc. Chang.* **2012**, *79*, 436–447. [CrossRef]
47. Skokan, K.; Pawliczek, A.; Piszczur, R. Strategic Planning and Business Performance of Micro, Small and Medium-Sized Enterprises. *J. Compet.* **2013**, *5*, 57–72. [CrossRef]
48. North, D.C. Institutions, institutional change, and economic performance. *Econ. Perspect.* **1990**, *5*, 97–112. [CrossRef]
49. Dilling, L.; Daly, M.E.; Travis, W.R.; Wilhelmi, O.V.; Klein, R.A. The dynamics of vulnerability: Why adapting to climate variability will not always prepare us for climate change. *Wiley Interdiscip. Rev. Clim. Chang.* **2015**, *6*, 413–425. [CrossRef]
50. Roberge, M.W.; Flaherty, J.C.; Almeida, R.M.; Boyd, A.C. Lengthening the Investment Horizon. 2014. Available online: http://conferences.pionline.com/uploads/conference_admin/mfse_time_wp_12_13.pdf (accessed on 4 January 2018).
51. WWF; Germanwatch. Klimarisiken für den Finanzsektor und Ihre Bearbeitung. 2015. Available online: https://germanwatch.org/de/download/13381.pdf (accessed on 4 January 2018).
52. Kelly, D.L.; Kolstad, C.D.; Mitchell, G.T. Adjustment costs from environmental change. *J. Environ. Econ. Manag.* **2005**, *50*, 468–495. [CrossRef]
53. European Commission. Climate Services. Serving a Society in Transition. 2017. Available online: https://ec.europa.eu/research/environment/index.cfm?pg=climate_services (accessed on 4 January 2018).
54. Santarius, T. Der Rebound-Effekt. Über die unerwünschten Folgen der erwünschten Energieeffizienz. Wuppertal Institute for Climate, Environment and Energy: Wuppertal, Germany, 2012. Available online: https://epub.wupperinst.org/frontdoor/deliver/index/docId/4219/file/ImpW5.pdf (accessed on 23 February 2018).
55. The Intergovernmental Panel on Climate Change (IPCC). Glossary. In *Climate Change 2014: Impacts, Adaptation, and Vulnerability. Part B: Regional Aspects. Contribution of Working Group II to the Fifth Assessment Report of the Intergovernmental Panel on Climate Change*; Cambridge University Press: Cambridge, UK, 2014; pp. 1757–1776.
56. Kates, R.W.; Travis, W.R.; Wilbanks, T.J. Transformational adaptation when incremental adaptations to climate change are insufficient. *Proc. Natl. Acad. Sci. USA* **2012**, *109*, 7156–7161. [CrossRef] [PubMed]
57. Kash, D.E.; Rycroft, R. Emerging patterns of complex technological innovation. *Technol. Forecast. Soc. Chang.* **2002**, *69*, 581–606. [CrossRef]

58. Kovach, J.J.; Hora, M.; Manikas, A.; Patel, P.C. Firm performance in dynamic environments: The role of operational slack and operational scope. *J. Oper. Manag.* **2015**, *37*, 1–12. [CrossRef]
59. Hartmann, J.; Moeller, S. Chain liability in multitier supply chains? Responsibility attributions for unsustainable supplier behavior. *J. Oper. Manag.* **2014**, *32*, 281–294. [CrossRef]
60. Huq, F.A.; Chowdhury, I.N.; Klassen, R.D. Social management capabilities of multinational buying firms and their emerging market suppliers: An exploratory study of the clothing industry. *J. Oper. Manag.* **2016**, *46*, 19–37. [CrossRef]
61. Akamp, M.; Müller, M. Supplier management in developing countries. *J. Clean. Prod.* **2013**, *56*, 54–62. [CrossRef]
62. Sridharan, R.; Simatupang, T.M. Power and trust in supply chain collaboration. *Int. J. Value Chain Manag.* **2013**, *7*, 76–96. [CrossRef]
63. Khurana, M.; Mishra, P.; Singh, A. Barriers to information sharing in supply chain of manufacturing industries. *Int. J. Manuf. Syst.* **2011**, *1*, 9–29. [CrossRef]
64. Kaluza, B.; Dullnig, H.; Malle, F. Principal-Agent-Probleme in der Supply Chain—Problemanalyse und Diskussion von Lösungsvorschlägen. 2003. Available online: https://pdfs.semanticscholar.org/c934/713884df2faf239d765464acdbab7418387b.pdf (accessed on 4 January 2018).

© 2018 by the authors. Licensee MDPI, Basel, Switzerland. This article is an open access article distributed under the terms and conditions of the Creative Commons Attribution (CC BY) license (http://creativecommons.org/licenses/by/4.0/).

Article

An Assessment of European Information Technology Tools to Support Industrial Symbiosis

Amtul Samie Maqbool *, Francisco Mendez Alva and Greet Van Eetvelde

Energy & Cluster Management, Faculty of Engineering and Architecture, Ghent University, B-9000 Ghent, Belgium; Francisco.MendezAlva@UGent.be (F.M.A.); Greet.VanEetvelde@UGent.be (G.V.E.)
* Correspondence: Samie.Maqbool@UGent.be

Received: 31 October 2018; Accepted: 21 December 2018; Published: 27 December 2018

Abstract: Industrial symbiosis (IS) has proven to bring collective benefits to multiple stakeholders by minimising underutilised resources, sharing knowledge and improving business and technical processes. In Europe alone, over €130 million have been invested since 2006 in research projects that enable IS by developing a methodology, tool, software, platform or network that facilitates the uptake of IS by different economic actors. This paper discusses and assesses information technology (IT) developments for supporting IS in Europe, following the five-stage methodology of Grant et al. (2010). It provides guidance to the applicants and reviewers of publicly funded research projects by listing the developments and gaps in the newly developed IT tools for IS. Content analysis of publicly available information on 20 IS supporting IT tools reveals a strong focus on synergy identification but a lack of support for the implementation stage of IS. The paper indicates that a vast quantity of IT tools and knowledge is created during the IT tool development stage and newer IT tools now also include implicit information for identifying IS. It was found that successfully operational IT tools are either part of a national or local IS programme or owned by a private company. The paper ends with the recommendation that better mechanisms are needed to ensure that publicly funded IS-supporting IT tools successfully reach the market.

Keywords: Industrial symbiosis; IT tools; research and innovation projects

1. Introduction

Optimisation of industrial sites through efficiency gains, carbon and energy savings and the use of renewable energy sources is a starting point to decouple economic growth from environmental degradation. However, the system boundaries can be widened to include other industries, process sectors and neighbouring municipalities to collectively strive for resource and energy efficiency and, ultimately, aim for a circular economy. This cooperative management of resource flows between businesses and engagement of traditionally separate entities in a collective approach to competitive advantage is termed as industrial symbiosis (IS) [1,2]. It involves physical exchanges of materials, energy, water and by-products, as well as sharing social tactics at the firm and multiorganisational level [3]. This interfirm cooperation or IS [2,4,5] enables businesses to strive for a collective economic and ecological benefit that is greater than the sum of the individual benefits each company can achieve [2,6,7]. IS is a crosscutting field that has relevance for policies relating to resource efficiency, the low carbon and circular economy, eco-innovation, green growth, regional economic development [8] and many more [9–14].

The political will to promote industrial symbiosis has grown over the last decade to the level of being fully integrated in Europe's long-term policies and strategies. Support is provided by all levels of governance in Europe in the form of the European Resource Efficiency Platform [15], Eco-Innovation Action Plan [16], Circular Economy Roadmap France [17], National Industrial Symbiosis Programme

(NISP) and West Midlands Industrial Symbiosis Programme (WISP) [18], etc. Europe invigorated its commitment to resource efficiency in 2011 by devising the "Roadmap to a Resource Efficient Europe" [19]. It is no surprise that IS was among the seven "top priority areas" outlined by the European Resource Efficiency Platform [15] in their 2012 Manifesto and Policy Recommendations. In 2015, the ambition to move away from a linear economy was endorsed with the release of the EU Circular Economy Package, revised in 2018 [9]. Industrial symbiosis is a means to support circularity and textbook examples of local industrial symbiosis cases like Kalundborg [6], Rotterdam [20], Tianjin [21] and many more [22–24] have repeatedly proven the potential of IS to enable progress towards the circular economy. Specifically, in Kalundborg, the symbiosis activities have resulted in reduction of CO_2 emissions, water savings, biofuel production from waste, reduction of imported primary materials, etc. [25]. It stands to reason that the material benefits in Kalundborg are coupled with social and economic benefits for all the parties involved.

Despite the proven benefits of IS, only 0.1% of the 26 million European enterprises are known to be active in IS [26] based on [27]. There is still a dire need for research and innovation at all levels of social, technological and commercial fronts to reshape policies, redesign products and processes and introduce new business models to show the feasibility of the circular economy [28] and of industrial symbiosis in particular. Starting from the launch of the Energy and the Innovation Unions under Europe's 2020 Strategy in 2011, symbiosis and circularity have become key priorities in political as well as scientific agendas [29,30]. In this regard, different funding opportunities have been announced to incentivise the move towards a low carbon, resource efficient and circular economy, the biggest of which is the H2020 research and innovation (R&I) funding scheme [31]. Cross-sectoral industrial symbiosis in process industries is a priority agenda for SPIRE cPPP (contractual Public Private Partnership on "Sustainable Process Industry through Resource and Energy Efficiency"), and several projects have received funding in this direction [32]. This paper helps to assess the effectiveness and impact of these initiatives in European countries, especially focusing on the development of information technology (IT) tools that support IS.

An interest in IT tools for IS has emerged on the research agenda in Europe, significantly supported by publicly funded R&I projects. It is evident that there is a need to avoid the pitfalls faced by earlier IT tools [33]. The literature shows that IT tools for IS have faced difficulties in remaining operational. In 2000, Chertow [2] presented a literature review of tools and approaches for industrial symbiosis which included three IT tools developed as part of the Designing Industrial Ecosystem Toolkit (DIET): DIET (Designing Industrial Ecosystems Tool), FAST (FAcility Synergy Tool) and REaLiTy (Regulatory Economic and Logistics Tool). The further development of the toolkit was cancelled due to changes in budget priorities [33]. In 2010, Grant et al. published an analysis of 13 IT tools for industrial symbiosis by applying a project lifecycle approach to identify the application of these tools in different project lifecycle stages (identification, assessment, barrier removal, implementation and follow up) of a symbiosis project [33].

Of the 13 IT tools that were discussed in the paper of Grant et al. (2010) [33], only four tools were reported to be operational, three of which were developed and implemented in Europe. Presteo, SymbioGis and CRISP were still operational in 2010 [33]. Of these three, only CRISP, succeeded by Synergie®, is still operational and constantly updated by the provider. In 2018, Benedict et al. identified four main barriers to IS and the corresponding IT support [34]. The first barrier is the lack of compatibility between the variety of required information from different sources and the underlying data-modelling framework of IT tools for IS creation. Second, technical feasibility and economic efficiency need to be accompanied by social aspects and mechanisms (willingness, trust, cooperation and reciprocity) that favour industrial symbiosis [35,36]. Third, the focus of most IT tools support IS identification (matchmaking), while the other project lifecycle stages of IS are often neglected. Fourth, the existing IT tools are difficult to access; often, there is no explicit mention of how and for whom the tool is available, and a gap is eminent in the literature about the management and development strategy of IT platforms [34].

This paper revisits the development of IT tools for IS by using the same five-stage methodology proposed by Grant et al. (2010) [33]. The aim is to objectively map the progress of IT tools for IS and conclude if the gaps in the development of IT tools for IS that have been identified in literature are being filled by newer IT tools. To reach this goal, content analysis of publicly available information on 20 IT tools was carried out and their focus on each of the five lifecycle stages of an IS project was evaluated. These five stages are, namely, synergy identification, symbiosis assessment, barrier removal, implementation and follow up [33]. To improve the robustness of the content analysis, the key performance indicators (KPIs) presented by Grant et al. (2010) for each stage were supplemented with the work of Van Eetvelde and colleagues (2005 and 2007) [5,36] and Maqbool et al. (2017) [37].

The five stages of any IS project lifecycle as outlined by Grant et al (2010) [33] are discussed below.

1.1. Synergy Identification

Synergy identification occurs through three primary means: new process discovery, resource (any underutilised materials, capacity, logistics, etc.) matching and relationship mimicking [33]. New process discovery refers to the identification of an industrial symbiosis enabled by technological development leading to value addition of a previously discarded by-product or waste through a novel transformation process [33]. Input–output resource matching refers to finding substitutes of resources among specific actors. Different models for synergy identification are deployed by public and private parties, such as IT-enabled identification by semantic matching [38,39], expert facilitated workshops [40] and integration of energy and material networks to achieve higher efficiency [41]. Relationship mimicking refers to the identification of an industrial symbiosis by making use of a documented case that resonates with the resources and industrial processes of the actors. However, relationship mimicking runs the risk of path dependence and, thus, research and innovation projects are crucial to the aim of innovative IS solutions.

1.2. Symbiosis Assessment

Symbiosis assessment evaluates the outcomes and challenges associated with IS. It is common practice to evaluate the environomic cost–benefit analysis of symbiotic activities between different partners [22]. However, other nontechnical aspects were also included in this study. This IS assessment stage was identified to be covered if the tools included:

- an assessment of compatibility of the IS activity with the national and local regulations;
- an evaluation of and distribution of economic gains between the IS partners;
- an assessment of spatial proximity between IS partners;
- a techno-environmental impact assessment of the industrial symbiosis; and
- the impact on job retention and creation under the symbiosis activity.

Focus on any one of the five aspects was considered sufficient to qualify the IT tool for achievement of this stage.

1.3. Barrier Removal

The barriers to industrial symbiosis have been enumerated by Van Eetvelde and colleagues [5,36], Lombardi [26] and Golev et al. [42]. Summarised by Golev et al., the barriers to IS are: lack of commitment to sustainable development, lack of information, difficulty in trust and cooperation between partners, technical infeasibility, uncertainty and inconvenience in regulatory compliance, lack of community awareness and, lastly, economic infeasibility [42]. Some of these barriers need to be removed within an individual organisation, some may need to be removed between organisations and still some are outside the bounds of the organisations, in which case, involvement of third parties to provide leverage and remove barriers is a common practice [43–45]. The barrier removal stage was considered to be covered if the IT tools focused on:

- removing legal barriers by providing a platform to jointly enrich legal expertise development [46];
- simplifying access to public investment funds;
- providing information on logistics for symbiotic transfers or the potential impact on existing material and energy networks;
- information to bring about IS-related emissions reduction and improved energy efficiency; and
- improving stakeholder interaction and overcoming information barriers between unrelated sectors.

It needs to be borne in mind that these barriers are interlinked and, hence, their solutions can have effects on each other.

1.4. Implementation

After barrier removal, decisions need to be made for implementing a symbiosis between industries. This entails the execution of the symbiosis activity, which cannot be decoupled from the selection of a management approach. Prior to any exchange, the decision on which approach will be useful to manage the symbiosis is to be made: either it will be self-organised by the participants of IS [7] or it will be facilitated or managed by a third party, such as a park manager or IS facilitator. In the latter case, the third party acts as an intermediary enabling cooperation [20,44]. The real-time handling of resource flows is the functionality that IT tools can provide to support the execution of the symbiosis. Regarding the selection of the management approach, IT tools can provide guidelines to businesses to support decision-making. Distribution of tasks and responsibilities between partners, which is defined by the clauses in the business contracts, also forms a part of the implementation stage. It was stressed in the work of Grant et al. (2010) that this stage is almost entirely handled by the participating organisations.

1.5. Follow Up (Review and Documentation)

There are two main functions covered in the final stage of the symbiosis cycle: thriving and propagating. Thriving is about continuous monitoring of impact and auditing to ensure stability of the activity and regular improvement of the symbiosis process. Disseminating is about publishing the results and lessons from the symbiosis activity at different levels of detail and diffusion, from own employees to the public. Documentation and dissemination within and outside the company help to replicate industrial symbiosis in the future; external outreach helps to generate value by improving the corporate image of the company and increasing the knowledge base of society. This creates the grounds for generalising symbiosis opportunities and thus generating IS mimicking by other businesses. IS tools that include a functionality to report the impact of the synergies or provide an IS case study database via publicly available online repositories cover this stage of the IS lifecycle. The five-stage IS lifecycle does not follow a linear pattern and, thus, follow up is a crucial stage for closing the loop between the implementation and synergy identification stages.

In this paper, the selected IT tools comprised existing and upcoming IT tools. These IT tools have been or are being developed as part of in-house research and development (R&D) projects by private companies or as a result of a publicly funded R&I project. To limit the scope of this study, only the IT tools developed in Europe were included in the assessment. The paper provides an answer to the following four questions: Is the combination of explicit and tacit knowledge used in identifying IS opportunities in the newer IT tools? Has the user base of newer IT tools expanded? Is the substantial focus on the IS identification stage still prevalent? Finally, what are the requisites to help IT tools for IS remain operational? The objective is to provide guidance for directing public funding and resources to projects that will help bridge the gap between IT tool development and the widespread application of IS in Europe.

The following section on methodology is added to define the data collection process and the KPIs that define the assessment criteria for each IS lifecycle stage. Then, the results and discussion section discusses the 20 IT tools with respect to their focus on the stages of the IS project lifecycle and insights are provided for future research and innovation endeavours. Finally, in the last section of the paper, conclusions are drawn.

2. Methodology

2.1. Data Collection

An inventory of 69 items was made to start the analysis by using an internet search for the terms industrial symbiosis, resource efficiency, resource and energy optimisation and circular economy. For each entry, relevant information was sourced from the publicly available online systems: websites, brochures and related academic literature. Twenty IT tools for IS were shortlisted from the set of data. These included 3 IT tools developed as part of in-house R&D projects by private companies; 16 IT tools developed or updated as part of a publicly funded project, 7 of which are still under development; and 1 IT tool that is being developed as part of independent academic endeavours.

Twelve IT tools that were privately owned and developed outside of Europe were excluded from the study. Also, 17 projects were discarded because of a lack of focus on industrial symbiosis. Twenty projects that focus on capacity building for industrial symbiosis were also excluded from the assessment because they do not deliver IT tools for IS. These 20 capacity-building projects and the 16 publicly funded projects for IT tool development received a funding of almost €137 million from the European Union (EU) since 2006.

One interview was carried out via Skype with the providers of the iNex platform. Specific information about the SymbioGis and Celero platforms was collected via email. Also, providers and researchers of the ZeroWin tool, Synergie®, Nova Light and SymbioSys were contacted via email to collect updates on their tools. Responses were received from the developers of Synergie®, Nova Light and SymbioSys. For the rest of the tools, publicly available information was obtained to carry out the assessment.

2.2. Key Performance Indicators for Content Analysis

To prepare the KPIs for the content analysis of the 20 IT tools for IS, the assessment criteria used by Grant et al. (2010) [33] was supplemented with some of the principles for park management by Van Eetvelde and colleagues (2005 and 2007) [5,36] and industrial symbiosis aspects used by Maqbool et al. (2017) [37]. The resulting assessment matrix is provided in Table 1.

Early in the analysis, it was found that the focus of the IT tools was not equally pointed at all of the five stages. Hence, a ranking with three levels was used to distinguish the level of focus of different IT tools on each of the five stages. These three levels are:

- 1 (no or low focus): minimal or no focus on the particular stage;
- 2 (moderate focus): supporting some aspects of the particular stage, but the main focus lies in another stage;
- 3 (strong focus): tools or projects with core objective and focus on this stage.

By introducing these three levels of measuring the focus on a given IS lifecycle stage, the desired flexibility for content analysis was introduced. An example of the IT tool developed in the EU funded H2020 project, EPOS (Enhanced energy and resource Efficiency and Performance in process industry Operations via onsite and cross-sectorial Symbiosis), is provided to elaborate how the three levels are used to reach a judgment about the focus of the IT tools. The EPOS IT tool optimises the generic models of industrial sectors, the so-called sector blueprints [47]. It does so by using techno-economic and environmental KPIs and identifies possible IS links between the unrelated industrial sectors, which is where the main focus of the EPOS tool lies. The EPOS User Club website also provides information on generic IS cases [48] that fall in the stage of follow up. The barrier of communication between unrelated sectors is overcome by anonymising data in the sector blueprints. Still, the stage where the main objective lies is synergy identification and thus received a score of 3. This is because the effort for anonymising data to provide techno-economic assessment and barrier removal (both stages scored as 2) are all done to achieve the identification of IS possibilities. The follow up stage received a score of 1 because, although the EPOS User Club is part of the project website, it is not an integrated part of the EPOS IT tool. Though no direct reference to the implementation stage was found in the publicly available information, this stage also received a score of 1 and not a 0 because the project is still running and this possibility cannot yet be conclusively eliminated. Hence, it was decided to include three levels of assessing the focus, which helps to avoid making unsupported harsh claims about the focus of the IT tools under study.

Table 1. Assessment criteria for content analysis of industrial symbiosis (IS)-related projects and information technology (IT) tools.

IS Project Lifecycle Stage	Assessment Criteria Presented by Grant et al. (2010)	Assessment Criteria Based on the Work of Van Eetvelde and Colleagues (2005 and 2007) Coupled with the Work of Grant et al. (2010) and Maqbool et al. (2017)
1. Synergy identification	New Process Discovery/Technology Innovation Input–Output Matching Case Study Mimicking	New process discovery/technology innovation Input–output matchmaking Relationship mimicking
2. Symbiosis assessment	Cost Benefit Analysis (CBA) Lifecycle Analysis (LCA) Economic Input–Output (EIO) Analysis and EIOLCA Barrier Assessment	Legal: Regulatory compliance Economic: CBA or best available techniques not entailing excessive costs (BATNEEC), business case assessment, lifecycle cost analysis (LCCA) Spatial: Distance between industrial symbiosis partners, land and transport availability Technical/Environmental: Process-based LCA, expert knowledge Social: Job retention and creation
3. Barrier removal	Technology Development Regulatory Approval Financing Business to Business Contractual Agreements Public Approval	Legal: Regulatory approval, business deals Economic/Financial: Information about public funds and unknown business opportunities Spatial: Optimisation of the network design, regional clustering of resources Technological: Technology/process development Social: Stakeholder workshops, community involvement, personnel (skill) training
4. Implementation	Commercialisation and Adaptive Management	Execution: real-time data handling of resource flows, contracts Management approach: self-organised, facilitated
5. Follow up	Documentation, Review and Publication	Thriving: External audits, standards, etc. Disseminating: Documentation platforms, wiki sites, marketing, etc.

Once the content analysis of the IT tools was finished, the scores were aggregated to find the trends in the development of these IT tools.

3. Results and Discussion

3.1. Content Analysis of the IT Tools

Of the 20 IT tools under study, 8 of these are available on the market. Two of these eight IT tools—Synergie®(by International Synergies) and iNEX platform (by iNEX Circular)—are products of private companies that couple their IT tool with IS facilitation services for the customers. Synergie®, an information and communication technology (ICT) resource management database and platform, is a web-based tool for IS facilitators. Additionally, the software allows companies to meet quality assurance protocol and audit requirements by offering database, project management and reporting functionalities to capture and store information about resources and to easily identify commercial opportunities for reuse [49]. Additionally, the R&I project SHAREBOX is currently running to include more functionalities that enhance the IT tool Synergie®. Now, Synergie®can be used, inter alia, by plant operators and production managers to effectively monitor and trade process resource streams in real time within their own supply chains or with other companies in a symbiotic industrial system [50]. These factors allowed for a high ranking of Synergie®in all the lifecycle stages of IS, with the highest focus on synergy identification. The iNEX platform aims to solve the recycling problem and address the gap in knowledge for waste producers and waste recyclers/users. The iNex platform has been active for five years and provides synergy identification and knowledge support about methodologies to their customers. Because of the focus on matchmaking, the iNex platform scored the highest in synergy identification.

Two of the available IT tools—IS DATA repository and CIRCULATOR—are self-service tools and both are freely available platforms for knowledge sharing. The IS DATA repository provides freely available information about industrial symbiosis in a structured way [51]. The repository is being developed as part of a project organised by the Eco-Industrial Development Council Section of the International Society of Industrial Ecology [52]. The aim of the project is to allow for and enable the construction of varied end-use applications for the research and facilitation of industrial symbiosis. Hence, the IS DATA repository scored high on knowledge and communication barrier removal, as well as follow up, because of the publication of IS case studies. CIRCULATOR is a project funded by EIT RawMaterials and provides customised information to the user in the form of existing cases of business strategies for circular businesses [53]. Since it is a passive platform which provides information to users about general IS possibilities, its objective was evaluated to lie in breaking information barriers and thus scored highest in the barrier removal stage.

Two more operational IT tools (SMILE Resource Exchange and the Italian Platform for Industrial Symbiosis) are part of national programmes to enhance industrial symbiosis. SMILE Resource Exchange, part of the Irish National Program, is a free online platform for businesses to connect and identify opportunities for resource exchange [54]. The platform is coupled with a service provided by local consultants in different regions of Ireland. SMILE Platform breaks the information and communication barrier between stakeholders to provide an opportunity for synergy identification. The Italian Platform for Industrial Symbiosis, managed by the Italian National Agency for New Technologies, Energy and Sustainable Economic Development (ENEA), provides a platform for local businesses to search for synergies between the registered companies [55]. The platform is part of the project "Eco-innovazione Sicilia Project". The IT tool that is used by ENEA uses geo-referenced data to identify opportunities between companies and helps establish a network between companies and stakeholders. Hence, it ranked higher on synergy identification and barrier removal.

At the local level in Belgium, the initiative SYMBIOSE (BE) by the region of Flanders brings together local actors and supports IS realisation to help reduce the environmental impact of economic activities [56,57]. Symbiosis 3.0 is the web-platform of this initiative, providing a matchmaking service to the users. The initiative also helps in providing further information to interested parties about realising IS. Because of the heavy focus of the platform on the matchmaking service, it scored highest in the first stage of the IS project lifecycle. The last of the operational IT tools, SymbioSyS, incorporates tacit knowledge to identify symbiosis matches. It is a freely available tool developed for a variety of users who do not require expert knowledge to manage it [58]. The literature suggests that SymbioSyS is a freely available tool [58], however, the IT tool is only provided under agreement with the Universidad de Cantabria. The IT tool achieved the maximum score of 3 points in synergy identification because of the use of implicit knowledge to identify IS opportunities. From the literature, it was found that the SymbioSyS tool also uses geo-referenced data and, hence, scored high in the symbiosis assessment stage because it provides spatial assessment for synergy.

Six out of 20 IT tools included in this study were inaccessible, of which e-symbiosis, ZeroWin and Locimap have been developed through European funding. Presteo and SymbioGIS are a result of a national programme, and Nova Light is a product of a private company. The nonoperational tools still provide a world of knowledge for future research, for example, the eSymbiosis project, which successfully integrated tacit knowledge in the process of IS identification with the use of ontologies [38,39]. It received a score of 3 for the synergy identification stage. There are examples of new IT tools that use knowledge embedded in nonoperational IT tools. For example, Presteo, based on the work of Adoue [59], preceded the development of SymbioGis. The web-based tool for input-output (IO) matching, Presteo, was made for end users to identify symbiosis opportunities [33] and was followed by SymbioGis, a web-based GIS tool to facilitate industrial symbiosis in Geneva, Switzerland.

The tool was developed as a result of a collaboration between the local government, university and the Sofies group [60]. Since the SymbioGis tool was developed to identify IO and service matches, it also provides a technical and geographical feasibility assessment and, hence, scored high on the first two stages of the IS lifecycle. Because of the functionality of the tool to provide information on locations for new facilities based on material flow in the region, it helps to remove the information barrier. Nova Light is a web-based platform developed by another private business. It was developed to provide matches between waste producers and consumers. Thus, a score of 3 was given for the synergy identification stage.

Six IT tools for IS that are under development are part of currently unfinished R&I projects. The IT tool of FISSAC scored the highest score of 3 in synergy assessment because it is planned to be able to respond to resource efficiency and environmental performance concerns (by the help of lifecycle analysis (LCA)) and scored a 2 on synergy identification because it provides matches for the users of the geo-referenced platform [61]. The set of tools developed by the Maestri project aims to enhance the overall efficiency of industrial processes [62]. The combination of Maestri front end tools (MSM and ecoPROSYS) and the Internet of Things platform focuses the most on the assessment stage (3) and also provides support for breaking barriers for information exchange between different software tools (scored 2). The BISEPS tool is being developed to bring improvement in energy efficiency for businesses in business parks by clustering individual energy needs and demands [63]. Since the focus is on match-making, the score (3) was the highest for the synergy identification stage. The synergy assessment stage of Symby-Net, the platform developed by the Symbioptima project [64], scored the highest (3) because this IT tool provides lifecycle sustainability assessments of symbiotic networks. Symby-Net was developed to be used by industry managers as well as IS facilitators or park managers [64]. The ERMAT project will provide a web tool to be used for matchmaking purposes. This IT tool will be freely available for use by the public, while more information will be provided for a fee [65]. The IT tool for ERMAT scores highest on the first stage of the IS project lifecycle. The last of the IT tools being developed as part of an R&I project is the EPOS toolbox, the content of which was discussed as an example in the methodology section.

3.2. Scoring the IT Tools against IS Project Lifecycle Stages

The results of the content analysis are summarised in Table 2. The acquired score gives the sum of the assigned scores for all IT tools for the respective lifecycle stage. The maximum score is calculated as 3×20, which would be the score had all the IT tools focused on the lifecycle stage with the highest emphasis.

Table 2. Results of the content analysis of the IT tools.

#	Name	Status	Funding Body and Funding Amount (in Million €)	Target Users	Available Since/From	Synergy Identification	Symbiosis Assessment	Barrier Removal	Implementation	Follow up
1	SymbioGIS	Completed, not operational	Geneva Govt.	Urban planners	2007	3	3	2	1	1
2	Presteo/LGCD	Completed, not operational	Not available	Industry	2010	3	1	1	1	1
3	Nova light	Completed, not operational	Private	Waste producers and users	2010	3	1	1	1	1
4	SMILE Resource Exchange platform	Operational	Public Private	IS facilitators	2013	3	1	2	1	1
5	eSymbiosis	Completed, not operational	EU 0.9, Others 0.9	Industry	2014	3	1	1	1	1
6	iNex platform	Operational	Private	IS facilitators, industry	2014	3	2	1	1	2
7	IT tool by project Locimap	Completed, not operational	EU 1.9, Others 0.6	Park managers	2014	1	3	2	1	2
8	Resource-eXange-Platform (project ZEROWIN)	Completed, not operational	EU 6.1, Others 3.3	General public	2015	2	1	1	1	2
9	Italian Platform for Industrial Symbiosis	Operational	Not available	Facilitators and industry	2015	2	1	2	1	1
10	CIRCULATOR (tool)	Operational	EIT RawMaterials	all users	2017	1	1	3	1	1
11	Symby-Net (project SYMBIOPTIMA)	In development	EU 6, Others 1.3	Industry, park managers	2018	2	3	1	1	1
12	SymbioSyS (tool)	Operational	Not available	General public	2018	3	2	1	1	1
13	EPOS toolbox by EPOS project	In development	EU 5.4, Others 0.5	Industry	2019	3	2	2	1	1
14	Synergie®(project SHAREBOX)	Operational and continuous update	EU 5.4, Other 0.5	Industry	2019	3	2	2	2	2
15	Maestri (Internet of Things) platform (project Maestri)	In development	EU 5.7	Industry	2019	1	3	2	2	1
16	IT tool by FISSAC project	In development	Not available	Cluster managers, industry	2020	2	3	1	1	1
17	BISEPS-tool (part of BISEPS project)	In development	EU	Park managers	2020	3	2	2	1	1
18	Industrial Symbiosis DATA repository	Operational	Not available	Researcher, IT developers	On going	1	1	2	1	2
19	Symbiosis 3.0 (project SYMBIOSE BE)	Operational	Public	General public	2016	3	1	2	1	1
20	ERMAT web tool	In development	EIT RawMaterials	General public	2018	3	1	1	1	1
	Acquired score					48	35	32	21	25
	Maximum score					60	60	60	60	60

The results show that 12 IT tools have their main objective in the stage of synergy identification, with a prevalent commitment to matchmaking (48 points out of 60). Symbiosis assessment received second priority when developing the IT tools (35 points). These results coincide with the findings of Grant et al. (2010). The reason for this trend lies in the fact that matchmaking tools and assessment methodologies hold the most promise for innovation for academia, IT tool developers and the facilitators of industrial symbiosis. Closely following the IS assessment stage was barrier removal (32 points). The focus on the follow up stage is due to the dissemination function of the IT tools. The implementation stage of the IS project lifecycle shows the least focus (21 points). This is also not an unexpected result, as once the synergies are identified and assessed and barriers are removed, then contractual details and commercialisation by the industry does not pose a major difference as compared to normal business practices [26], resulting in the lowest focus being on the implementation stage [33] by IT tools for IS.

3.3. Discrepancies Resolved by Newer IT Tools for IS

Similar to the observations made by Grant et al. (2010), the assessment of newer IS-enabling IT tools still shows a heightened focus on the identification of symbiosis opportunities. This trend raises the legitimate question of if future research should still focus on developing more tools in the same trend, or should the focus be shifted to customising the existing tools to embed a management functionality to provide specific solutions to symbiosis partners. Learning from the available IT tools, it is evident that there is a lack of IT tools that provide support to implementation and management of the symbiosis activities, and more research and development efforts are required in that area.

The nontechnical information or tacit knowledge regarding industrial processes, business interactions, regulatory compliance, etc., is crucial for successful identification and application of IS. The older IT tools had a limited focus on the inclusion of nontechnical information in the development framework of the IT. Academia and industry have understood and expressed the importance of nontechnical barriers and drivers to IS [25,26] and the inclusion of tacit knowledge in the process of opportunity identification. The newer IT tools developed under R&I projects have focused on the inclusion of tacit knowledge for symbiosis identification and assessment, be it by the use of ontologies (projects e-symbiosis and SymbioSys) or by use of a recommender which identifies opportunities based on machine learning (AI) algorithms [66] (project SHAREBOX).

The other discrepancy in the older IT tools was the limited user profile, mainly targeted at the IT developer or engineer. This paper shows that more IT tools are being developed for use by industries that are referred to as participants by Grant et al. (2010) [33]. As mentioned by Van Capelleveen et al. (2018), organisations need to justify the time and resources invested in exploration of potential opportunities, the benefits of which are not certain [66]. This becomes even more difficult when an outsider requests information that may require time and resources and a risk of breach of confidentiality. IS opportunity exploration via IT tools can help industries control the information flow outside of the organisation [47]. As more IT tools are being developed for participants (EPOS toolbox, updates in Synergie under the project SHAREBOX, Maestri IoT, e-symbiosis platform), this should enhance the application of IS in various industries. The multidisciplinary and cross-sectoral nature of IS requires the IT tools to be used by a myriad of users. The knowledge that these users have also differs based on the sector to which they belong. Currently, the existing IT tools are specialised in solving sector-specific problems, while IS-enabling IT tools should address users from different sectors to solve cross-sectoral problems, which may result in the development of complex IT tools that require a high level of expertise to use. This dichotomy is shown in Figure 1. Passive online tools for matchmaking occupy the block of "nonspecific and easy to use IT tools"; however, these tools have shown to be less effective [26]. IT developers must ensure to avoid the effort invested to build tools that fit in the lower half of Figure 1.

Figure 1. The relationship between specificity and user friendliness of IT tools for IS.

This figure, combined with the findings of this paper, will help researchers, industries and funding agencies to direct future endeavours in a more efficient and effective manner in order to pave Europe's way for resource efficiency and circularity.

The last research question to answer relates to the features of the successfully operational IT tools. The operational IT tools Synergie and the iNex platform are being developed by IS facilitators whose core business is coupled with the development of such a tool. The other operational IT tools for IS are part of local or national symbiosis programmes, such as Symbiose 3.0 and the SMILE platform, respectively. In both cases, the IT tools have a clear ownership and undergo continuous improvement. These are the two main features of successfully operational IT tools for IS. Another supporting mechanism to ensure that IT tools for IS find their way to the market and effectively support the uptake of IS in industry is through a variety of R&I projects and local initiatives. These projects and programmes help to build capacity for IS in industry, small- and medium-sized enterprises, academia and local and national administrative bodies. These capacity-building efforts coupled with IT tools provide the advantage of focusing on all the stages of an IS project lifecycle that might not be fully supported by IT tools alone.

The research also showed that more IT tools are being developed with public funding. Five out of the six nonoperational IT tools included in this study were funded by some public organisation. It is crucial to use the knowledge created by these nonoperational tools and develop new and improved IT tools for IS. For example, as a follow up of SymbioGIS, since 2014, a consortium including the Sofies group has been busy developing a Celero platform to identify and facilitate industrial symbiosis and cleaner production among companies [67]. Effective mechanisms to support revitalising and improving publicly funded IT tools for IS need to be put in place in order to disseminate the benefits to the wider society.

4. Conclusions

This paper provided findings of the assessment and quantification of IS-enabling IT tools by focusing on the lifecycle stages of an IS project. The study enumerated the strengths and weaknesses of IT tools for IS; therefore, the KPIs and findings presented in this paper can be used as reference for self-assessment by applicants of R&I projects for development of new IT tools for IS. The assessment matrix followed in this paper will also prove useful for funding agents and evaluators.

From the results, it can be concluded that the gaps identified by Grant et al. (2010) are being eliminated by the newer IT tools developed in Europe. Although the focus of IT tools still mainly centres on the identification of IS opportunities, the stage of implementation and management of a symbiosis activity is overlooked by IT development and research efforts. This gap can be easily bridged by updating existing and nonoperational IT tools for IS.

It was observed that IT tools have a higher chance of remaining operational when the companies responsible for developing them can improve their core business by using it. Hence, coupling IT tools with the services of an IS facilitator or a local or national IS programme increases the chances of the tools remaining functional and economically viable. It was also observed that newer IT tools utilise tacit knowledge when identifying symbiosis opportunities and thus attempt to fill the gap in the older IS-supporting IT tools.

To conclude, the IT tools being developed in Europe are well on their way to proving effective for wider application. However, better mechanisms are still needed to ensure that IS-supporting IT tools developed with public funding reach the market and that the capacity developed from successful R&I projects is made available to peers, industry and the general public for successful application of IS in Europe.

Author Contributions: Conceptualization, Writing-Original Draft Preparation, A.S.M.; Writing-Review & Editing, A.S.M. and G.V.E.; Methodology, Formal Analysis, Investigation, A.S.M. and F.M.A.; Supervision, G.V.E.

Funding: The research leading to these results has received funding from the European Union's Horizon 2020 research and innovation programme under grant agreement no. 679386, EPOS project (Enhanced energy and resource Efficiency and Performance in process industry Operations via onsite and cross-sectorial Symbiosis). The sole responsibility of this publication lies with the authors. The European Union is not responsible for any use that may be made of the information contained herein.

Conflicts of Interest: The authors declare no conflict of interest.

References

1. Lowe, E.A.; Evans, L.K. Industrial ecology and industrial ecosystems. *J. Clean. Prod.* **1995**, *3*, 47–53. [CrossRef]
2. Chertow, M.R. Industrial Symbiosis: Literature and Taxonomy. *Annu. Rev. Energy Environ.* **2000**, *25*, 313–337. [CrossRef]
3. Puente, M.C.R.; Arozamena, E.R.; Evans, S. Industrial symbiosis opportunities for small and medium sized enterprises: Preliminary study in the Besaya region (Cantabria, Northern Spain). *J. Clean. Prod.* **2015**, *87*, 357–374. [CrossRef]
4. Lombardi, D.R.; Laybourn, P. Redefining Industrial Symbiosis. *J. Ind. Ecol.* **2012**, *16*, 28–37. [CrossRef]
5. Van Eetvelde, G.; Delange, E.; De Zutter, B.; Matthyssen, D.; Gevaert, L.; Schram, A.; Verstraeten, B.; Dierick, B.; Allaert, G.; Vanden Abeele, P.; et al. *Groeiboeken Duurzame BedrijvenTerreinen Juridisch, Economisch, Ruimtelijk, Technisch Bekeken*; Vanden Broele Grafische Groep: Brugge, Belgium, 2005.
6. Jacobsen, N.B. Industrial Symbiosis in Kalundborg, Denmark: A Quantitative Assessment of Economic and Environmental Aspects. *J. Ind. Ecol.* **2008**, *10*, 239–255. [CrossRef]
7. Chertow, M.; Ehrenfeld, J. Organizing Self-Organizing Systems. *J. Ind. Ecol.* **2012**, *16*, 13–27. [CrossRef]
8. Laybourn, P.; Lombardi, D.R. Industrial Symbiosis in European Policy. *J. Ind. Ecol.* **2012**, *16*, 11–12. [CrossRef]
9. European Commission. Circular Economy Package. January 2018. Available online: http://ec.europa.eu/environment/circular-economy/index_en.htm (accessed on 21 December 2018).
10. European Commission. *A Sustainable Bioeconomy for Europe: Strengthening the Connection between Economy, Society and the Environment*; Directorate-General for Research and Innovation; Publications Office of the European Union: Luxembourg, 2018.
11. Publications Office of the European Union. *Innovating for Sustainable Growth: A Bioeconomy for Europe*; Publications Office of the European Union: Brussels, Belgium, 2012.
12. European Commission. *A Roadmap for Moving to a Competitive Low Carbon Economy in 2050*; Europese Commissie: Brussels, Belgium, 2011.
13. Directorate-General for Research and Innovation (European Commission). *Pathways to Sustainable Industries: Energy Efficiency and CO2 Utilisation*; European Commission: Brussels, Belgium, 2018.
14. European Commission. *A European Strategy for Plastics in a Circular Economy*; European Commission COM(2018) 28; European Commission: Brussels, Belgium, 2018.
15. EREP. *European Resource Efficiency Platform—Manifesto & Policy Recommendations*; Policy Recommendations; European Commission: Brussels, Belgium, 2012.

16. Funding Programmes—Eco-Innovation Action Plan—European Commission. Plan—European Commission. 2011. Available online: https://ec.europa.eu/environment/ecoap/about-action-plan/union-funding-programmes_en (accessed on 11 June 2018).
17. Davies, P.A. France Unveils Circular Economy Roadmap. 2018. Available online: https://www.globalelr.com/2018/04/france-unveils-circular-economy-roadmap/ (accessed on 11 Jun2018).
18. Mirata, M. Experiences from early stages of a national industrial symbiosis programme in the UK: determinants and coordination challenges. *J. Clean. Prod.* **2004**, *12*, 967–983. [CrossRef]
19. European Commission. *Roadmap to Resource Efficient Europe*; European Commission COM/2011/571; European Commission: Brussels, Belgium, 2011.
20. Baas, L. Planning and Uncovering Industrial Symbiosis: Comparing the Rotterdam and Östergötland regions. *Bus. Strategy Environ.* **2011**, *20*, 428–440. [CrossRef]
21. Shi, H.; Chertow, M.; Song, Y. Developing country experience with eco-industrial parks: A case study of the Tianjin Economic-Technological Development Area in China. *J. Clean. Prod.* **2010**, *18*, 191–199. [CrossRef]
22. Chertow, M.R.; Lombardi, D.R. Quantifying economic and environmental benefits of co-located firms. *Environ. Sci. Technol.* **2005**, *39*, 6535–6541. [CrossRef] [PubMed]
23. Li, Y.; Shi, L. The Resilience of Interdependent Industrial Symbiosis Networks: A Case of Yixing Economic and Technological Development Zone. *J. Ind. Ecol.* **2015**, *19*, 264–273. [CrossRef]
24. Sharib, S.; Halog, A. Enhancing value chains by applying industrial symbiosis concept to the Rubber City in Kedah, Malaysia. *J. Clean. Prod.* **2017**, *41*, 1095–1108. [CrossRef]
25. Van Eetvelde, G. Industrial Symbiosis. In *Resource Efficiency of Processing Plants: Monitoring and Improvement*; Krämer, S., Engell, S., Eds.; John Wiley & Sons: Hoboken, NJ, USA, 2017.
26. Lombardi, R. Non-technical barriers to (and drivers for) the circular economy through industrial symbiosis: A practical input. *Econ. Policy Energy Environ.* **2017**, *1*, 171–189. [CrossRef]
27. Statistics Explained. Eurostat Business Demography. 2017. Available online: http://ec.europa.eu/eurostat/statistics-explained/index.php/Business_demography_statistics (accessed on 18 June 2018).
28. Stahel, W.R. The circular economy. *Nature* **2016**, *531*, 435–438. [CrossRef]
29. European Commission. Strategic Energy Technology Plan—Energy. Available online: https://ec.europa.eu/energy/en/topics/technology-and-innovation/strategic-energy-technology-plan (accessed on 11 June 2018).
30. European Commission. *Erasmus+ Programme Guide*; European Commission: Brussels, Belgium, 2018.
31. European Commission. Horizon 2020—European Commission. Horizon 2020. Available online: https://ec.europa.eu/programmes/horizon2020/en/ (accessed on 11 June 2018).
32. SPIRE. Sustainable Process Industry through Resources an Energy Efficiency. Available online: https://www.spire2030.eu/ (accessed on 13 June 2018).
33. Grant, G.B.; Seager, T.P.; Massard, G.; Nies, L. Information and Communication Technology for Industrial Symbiosis. *J. Ind. Ecol.* **2010**, *14*, 740–753. [CrossRef]
34. Benedict, M.; Kosmol, L.; Esswein, W. Designing Industrial Symbiosis Platforms—From Platform Ecosystems to Industrial Ecosystems. In Proceeding of the Pacific Asia Conference on Information Systems, Yokohama, Japan, 26–30 June 2018.
35. Van Capelleveen, G.; Amrit, C.; Yazan, D.M. A Literature Survey of Information Systems Facilitating the Identification of Industrial Symbiosis. In *From Science to Society*; Springer: Cham, Switzerland, 2018; pp. 155–169.
36. Van Eetvelde, G.; Deridder, K.; Segers, S.; Maes, T.; Crivits, M. Sustainability scanning of eco-industrial parks. In Proceeding of the European Roundtable for Sustainable Consumption and Production (ERSCP), Basel, Switzerland, 20–22 June 2007; p. 20.
37. Maqbool, A.S.; Piccolo, G.E.; Zwaenepoel, B.; van Eetvelde, G. A Heuristic Approach to Cultivate Symbiosis in Industrial Clusters Led by Process Industry. In Proceeding of the Sustainable Design and Manufacturing (SDM-17), Bologna, Italy, 6–28 April 2017; pp. 579–588.
38. Trokanas, N.; Cecelja, F.; Raafat, T. Semantic input/output matching for waste processing in industrial symbiosis. *Comput. Chem. Eng.* **2014**, *66*, 259–268. [CrossRef]
39. Raafat, T.; Trokanas, N.; Cecelja, F.; Bimi, X. An ontological approach towards enabling processing technologies participation in industrial symbiosis. *Comput. Chem. Eng.* **2013**, *59*, 33–46. [CrossRef]

40. Jensen, P.D.; Basson, L.; Hellawell, E.E.; Bailey, M.R.; Leach, M. Quantifying 'geographic proximity': Experiences from the United Kingdom's National Industrial Symbiosis Programme. *Resour. Conserv. Recycl.* **2011**, *55*, 703–712. [CrossRef]
41. Hackl, R.; Andersson, E.; Harvey, S. Targeting for energy efficiency and improved energy collaboration between different companies using total site analysis (TSA). *Energy* **2011**, *36*, 4609–4615. [CrossRef]
42. Golev, A.; Corder, G.D.; Giurco, D.P. Barriers to Industrial Symbiosis: Insights from the Use of a Maturity Grid. *J. Ind. Ecol.* **2015**, *19*, 141–153. [CrossRef]
43. Van Eetvelde, G.; de Zutter, B.; Deridder, K.; de Roeck, V.; Devos, D. *Groeiboek Duurzame Bedrijventerreinen, Juridisch Bekeken*; Vanden Broele: Brugge, Belgium, 2005.
44. Siskos, I.; van Wassenhove, L.N. Synergy Management Services Companies: A New Business Model for Industrial Park Operators. *J. Ind. Ecol.* **2016**, *21*, 802–814. [CrossRef]
45. Paquin, R.L.; Howard-Grenville, J. The Evolution of Facilitated Industrial Symbiosis. *J. Ind. Ecol.* **2012**, *16*, 83–93. [CrossRef]
46. Harmonised Assessment of Regulatory Bottlenecks and Standardisation Needs for the Process Industry I SPIRE. 2019–2017. Available online: https://www.spire2030.eu/harmoni (accessed on 4 July 2018).
47. Cervo, H.; Bungener, S.; Méchaussie, E.; Kantor, I.; Zwaenepoel, B.; Maréchal, F.; van Eetvelde, G. Virtual Sector Profiles for Innovation Sharing in Process Industry—Sector 01: Chemicals. In Proceeding of the Sustainable Design and Manufacturing (SDM-17), Bologna, Italy, 6–28 April 2017; pp. 569–578.
48. Generic IS Cases/EPOS. The EPOS User Club. Available online: https://epos.userecho.com/knowledge-bases/10-generic-is-cases (accessed on 28 November 2018).
49. SYNERGie®Software. International Synergies. 2018. Available online: https://www.international-synergies.com/software/ (accessed on 28 June 2018).
50. Introducing SHAREBOX—A Systemic Leap Forward for Industrial Symbiosis. International Synergies. 22 March 2016. Available online: https://www.international-synergies.com/news/introducing-sharebox-a-systemic-leap-forward-for-industrial-symbiosis/ (accessed on 28 June 2018).
51. What Is ISDATA I ISDATA. Available online: http://isdata.org (accessed on 31 July 2018).
52. Davis, C. The Industrial Symbiosis Data Repository/ Primary Database Project. In Proceeding of the 2013 Industrial Symbiosis Research Symposium, Ulsan, Korea, 23–24 June 2013. Database Project Handout.
53. Mix Your Strategies—Circulator. Available online: http://circulator.eu//mix-your-strategies (accessed on 31 July 2018).
54. Macroom E Enterprise Centre. About SMILE Resource Exchange—Save Money, Divert Waste from Landfill—Smile Resource Exchange. Available online: http://www.smileexchange.ie/about-us (accessed on 31 July 2018).
55. Cutaia, L.; Luciano, A.; Barberio, G.; Sbaffoni, S.; Mancuso, E.; Scagliarino, C.; La Monica, M. The experience of the first industrial symbiosis platform in Italy. *Environ. Eng. Manag. J.* **2015**, *14*, 1521–1533. [CrossRef]
56. OVAM. Symbiose Verzilvert Reststromen (Symbiosis to Valorise Waste). Available online: https://ovam.be/symbiose-verzilvert-reststromen (accessed on 2 August 2018).
57. Symbiose—Catalisti. CATALISTI. December 2016. Available online: http://catalisti.be/project/symbiose/ (accessed on 30 November 2018).
58. Álvarez, R.; Ruiz-Puente, C. Development of the Tool SymbioSyS to Support the Transition Towards a Circular Economy Based on Industrial Symbiosis Strategies. *Waste Biomass Valorization* **2017**, *8*, 1521–1530. [CrossRef]
59. Adoue, C. Méthodologie d'identification de Synergies éco-Industrielles Réalisables Entre Entreprises sur le Territoire Français. Ph.D. Thesis, Université de Technologie de Troyes, Troyes, France, 2004.
60. Massard, G.; Erkman, S. A regional industrial symbiosis methodology and its implementation in Geneva, Switzerland. In Proceeding of the 3rd International Conference on Life Cycle Management, Zurich, Switzerland, 27–29 August 2007; Volume 27.
61. Sánchez, B.J.; Hiniesto, D.; Guedella, E. D1.8: Initial outline of FISSAC Industrial Symbiosis Model and Methodology/Report. 2017. Available online: http://fissacproject.eu/wp-content/uploads/2017/09/FISSAC-D1.8-Initial-outline-of-FISSAC-IS-Model-Methodology.pdf (accessed on 21 December 2018).
62. Baptista, A.J.; Lourenco, E.J.; Pecas, P.; Silva, E.J.; Estrela, M.A.; Holgado Granados, M.; Benedetti, M.; Evans, S. MAESTRI Efficiency Framework as a support tool for Industrial Symbiosis implementation. 2017. Available online: https://www.repository.cam.ac.uk/handle/1810/267299 (accessed on 21 December 2018).

63. BISEPS Tool—Biseps. BISEPS Project Website. Available online: http://www.biseps.eu/biseps-tool/ (accessed on 30 November 2018).
64. Brondi, C.; Cornago, S.; Ballarino, A.; Avai, A.; Pietraroia, D.; Dellepiane, U.; Niero, M. Sustainability-based Optimization Criteria for Industrial Symbiosis: The Symbioptima Case. *Procedia CIRP* **2018**, *69*, 855–860. [CrossRef]
65. Grazia, B.; Jacobsson, E.; Björn, H.; Karhu, M.; Meneve, J.; Kinnunen, P.; Karu, V. Sustainability and business for residual materials: The ERMAT project approach. 2017. Available online: http://www.enea.it/it/seguici/events/sun_simbiosiindustriale_25ott17/ERMAT_ENEA_SUN.pdf (accessed on 28 June 2018).
66. Van Capelleveen, G.; Amrit, C.; Yazan, D.M.; Zijm, H. The influence of knowledge in the design of a recommender system to facilitate industrial symbiosis markets. *Environ. Model. Softw.* **2018**, *110*, 139–152. [CrossRef]
67. The Celero Platform: Detection of Industrial Symbiosis and Cleaner Production Opportunities. Sofies. Available online: https://sofiesgroup.com/en/projects/ecoman-platform-detection-industrial-symbiosis-cleaner-production-opportunities/ (accessed on 28 June 2018).

© 2018 by the authors. Licensee MDPI, Basel, Switzerland. This article is an open access article distributed under the terms and conditions of the Creative Commons Attribution (CC BY) license (http://creativecommons.org/licenses/by/4.0/).

Article

A Framework for Implementing and Tracking Circular Economy in Cities: The Case of Porto

António Cavaleiro de Ferreira [1,*] and Francesco Fuso-Nerini [1,2]

1   Unit of Energy Systems Analysis (dESA), KTH Royal Institute of Technology, Brinellvägen 68, SE-100 44 Stockholm, Sweden; francesco.fusonerini@energy.kth.se
2   Payne Institute, Colorado School of Mines, Golden, CO 80401, USA
*   Correspondence: antonio57_cf@hotmail.com

Received: 28 January 2019; Accepted: 16 March 2019; Published: 26 March 2019

Abstract: Circular economy (CE) is an emerging concept that contrasts the linear economic system. This concept is particularly relevant for cities, currently hosting approximately 50% of the world's population. Research gaps in the analysis and implementation of circular economy in cities are a significant barrier to its implementation. This paper presents a multi-sectorial and macro-meso level framework to monitor (and set goals for) circular economy implementation in cities. Based on literature and case studies, it encompasses CE key concepts, such as flexibility, modularity, and transparency. It is structured to include all sectors in which circular economy could be adopted in a city. The framework is then tested in Porto, Portugal, monitoring the circularity of the city and considering its different sectors.

Keywords: circular economy; urbanization; framework; indicators; circular city

1. Introduction

Circular economy (CE) is an emerging concept which is seen as the alternative to the current linear economy [1]. Its impacts are seen as relevant—now more than ever. Given the present environmental crisis, alongside economic uncertainty, governments have started seeking alternatives. China has made successful use of CE to tackle urban pollution and a wasteful system [2]. Its results have also motivated the European Union (EU) to promote it.

The EU has a variety of international environmental, technological, economical, and social goals. Particularly, it looks to follow the United Nations (UN) 17 Sustainable Development Goals (SDGs). A CE can be the answer to a systemic approach that eases the completion of several goals, specifically Goal 12 [3].

Alongside a CE comes urbanization, so the need to define and understand circular cities. This gives rise to a new set of research gaps, as the definition of a circular city, how to monitor the city's circularity, how to implement this circularity, which indicators to use, and what data to collect [4]. This paper explores these different research gaps, with a focus on developing a tool to measure and set goals of circularity in a city.

With this tool, the Circular City Analysis Framework (CCAF), the main goal is to help municipalities and cooperating actors understand a city's circularity. It is based on the city's intrinsic properties and sectors, as well as the circular economy and the circular city's characteristics. Moreover, it is structured to adapt to different cities, to other tools and future indicators. It was tested in the Portuguese city of Porto, as a case study.

The introduction of this paper is subdivided into the present global context, followed by CE key concepts. This is then bridged to the circular city definition, which is essential to the framework development. Finally, the Portuguese context is introduced, together with the northern city of Porto.

Second, it explores the methodology used to develop the framework. This section is subdivided into a literature review timeline, followed by a description of the framework's general characteristics. Relevance of the indicators of the methodology is explained, and a field-by-field description follows.

The case study of Porto is the Section 3, where the framework is implemented to monitor and reflect Porto's circularity. This is followed by the Section 4 where the case study results and the framework functionality are further discussed.

The paper ends with a conclusion. Extra information can be found in the appendixes, such as the complementary tables of the framework and the table of interviewed experts.

1.1. International Contextualization for Cicular Economy

Internationally, there is a notable effort to tackle environmental challenges such as climate change and air pollution. Technological disruptions and shifts are also emerging, mainly in the form of digitalization and new power and storage sources. These new technologies can be used to tackle the issues in innovative ways, allowing for ambitious goals.

Many countries embrace these goals, believing that the shift towards a more sustainable system can empower a country, while increasing the lifestyle of their areas and reducing international environmental stresses.

One of the most relevant sets of goals is that of the SDGs. They encompass different areas, acknowledging the different possible technologies and trends of the 21st century. These goals have holistic influence, focusing on social, environmental and economical progression, and are highly interconnected [5].

The EU also embraced a more sustainable path, reflected in several goals to reduce materials and energy use. Its goals include, by 2030, increasing municipal waste prepared to reused and recycling to 65%, packaging waste to reuse and recycling to 75%, limit municipal landfill waste to 10% and reduce marine litter by 30% [6]. Furthermore, the EU strives to meet global environmental goals, such as the below 2 °C target embedded in the Paris agreement, and halve the food waste per capita [7].

This is in line with CE impacts, and the EU has developed an action plan to promote it. The EU, thus, acknowledges this and promotes CE implementation. It dedicates funds towards it, such as the Horizon 2020 and LIFE [7], and has also dedicated a set of documents, the BRIEFs, that, in different sectors, share behaviors and technologies that influence the CE.

1.2. Circular Economy Key Concepts and Impacts

The CE has no strict definition. However, it is commonly understood as a "system that is restorative or regenerative by intention and design. It replaces the 'end-of-life' concept with restoration, shifts towards the use of renewable energy, eliminates the use of toxic chemicals, which impair reuse, and aims for the elimination of waste through the superior design of materials, products, systems, and, within this, business models" [8].

International actors such as the EU, China, Google, and the Ellen MacArthur Foundation (EMF) are intensifying legislation, technology, and research that promote CE. This allocates the CE into different levels and sectors, as is expected of an economic system, and its definition adapts to each one. Therefore, one can expect a different, but not contradictory, definition of a CE for different purposes, for instance from a business perspective, or a political perspective.

Today, a CE can be commonly understood as a system with a holistic impact that works in loops, at different levels, which mimic the loops seen in nature [9]. At its core there is the design for second usage, the goal to eliminate waste and to avoid toxic materials, the importance of waste management, and the implementation of the 9Rs (reduce, reuse, recycle, recover, refuse, repair, refurbish, remanufacture, and repurpose) [10].

Moreover, it is acknowledged that the outcome of the CE is a more sustainable economy, referenced as growth from within. However, to achieve it, businesses need to be transparent and cooperative [11].

This will require a more active, aware, and skilled society. Being flexible, modular, and resilient, it needs to take innovation into account [12].

CE implementation can have significant economic and environmental impacts. Through a systemic implementation of a CE, it is estimated that European gross domestic product (GDP) can increase by 7%, reflecting annual savings of 600 billion EUR, benefits of 1.8 trillion EUR per year, and the generation of 170,000 jobs by 2035 [6]. Furthermore CE can reduce carbon dioxide emission by 48% by 2030 and 83% by 2050 [12]. During the same period, primary material consumption can be reduced 32% and 53% [13].

Today, different tools for assessing circularity exist, such as the Butterfly Diagram, the RESOLVE, mass flow analysis (MFA), and life cycle analysis (LCA) [13–16]. These frameworks were not designed specifically for a CE in cities but incorporate some of its features. To date, CE research has focused on the business side and on a global perspective [2]. This framework brings the analysis to the city level, while discretizing it in sectors.

1.3. Circular City Definiton

Cities host 50% of the world's population. They are responsible for 85% of the total GDP, 75% of natural resources consumption, 50% of waste generation, and 60–80% of total greenhouse gas (GHG) emissions [17]. By 2050, the share of the population living in cities is expected to increase to 70%, increasing the weight of these areas in economic, environment, and social matters [18]. Therefore, there is a need to explore circularity in cities, merging these two holistic and impactful trends. With increasing urbanization, cities are an ideal location to implement circular changes, originating the circular city concept. There are some cities already embracing this challenge, such as Amsterdam [14], Glasgow [19] and Barcelona [20], contributing with case studies to the literature [21]. The research in this specific area is in progress, with a need for more case studies. Moreover, there is a need to monitor circularity in an urban context, but there is a lack of tools to do so, as well as indicators [16].

However, circular cities face now what CEs faced before: the need of a definition. This definition still has to embrace the relevant aspects of CEs, but it needs to shift them to the city perspective. This results in a merging of CE and city dynamics and fundamental structures. The city is a complex system that involves areas beyond the economy and the CE [22].

Therefore, tools that are developed or are adequate for circular business analysis—such as LCA, MFA, or RESOLVE—fail to capture the circularity of a city. They are generally too specific or too broad, since they were not developed specifically for circular cities. However, these tools reflect CEs and should partially represent the city, i.e., they can still have a role in city monitorization and a definition [22]. For instance, RESOLVE focuses, among other particularities, on the consumption of materials and energy. Consumption is still relevant in a city context but is complemented with the production, as well as the flow, of materials and possible synergies [22]. Being a source of resources, instead of a drain, is a characteristic of a circular city, as is the capacity to enable connections between sectors [22]. The resources of a city, focusing on locality, together with its demography, are relevant [22]. They transmit the city context and identity. The infrastructures of the city, concerning mobility, industry, housing, and offices determine a city's dynamics. These have a heavy impact in circular terms within the city context, and must be in the definition of a circular city. Moreover, the city should be adaptable, embracing new technologies to come (i.e., digitalization, shared mobility, renewable energy, and 3D printing, inter alia) and, again, allowing synergies and material flows [22].

Finally, it is important to point out that this new approach to monitor a city, as circular and has a whole, comes with challenges. The city is composed of sectors, which can be represented in the framework to illustrate the city. These sectors need to be monitored, as well as their interactions, in terms of circularity, translating the complexity of a multi-sectorial system, which is characteristic of a city [23]. Despite the recent efforts and progress in this area, there is still a lack of data and indicators in the different sectors and in the city as a whole for implementation. Besides the lack of circular data and standardization, these challenges are burdened with a lack of circular city case studies [22].

1.4. Portugal and Porto Context

The Portuguese CE action plan is aligned with EU directives. It targets parallel goals on air pollution reduction and energetic dependency, among others. Several programmes support the CE action plan implementation, such as the Innovation, Technology and CE Fund, Portugal 2020 and ECO.NOMIA. By implementing a CE, Portugal expects to reduce raw material dependency by 30%, with a gross value-added (GVA) increase of 3.3 billion EUR [24].

Porto is a northern city of Portugal and was selected for this study given its goal to be a circular city by 2030. It focuses on industrial symbioses within the Eurocities programme [25]. Symbioses are important for a CE, as they allow synergies between entities that promote waste reduction and an increase in efficiency. Industrial symbioses focus these synergies within and between industries.

Porto is geographically part of the Great Porto—responsible for 12% of national wealth—and the Metropolitan Area of Porto (AMP). The average waste generation in AMP is 588.2 kg per capita every year [26]. Finally, Porto is culturally and economically defined by different sectors, such as the wine industry, the cork industry, the sea economy, the agroindustry, furniture technology, and sustainability [27].

2. Materials and Methods

The development of this paper is based on a literature review, complemented by data sets and semi-structured interviews of experts on the circular economy field and on the respective sectors of a city.

During this process—which was carried out throughout the entire development of the paper—the biggest challenge faced was the gathering of data and indicators, due to the lack of support and transparency of important entities.

However, other entities were also critical to its development. For instance, literature was gathered from the EU and the EMF. The interviewed experts contributed to the lack of data and indicators, and provided a more intuitive and holistic interpretation of circularity in a city.

Finally, different data sets were used. Between them are the Eurostat, the Diário da República Electrónico (DRE), the Instituto Nacional de Estatística (INE), the PORDATA—mostly supported by INE—and IRENA data sets. Alongside them were data sets provided by private companies, namely Cork Amorim and Águas do Porto.

2.1. Literature Review

This paper started with the intent of understanding the impacts of CE in cities. Therefore, a literature review was carried out, involving academic papers and reports that would be related to a circular economy.

The EMF is a very active actor in the field [1,12,13,15,27]. With different reports on the subject, also in collaboration with Google [17] or the World Economic Forum [8], it described the role of a CE, its economic and environmental impacts, and the lack of research. This was supported by academic papers confirming reports and providing a more scientific insight. CE key concepts, technologies, and behaviors result from these reports.

Past case studies regarding the CE in Amsterdam [14] and Glasgow [19] were insightful in how a CE is implemented and monitored today in a city. Other cities, such as Barcelona [20], London [28], Lisbon [29], and Stockholm [30], inter alia, were also studied through other entities and academic research. Alongside it came the definition of a circular city, the tools to monitor it, the in-use indicators, and sector-specific circular technologies and synergies. After analyzing different cities, the common relevant sectors and circular city dynamics were sketched.

The political framework, together with international and national goals, is important to understand where a CE is heading and the EU's commitment towards it. Different EU action plans and directives were part of the literature review to understand this [31–33]. They were then narrowed down to Portugal's different action plans and active legislation [24,29,34–44]. Finally, the Porto Circular Economy Roadmap gave insight into the Porto context, together with regional legislation [25–27,45].

When the framework was already defined, a variety of interviews were carried out with experts of the different analyzed sectors to gain insight. These interviews are available in Appendix B. A literature review of each specific subject was carried out at the same time, with the aim to better understand each sector and to collect data and possible indicators.

2.2. The Circular City Analysis Framework (CCAF)

The created CCAF framework aims to keep the key concepts of a CE adapted to a city perspective and to capture the circularity of any analyzed city. It is in line with different CE interpretations as well as the circular city.

Besides the aim to analyze circularity in cities, focus was placed on the framework being simple and intuitive. This will increase understanding between the city's many agents, from municipalities, to academics and enterprises [3]. Therefore, a multi-sector analysis was adopted, representing the different sectors of a circular city.

The CCAF is composed of the Circular City Diagram (CCD), shown in Figure 1, and three tables. This display aims to picture the holistic perspective of a city, as well as its multi-sectorial aspects. The combination and labeling of the fields bring flexibility, adapting to different cities, and modularity, enabling the integration of different tools and future multi-level analysis. The sectors themselves showcase the different aspects of the circular city.

The three tables complement the CCD. Firstly, the fields table describes the fields, the relevant agents, technologies, and behaviors of these fields, its indicators, its goals, and the current situation. Secondly, the synergies table describes illustrated synergies, the sectors they compromise, the goals, and the current situation. Thirdly, the policies table lists the policies, its level (regional, national, or international), and the fields it affects, describes the policy, and presents alternatives. These tables are present in Appendix A.

The CCD is organized into three areas: the inner circle, the intermediate circle, and the outer circle. The inner circle gives CE information regarding the city, as well as the source of different businesses, materials, and energy flows. The intermediate circle focuses on the industries and sectors that characterize every city. However, it does not reflect every relevant aspect of a city. It is the outer circle's purpose to capture aspects with broader fields.

Each field is composed of one or more indicators that aim to reflect the city's circularity level. Therefore, not only indicators have to be selected but also goals. These goals are chosen by identifying realistic circularity levels within a city or go alongside a certified objective set by the EU, country, or region. For each of the indicators, the % of the completion of such goals is represented, with the completion of the bar representing the indicator measuring each goal.

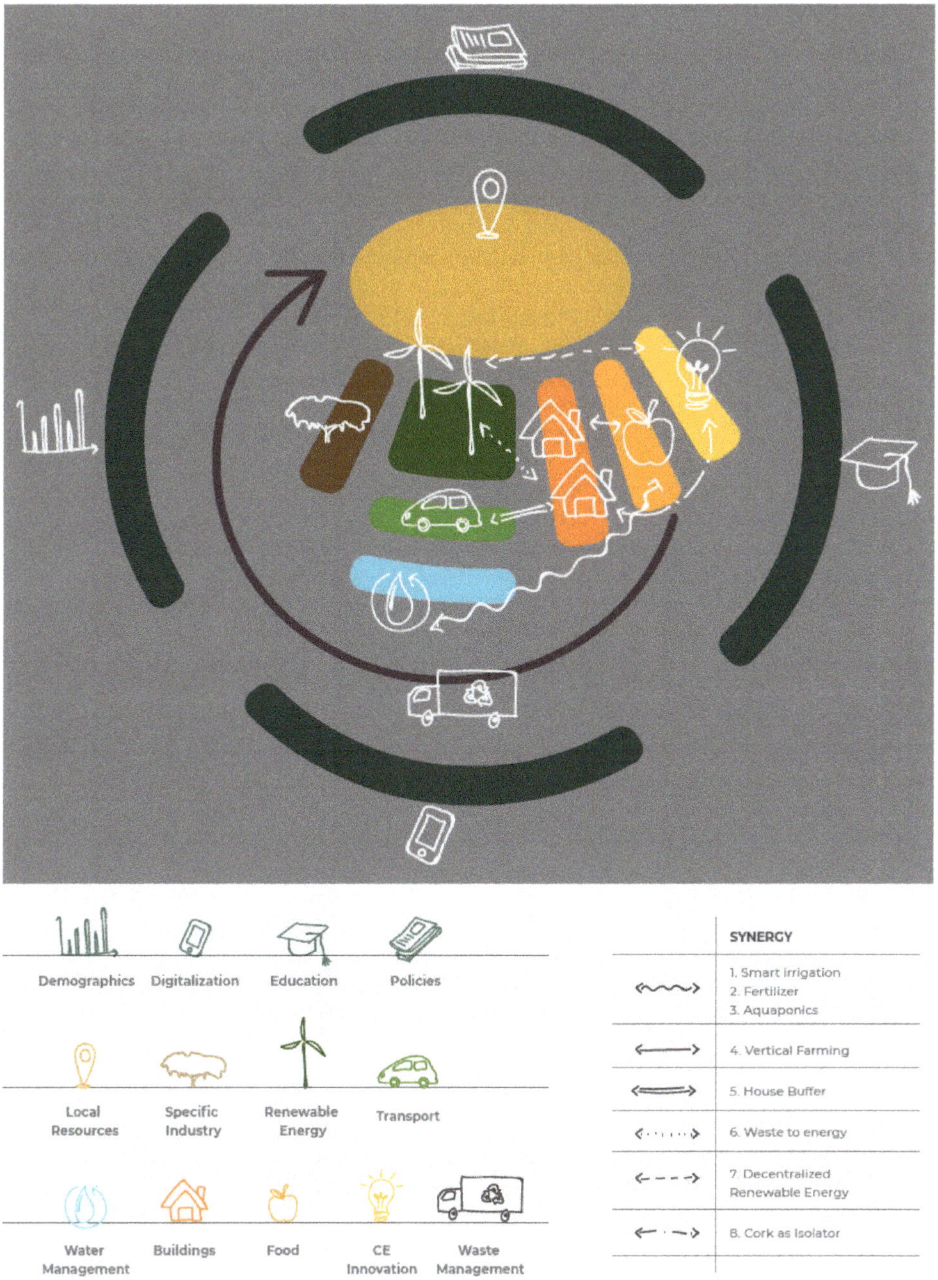

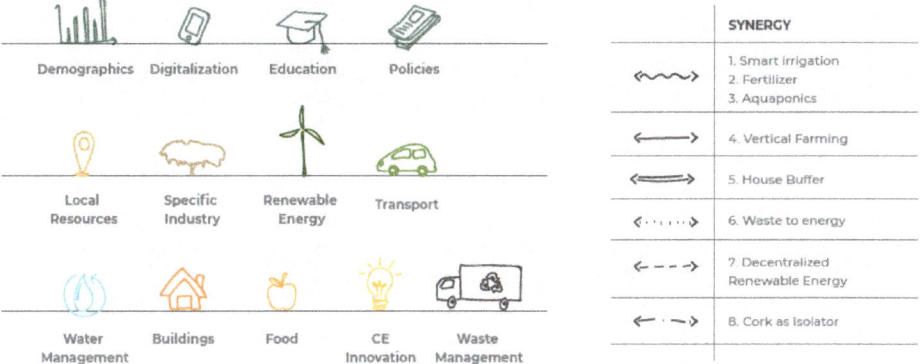

Figure 1. Circular City Diagram (CCD).

2.3. Circular Economy Indicators

Indicators are widely discussed in CE. The lack of established indicators and data are an acknowledged barrier to CE´s implementation and are currently under development. In circular cities, there is the need to define what is relevant and take into account what can be measured [46].

Linear economy indicators are no longer applicable to CEs, and this brings about the need to develop a new set of indicators to power CEs and circular cities. Despite the recent progress in this field, there is still room for improvement, especially in the standardization of indicators and data collection [47].

Academic literature is constantly proposing new indicators for CEs. Even for circular cities, some indicators are available [3]. However, generally these indicators are not applicable due to the lack of data. Moreover, different tools use different indicators that generate incomparable monitorization [48].

The CCAF aims to capture the relevant aspects of a city through the different sectors. To do so, it requires different indicators per sector as well as indicators that allow for comparison of the different sectors. Some of these indicators are available, but the data to feed them is not. Therefore, for testing in the Porto case study, a more pragmatic approach was taken, using available data and available indicators.

Hence, the Porto case study can be analyzed in its different sectors, whereby the functionality of the framework is shown, but these results cannot be considered definitive or conclusive. This means that further work must be done in this area to implement the CCAF to its full potential.

Different data sets of indicators were reviewed, and possible indicators for the sectors are already in the literature review. However, due to the innovative nature of this framework its indicators need to be tailor-made. This does not mean that other circular, sustainable, or sectorial indicators and data sets are negligible. These are present in the literature review and are the origin of the CCAF current indicators. They will also be the foundations of future indicators if further work is done in this framework.

A starting point for circular indicators is [49], where the SDGs are related to different indicators. The diverse nature of these indicators goes in line with the CE holistic impact, as well as its goals. Similar information is gathered in [46], with its proposed indicators more limited to CEs. Supporting and providing better insight into these indicators is [50], followed by [51,52], where diverse indicators are applied to EU country comparisons and United States of America (US) city comparisons, respectively.

An extensive country comparison analysis of sustainable and environmental indicators was performed in [53]. These indicators become more circular city-, business-, and economy-focused in [3,4,16].

Regardless of the relevance of this literature review, no data were available for Porto through these indicators. Furthermore, many of the indicators lacked a sectorial identity. Therefore, available data sets for Porto, generally found in INE or PORDATA, were used, leveraging on precision, but being outdated and lacking in circularity. Other indicators for Porto were considered, through sector reports and interviews. Acknowledgement of different tools such as RESOLVE, MFA, and LCA, together with city case studies [21], completed the indicator literature review.

An immense number of different indicators have been emerging in spite of the lack of data and even of their utility [4]. The CCAF requires different types of indicators. On the one hand, sector-specific indicators that capture the reality of each sector are used; on the other hand, standard indicators are used, and they apply to every city and allow for sector and city comparisons. This is a bottleneck of the framework due to the circular city indicators state-of-the-art.

2.4. Field-by-Field Development

To identify the circularity of the city, 13 different sectors were identified and split into three different levels: inner, intermediate, and outer circles. Fields with more possible synergies were placed closer to each other, with such synergies mostly happening in the intermediate circle.

Each field is followed by a list of indicators found in the literature review and proposed by the authors that can reflect circularity in the sector. Two sets of indicators are identified: possible indicators and used indicators. The description of the different fields is presented below:

Inner Circle:

- **Local resources:** The inner circle is solely composed of local resources, allowing this definition to be flexible enough to embrace different local aspects such as energy, food, material, and cultural sources. It is connected, at its core, to most of the intermediate circle sectors.

    ○ Indicators used: wind potential (m/s); solar potential (W/m^2); green roofs (%); imports/exports (€/€).

Intermediate circle:

- **Renewable energy** is in the most central position of the intermediate circle due to its overall impact. Energy connects to virtually every sector, enabling many inter-sector synergies [5]. Renewable technologies also enable waste reduction, foster efficiency, and bring a diverse and clean identity to the city and to circularity from the beginning [48].

    ○ Indicators used: renewable penetration (%); access to electricity (%); energy intensity (GWh/M€).

- **Transport sector:** A major component of renewable energy is the transport sector, playing a central role and close to different sectors. It allows synergies and is a buffer to buildings or renewable energy storage. This sector, as well as the building sector (next to the renewable energy) and the food sector (next to the building sector), is integral to every city [12]. It faces structural challenges, with cars being parked 92% of the time and only 1–2% of the total energy used to move people. Moreover, it accounts for 24.3% of GHG emissions, but can be shifted towards a more sustainable pathway through sharing systems and electrification [12].

    ○ Indicators used: public transport usage (%); electrical energy consumed in the transport sector (%).

- **Building sector:** This central sector in cities generates 14 million jobs in Europe, representing 8.8% of its GDP. It is responsible for 32% of GHG emissions and 40% of the energy consumption. This inefficient sector can be influenced by CEs, through material passports and banks, digitalization, and decentralized renewable power. For instance, 3D printing can reduce material waste and cut costs by 30% and delivery time by 50% [12].

    ○ Indicators used: retrofitting (%); very degraded buildings (%).

- **Food sector:** Besides being responsible for 40% of EU land, the food sector accounts for 19% of the European average household and 45% of the EU Commission budget. However, on average 20% of the food value is wasted through its value chain, and 11% is due to consumers [12]. Digitalization, as well as new technologies (e.g., aquaponic, urban farming, and precision farming) and circular behaviors, can transform this sector, increasing irrigation efficiency by 20–30% and reducing pesticide use by 10–20% and fertilizer consumption by 70–90% [12].

    ○ Indicators used: food waste treated (%); food waste treated in small and medium enterprises (SMEs) (%).

- **Water management** plays a central role due to its necessity. In many cities, it is an inefficient sector that can be upgraded through new monitoring technologies and smarter networks [12].

    ○ Indicators used: safe water accessibility (%); water efficiency (%).

- **Waste management** is critical to a circular society. This sector's responsibility is to collect the waste of different industries and assure for it a second life. This means planning longevity and designing waste-avoiding toxic materials, while keeping assets in the market at high value through tight loops [28]. This field limits the intermediate ring, starting in the food sector and ending in local resources.

    ○ Indicators used: landfilled waste (%); separated waste (Kg/capita*year).

- **CE innovation:** This field represents the motor of modern business creation, here with a focus towards a CE [28]. Activities such as reverse logistics can end up being incredibly complex, as can synergies [54]. An innovative business plan (or disruptive technology) can overcome this complexity, bringing about a competitive advantage for those who are implementing the CE, and enable an improved CE scenario. CE innovation is situated on the right side of the intermediate ring, close to education, due to its natural symbiosis with academic R&D.

    ○ Indicators used: CE innovation budget (%).

- **Specific industries** are on the left side of the intermediate circle. Their purpose in the framework is to highlight relevant sectors for the CE and the city. These industries are economically representative of the city. Moreover, they bring flexibility and singularity to the framework, allowing cities to monitor their different impactful sectors.

    ○ Indicators used: recycling rate (%); synergies (%).

Outer circle:

- **Education** is close to CE innovation and is the backbone of every society that seeks progress. It is an important sector, since CE requires a set of skills that today societies are still lacking [1].

    ○ Indicators used: basic education quitting (%); superior course (%).

- **Digitalization** is at the bottom, influencing almost every sector. As written in [17], "[p]owering the circular economy by providing digital solutions and closing the information gap is probably the best investment that technology companies of our time can make."

    ○ Indicators used: accessibility to smartphones (%).

- **Demographics** illustrate the society of a city and, therefore, are next to specific industries that can be affected by them. These industries are representative of the characteristics and identity that make every city different.

    ○ Indicators used: balance between men & women (%); heaviest age group (years); active population (%).

- **Policies**, being on the top, represent the top-down approach of policies and reflect the legal framework in which a city is inserted.

    ○ Indicators used: man–woman balance in politics (%).

3. Porto Case Study

This framework was then tested in the city of Porto. The framework´s indicators were quantified with the best available reports and data, combined with semi-structured interviews with experts in the field. When quantitative data were not available, only a qualitative analysis was performed. One first key result is that data paucity is an issue, and more effort will need to be dedicated to data collection.

However, the framework could help cities decide which indicators to use to start monitoring CE. The boundaries of this analysis are the Porto geographic limits, with exceptions of expansion to the AMP or Portugal.

The CCD for Porto is represented in Figure 2, with the fields (Table A1), synergies (Table A2), and policies (Table A3) tables in Appendix A. The field table lists the used indicators. Below is an explanation of the different sectors.

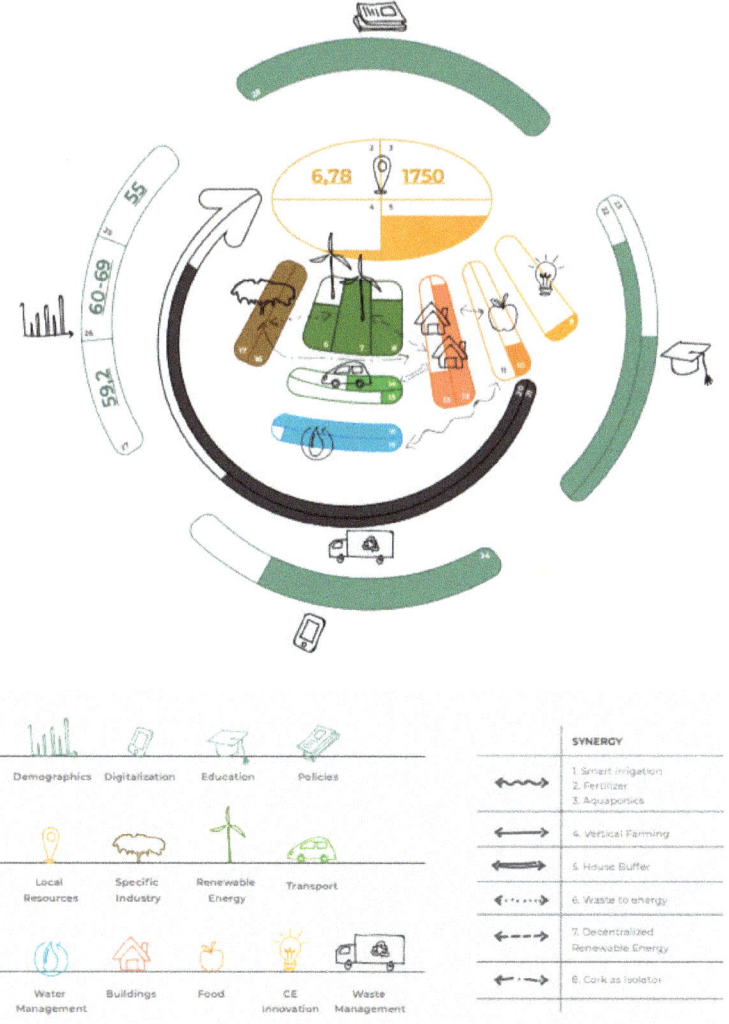

Figure 2. The CCD of Porto.

Inner Circle:

- **Local resources:** Porto has good wind and solar potential [55,56]. Porto wants to increase its green space areas, which will promote a better lifestyle and allow circular synergies, for instance, through water waste usage. Currently, it has around 0.5% of its area covered by green rooftops, an initiative to promote green spaces [45]. Moreover, its import/export is higher than 1; in a circular city, it should be below 1, i.e., it should generate more than it consumes [57].

Intermediate Circle:

- **Renewable energy** is taking advantage of the wind and solar potential, representing, in Portugal, 63% of the power generated [58]. The Portuguese grid gives access to, virtually, every citizen [59]. This means that Porto is in line with the national renewable energy values. However, it has an energy intensity of 1.56 GWh/M€, above the EU average of 1.4 GWh/M€ [58].
- **CE innovation:** Porto has different innovation hubs with diverse entrepreneurial fronts. Many of these fronts can have a positive circular impact. Porto Digital, together with OPO Lab and the innovation hub, are relevant platforms that promote circularity [26]. The budget dedicated for innovation is 0.009% of the total municipal budget [60,61].
- **The food sector** is difficult to measure, due to the many stakeholders involved and the difficulty of keeping track of energy and material flow. In Portugal, Continente, a wholesale market company, embraced circular practices, making them pioneers among its competitors. Except for this market, little evidence of steps towards circularity has been found. Continente has seen positive outcomes due to its efforts in savings, sustainability, and customer engagement. Therefore, it was selected as a reference, having close to zero food waste, with the rest of the wholesale market making no contribution. This gives a share of around 21% of food waste being recovered in the wholesale market [62]. However, the SMEs in the food industry, show no evidence of circularity, and it was considered 0% [63].
- **The building sector** is representative of data paucity in the field. It indicates extreme building degradation of 1.7%, which is not the perspective Porto itself transmits. Porto's buildings are usually old, degraded, unoccupied, and not monitored. They are not prepared to embrace synergies as renewable energies. Nevertheless, there is opportunity to remodel the city, translating to a retrofitting percentage of 13.6% among all work buildings in the city [64].
- **The transport sector** does not yet indicate any positive progress. This sector aims for electrification, shared mobility, and increased usage of public transportation. A percentage of 19.6% of mobility is indeed through public transport, but electrification is blocked [65]. This can be attributed to the low incentives for this technology, alongside a monopolized charging infrastructure that is now stagnating and bottlenecking electric mobility [41]. Representative of this is the electrical energy consumed in the transport sector: only 0.6%, far below the 10% mark set by the EU [66].
- **Specific industries:** Only one was selected for Porto: the cork industry. This industry is a model of circularity in Porto. Its business relies strongly on cork as a raw material, and it is led by Cork Amorim. By verticalizing its business model, Cork Amorim expanded from raw materials to different products—from crop stoppers to space composites [67]. Cork characteristics, together with this vertical approach, allow Cork Amorim to recycle and reuse its material, allocating it for different purposes while retaining its value. Its vision and innovative perspective leads to new market opportunities for cork, bringing possible synergies with every sector of the intermediate ring of this diagram [68].
- **Water management** in Porto is an old network and lacks monitoring and nutrient extraction. It has plans in motion, oriented by Águas do Porto, to create stations to treat wastewater and generate fertilizer [12]. It features, as expected of Portugal, 100% safe water access, together with an efficiency of 81% [69]. Its efficiency is blocked by the networks and their monitorization, and upgrades to increase efficiency—at this level—are expensive and even considered economically unfeasible [26].
- **Waste management:** Lipor is responsible for waste management and is a good example in Portugal and the EU. Only 1% of the waste ends in a landfill; the rest is treated or energetically valued [70]. Lipor also organizes sensitizing campaigns, with a recent one promoting the use of

combustors by citizens, in households, generating their own fertilizer. This is reflected in a waste separation of 46.54 kg/year per capita [71].

Outer Circle:

- **Education** in Portugal is mandatory between the age of 6 and 18, or until the 12th grade is finished [40]. This policy promotes an educated society but has yet to translate into impactful results. Still, 11% of the students in Porto quit basic education, and only 25% of the population have good grades [72,73]. Despite the efforts to incorporate more students and to provide a better education—including circular and sustainable behaviors—there is still a lack of skills that CEs require to thrive, since they rely on a skilled labor force, which can lead a holistic, circular shift through large enterprises and SMEs [26].
- **Digitalization** is present in Portugal and Porto. Companies embrace new technologies, programming is a course gaining more spotlight, and smart metering is a discussed and aimed solution by large utilities, such as EDP and ENDESA [59]. It is a process that is already underway, combined with an equipped society and connected to the world through smartphones. In Portugal, 71.6% of the population has access to a smartphone, and this is reflected in Porto [74].
- **Demographics** represent a gender-balanced society in Porto, with up to 60% being able to work. It also translates the Portuguese aging trend, with the heaviest age group being 60–69 [75].
- **Policies**: Portugal, and consequently Porto, is in line with the EU CE action plan and consequent directives. A more social indicator was used to analyze this field, showcasing the framework adaptability to embrace different data and indicators. This also showcases the lack of data in some of the fields. The indicator consists of the percentage of women present in the municipality directive board. A cap of 30% was selected [76], and Porto overcame this with 36% [77]. In the framework, this is represented as 100%, indicating that this field has reached its goal.

4. Discussion

Porto is still in the initial stages of becoming a circular city. Nevertheless, the city is part of the Eurocities group and has the objective of becoming fully circular by 2030 [25]. Evaluating the city circularity with CCAF shows how the city is doing in some sectors, while still lagging in others.

The different sectors of Porto demonstrate initiative towards circularity. Most significantly, waste management, water management, and the cork industry are successful examples. CE innovation and the food and transport sectors require a critical shift towards circularity. The remaining sectors are already in a circular path and require further measures.

A higher interconnectivity between sectors—synergies that result in less waste and higher efficiency—is desirable. This can be achieved if the different sectors adopt a circular perception, together with transparency and cooperation [8].

4.1. Case Study Discusion

In this section, each sector will be discussed. This discussion relies mostly on Porto's performance, and lessons can be learned for other cities aiming to shift towards circularity.

Looking at Porto's local resources, the city has the potential to increase energetic independency through solar and wind. Economically, a more export-oriented philosophy must be adopted, making more use of the local resources and recirculating some of Porto's products. This is in line with the CE characteristic of being a generator instead of a consumer [15]. Finally, there is a push from Porto towards the installation of more green areas, which result in a higher life quality and enable synergies [45].

In the renewable energy field, Portugal is on a promising path, supported by a grid network that connects the entire country. A next step could be decentralized power adoption that is incentivized by the government; this would provide more energetic independence and resilience to cities and buildings.

Porto is increasing its innovation in general, with a higher focus on hubs and cooperation with universities, and benefits from a platform of entrepreneurs and innovation that promotes CEs. The actions of enrolling in international projects are a positive aspect of Porto innovation centers and will lead this transition towards a circular city.

The food sector requires major improvements, especially on the two fronts analyzed by the indicators. On one hand, the retail sector has yet to increase food waste allocated to a second life. Continente is an example in this area. On the other hand, food SMEs need to decrease their waste flow, adopting sharing schemes such as OLIO that allocate flows to citizens who want it [78].

Porto's housing must both be retrofitted and adopt circular materials and technologies. It must adopt smarter electrical networks and metering, allowing synergies from decentralized power in buildings and the integration of electrical vehicles (EVs) as buffers. Smart housing and offices should be further explored, implementing sharing behaviors and connecting through IoT [12]. Moreover, retrofit must increase its share in building works and be complemented by material passports and material banks. For these three actions to happen, legislation is required, focusing on heavier penalties for the non-registration of materials in buildings, the allocation of local banks of building materials, and a push to shift retrofitting based on circularity as a viable alternative [79].

Individual transport is the main source of transportation in Porto and Portugal, mostly composed of fossil fuel powered technologies, despite a good public transport sector [80]. This goes alongside a political panorama that only slightly promotes the shift to EVs and a very weak charging network for EVs, with its development blocked by the monopoly of MOBI.E [36,66]. To promote a shift towards public transportation, Porto can increase the area of prohibited zones for individual transportation, complemented by restrictions for fossil fuel transportation. More than just a charging network for EVs, the infrastructure of Porto needs to be rebuilt, focused on the future of mobility and making it flexible for technologies that will most likely thrive—automation and IoT, for example—in the mobility system [17].

It is understood that a shift to circularity can happen by connecting specific industries through synergies, shaping the economic and cultural panorama of the city [12]. In the case of Porto, other industries to join the cork industry may be the textile, furniture, shoemaking, plastic, rubber, metallurgic manufacturing, and wine industries [27]. These industries are all present in Porto and all share opportunities—some due to the ease of implementing circular business models, others due to the substantial potential and positive impacts that would occur if these circular business models were implemented [27]. The cork industry is a great example of how circularity can thrive in companies, with positive impacts on the environment and finances of the enterprise. The ease, compared to most other sectors, of recovering data in Cork Amorim must be highlighted. Companies such as these have the opportunity to monitor their business closely and focus on circularity, and to use it as leverage against competition. This is a sector that is expected to keep expanding, which can use the CE trend to lead the circular transition.

The Porto municipality and Águas do Porto have a high focus on water management and on how to upgrade it, especially in matters of losses [26]. The indicators used in this work do not reflect this situation. It shows the liability of current indicators, and the need for information on relevant and accessible entities. The political framework, although lacking a set of objectives and oversight, is organized to tackle the leakage issues, increase synergies by the recovery of nutrients, and upgrade the digitalization of the network [26].

The waste management sector is well developed, mostly due to Lipor. The greatest achievement is the almost zero landfilled waste, which should place Lipor as the reference to Portuguese cities. Nevertheless, many of this waste ends up incinerated, and such solution does not recover most of the value of the material. Hence, in a circular model, should be a final solution [10]. Lipor is already promoting a better second-life usage of the waste separation of materials, but it is the company's responsibility to design waste in a smart way until disassembly, using materials that are durable and circularity-friendly [13]. For that to happen, reverse logistics business models need to be explored and supported, together with second-life material markets. Moreover, Lipor is asking the citizens of Porto (and other municipalities) to separate their waste and is educating them on how to directly reuse it at home. This is being described, by Lipor, as a success programme.

Portugal has a good political framework that promotes education. Porto follows this profile with a strong college presence, backed by some relevant schools and leads in scholar rankings [81,82]. Moreover, there is cooperation between innovation hubs and Porto University, together with the creation of a college course of circular economies. However, Portugal still has to invest in the population's basic education and the elderly, who will be a big share of the population in the next several decades. To achieve the potential of this age group as an example, education programmes focusing on CEs must be available to elder citizens.

Digitalization is the most impactful field of Porto, with the potential to impact all sectors, bringing Porto to a circularity-friendly position. It is a trend that has gained traction in past years and is supported by big enterprises— mainly Google in the circular economy—in every sector. It is reflected in population access, for instance, smartphone accessibility [17]. All sectors need to have better monitorization. The cork industry can be a role model in this matter. Allowing the gathering and treatment of data will foster more reliable indicators that will reflect the current situation of the city in circular terms more accurately.

According to policies, many incentives are in progress or planned, with the EU being responsible for many of them. Legislation must be reviewed, with a national effort to focus it on CEs. It must be accessible and organized to the public, making it easier for individuals and SMEs to understand the political framework they are navigating in while also reducing the uncertainty of investors by showing commitment to this path. The indicator of this sector, together with the demographic indicators, showcases a balanced society without gender discrimination. It is the role of Porto, Portugal, and the EU to be generally non-discriminatory, not just with respect to gender. This translates into a balanced and progressive society that will foster awareness, acceptance, transparency, and a sense of community. This can then foster circular, social, and economic development [83].

4.2. Framework Discussion

The framework fulfilled its purpose of supporting a structured reflection of what Porto's progress is in circularity. The relevant fields were identified, and each one was analysed individually. It provided an idea of overall progress in CE as a city.

The local resources and demographics illustrate Porto's characteristics, allowing for an understanding of the intrinsic strengths and weaknesses of the city. The other sectors reflect how far Porto is from reaching the desired CE goals.

The specific sectors successfully bring identity to the city. It demonstrates that Porto has an influential cork industry that is an example of circular business. This is an area that adapts from city to city, and with more industries analyzed in this side of the framework, the city's economic and industrial character can be better understood.

Moreover, the fields interact well between each other. They have a mix of individuality and connectivity that is ultimately showcased by the synergies. These synergies will be further explored and tested. However, it is already possible to identify possible and ongoing synergies, when complemented by the tables.

These synergies have even fewer indicators and data support than the fields though. A better design of these synergies can allow MFAs to be implemented between fields (and within fields), solidifying the connection between fields and allowing a holistic overview of the city material and energy flows. This can be achieved with further computational work and data gathering.

Data paucity is a concern for the framework and more broadly to the implementation of CE in cities. Most cities do not track circularity and are not equipped to gather this data. A good starting point would be to define the indicators, so that the data can be targeted. A framework like the one presented in this paper can support the discovery of such indicators and support monitoring efforts.

The CCAF fulfils the objective of monitoring the complexity of city circularity and setting goals. However, it misses a mapping feature, which would allow the identification of physical synergies and ease urban planning. This can be tackled with further computational work that connects the CCAF to intuitive maps, already in use in other analyses [14].

Cities such as Paris, London, Milan, and Amsterdam [23] are already taking steps towards monitoring CE in cities. The tools usually used are the MFA, the LCA, intuitive mapping, and RESOLVE [10]. Connecting the first three tools to this framework (even a simplified version that does not complement all the fields) can bring about a precise understanding of the progress of these cities towards circularity. Even more, it can bring about the standardization of tools and indicators, pushing cities to work together towards circularity and enabling comparisons.

Portugal has the opportunity, due to Porto's and Lisbon's circularity initiatives, to develop an environment where CEs in cities can thrive. For that, standard indicators that reflect each city but also allow one to compare them on a national level can be implemented. National regulators can take an important role here, making connections between cities and identifying, together with municipalities, potentials and where each can be improved. Furthermore, intercity synergies can be found through national level meetings. This can be extrapolated to the EU level.

5. Conclusions

Cities are a hotspot of economic, environmental, technological, and social development [20]. With an international push towards circularity in cities, a framework that can support municipalities in the pursuit of CEs is required. This work presents a framework that comes from a holistic definition of CEs in a city context, and the framework was developed to be modular, flexible, and transparent. The framework strives to represent the most relevant fields in a circular city and their interactions. It was developed to be modular enough to be applied in different areas and at different regional scales.

For its improvement, the framework should be applied to new cases and could be upgraded in several ways. A set of standard indicators that could be used in all cities should be set, and data should be collected for it. The framework could be coupled with multi-criterion analysis to reflect the weights of different indicators in each sector and enable city comparisons. It also aims to be flexible so that different levels of analysis can occur. LCA and MFA of companies from different sectors can be an extension of this framework, reaching a meso level. This can then be repeated at the product level, reaching a multi-level analysis characteristic of CEs, achieving the holistic requirement of its implementation [84].

As stated before, the target users of this framework are municipalities, alongside other agents promoting and monitoring circularity in a city. It is understood that the framework is intuitive enough to be used by non-specialist and non-scientific personnel, leading to a reflection of a city's circularity from a multi-sectorial perspective. Furthermore, matching the CCAF with an intuitive city map can create an even better understanding of where to act, who is acting, and the local synergies, and this can ease urban planning at the same time.

This study could be enriched by (and can inform) the work being done by the European Commission, together with relevant entities in the CE subject, such as EMF and ESAC, as they identify indicators to monitor circularity. While all cities are different, some standard aspects exist. Therefore, modularity and flexibility need to be present in the development of indicators. They need to be balanced with a standardization and simplicity perspective that allows for city comparisons and rankings. Consequently, the integration of multi-criterion analysis indicators is achieved, and the circularity of a city is concisely reflected.

Finally, there is a need, in terms of the analysis of a city in its circularity, for a broader involvement of agents, reducing the needed assumptions and therefore allowing a reflective understanding of a city standpoint as well as its future objectives and pathways. Circularity in a city needs to be harmonised and understood together with other action on decarbonisation [85,86]. With a broad enough work group, deeper analysis can become a reality, as this framework with economic, technical, social, and environmental analyses of possible paths will be complemented.

Author Contributions: This article was co-developed by the two authors. A.C.F. developed the core of the paper and underlying analysis and F.F.N. supervised the work and co-wrote & reviewed the paper.

Funding: The work presented in this paper was funded by the European Commission under contract number no. 642242 (https://deeds.eu/).

Acknowledgments: The graphical design of the framework was developed together with Teresa Segismundo. English revision was kindly provided by David Hughes and Charlotte Wragg.

Conflicts of Interest: The authors declare no conflict of interest.

Appendix A Framework Tables, for Porto

To complement the framework, three tables were developed. These tables work as the backbone of the CCD and provide more insight to each sector. They are the starting point to develop the CCAF, since it is in there that most of the information is stored. However, in this case study, it is important to remember the lack of data and support. This led to less insightful tables. Nevertheless, as well as the CCD, the tables need to be updated once more recent and reliable information is gathered.

The first table analyzes each field in more depth. It is divided into different columns, with the left column listing the different fields. First, it provides a description of each field, contextualizing the results and providing qualitative information that cannot be reflected by the indicators. Second, it identifies the principal agents influencing that sector. This can be extremely useful if the CCAF is merged with an intuitive map, since it indicates which entities should be highlighted. Third, it lists the technologies in use in each field that promote circularity. This column can be further explored by matching it with a list of all possible technologies and behaviors that can promote CEs. Again, this is useful when merging CCAF with other tools, because it determines the context of that field. The last three columns refer to the indicators that are in use, the current value of that indicator, and the desired goal.

The second table analyzes the synergies in action in the city. This table is extremely useful, since an extensive visualization of these synergies in the CCD would increase the framework complexity and hinder its understanding. As a side note, synergies could be further explored in the CCD if this framework is adapted into a computational framework. After listing the different synergies in the first column, the second column indicates the fields that the synergy involves. The next column describes the synergy itself and contextualizes it in the analyzed city (in this case, Porto). The following two columns describe the current situation of the synergy, followed by the aimed goal. These two columns can be further explored, as they apply quantitative goals to the different synergies as well as indicators.

The third table focus on policies. Policies have a heavy impact on city dynamics and its future. Moreover, due to its holistic and qualitative characteristics, they are difficult to translate into indicators in the CCD. Hence, this table complements the policies field. It lists different policies, on different levels, that impact and define the city circularity. Mirroring the two other tables, the policies are listed in the first column. This is followed by a column identifying the level of implementation of the policy (regional, national, EU, global, etc.) and then the fields affected by it, similar to the synergies table. Another column describes the most relevant points of the policy, followed by another with recommendations on that policy, so that it better aligns with the CE objectives and context of the city.

Table A1. Fields table.

Field	Description	Agents	Technologies/Behaviors	Indicator	Current	Goals
Local Resources	Source of energy in Porto, its macro-economic profile and life quality	AMP, Câmara Municipal do Porto	Green spaces, energy source data, air pollution levels	1. Wind Potential (m/s)	6.78	-
				2. Solar Potential (W/m^2)	1750	-
				3. Green Roofs (%)	0.5	10
				4. Imports/Exports (€/€)	1.5	1
Renewable Energy	A broad analysis considering the Portuguese grid and local production	EDP, REN, Endesa, DGEG	Decentralized production, PVs, wind kit-based, biomass, waste-to-energy	5. Renewable Penetration (%)	63	100
				6. Access to Electricity (%)	100	100
				7. Energy Intensity (GWh/M€)	1.56	1.4
CE Innovation	Platforms and business that lead to innovation in CE subjects	Innovation Hub, OPO Lab, ScaleUp, Porto Digital	Platforms connecting academics, companies and entrepreneurs, public incentives, hubs	8. CE Innovation Budget (%)	0.009	0.5
Food	Food value chain focused on retailers and SMEs embracing urban production	Continente, Canal Horeca, Pingo Doce, Intermarche	Aquaponics, hydroponics, urban and peri urban farming, smart irrigation; vertical and community farming	9. Food Waste Treated (%)	21	100
				10. Food Waste Treated in SMEs (%)	0	30
Buildings	Buildings profile, relating housing and abandoned buildings	Câmara Municipal do Porto, OASRN	Housing sharing, office sharing, retrofitting, 3D printing, industrial building work, material passport, bank of materials	11. Retrofitting (%)	13.6	50
				12. Very Degraded Buildings (%)	1.7	0
Transport	The mobility within Porto, regarding the shift towards EM	MOBI.E, STCP, UBER, Endesa	Shared mobility, smart transport infrastructure, EVs, automation	13. Public Transport Usage (%)	19.6	50
				14. Electrical Energy Consumed in the Transport Sector (%)	0.6	10
Specific Industry—Cork	Overview of the cork industry labeled as circular and a world leader in its area	Amorim	Cork composites, recycling	15. Recycling Rate (%)	100	100
				16. Synergies (%)	100	100
Water Management	Water issues regarding its treatment and distribution	Águas do Porto	Nutrients recuperation, leakage monitoring, recirculation	17. Safe Water Accessibility (%)	100	100
				18. Water Efficiency (%)	81	85
Waste Management	Recovery and treatment of waste generated in Porto, as well as the actions of the principal agents	Lipor	Ecopontos, house waste treatment, incineration, digitalization of the separation system	19. Landfilled Waste (%)	1	0
				20. Separated Waste (Kg/capita*year)	46.54	70

Table A1. Cont.

Field	Description	Agents	Technologies/Behaviors	Indicator	Current	Goals
Education	Levels of overall education in Porto, including college and its embracing of CE	Ministério da Educação, Câmara Municipal do Porto	CE Schools, programmes in universities, sensitizing projects to overall citizens	21. Basic Education Quitting (%)	11	0
				22. Superior Course (%)	25	50
Digitalization	The digital overview of citizens combined with the digital platforms and infrastructures that lead to CE	Google, INESC, EDP, REN, Endesa	Smart metering, asset tagging, geospatial information, big data management, connectivity	23. Accessibility to Smartphones (%)	71.6	100
Demographics	The demographic profile of Porto, showing weakness and potentials	INE	Main data collection from Censos	24. Balance between Men & Women (%)	55	-
				25. Heaviest Age Group (years)	60–69	-
Policies	An overview of the commitment of Porto political environment towards CE	Governo de Portugal, EU, EC, Câmara Municipal do Porto, AMP	Incentives, tax penalties, transparency, municipalities autonomy	26. Active Population (%)	59.2	-
				27. Man–Woman Balance in Politics (%)	38	>30

Table A2. Synergies table.

Synergy	Fields	Description	Current	Goals
Smart Irrigation	F + WatM	Usage of the wastewater as input to irrigation system	Already implemented, but small scale	After feasibility study, if positive, increase the smart irrigation network
Fertilizer	F + WatM	Collection of nutrients from wastewater and transformation into fertilizer for food production	Águas do Porto investing to make it real	Extract most of phosphorus and cellulose fiber, reusing it in fertilizers, reducing dependency on it
Aquaponics	F + WatM	Combination of water treatment and food production through aquaponics	No implementation yet, only referred as a possibility	Exponential growth, using wastewater and the river to develop fisheries and food production and to tackle water waste
Vertical Farming	B + F	Implementation of vertical farms in unused building areas	Close to zero presence	Citizen taking care of this business model, exploring self-production and community farms
House Buffer	T + B	Remodeling of house systems to include EVs as buffers and ESS.	Still to implement due to lack of technology and infrastructure	Product available in the market and leveraged by ESCOs
Cork as Isolator	C + B	Usage of waste cork as wall and floor isolation (sound and heat)	Already part of Amorim strong ramifications of business	Besides other synergies to Amorim, more companies to follow the lead in this type of synergies
Decentralized Renewable Energy	B + RE	Leverage building heights to gather solar, wind, or rain energy through decentralized technologies	Installations too small to be considered	The citizen embraces self-production, buildings, in conditions, having at least one decentralized technology promoting sustainability

Table A3. Policies table.

Policies	Level	Fields	Description	Recommendation
Circular Economy Action Plan (CEAP)	EU	All	Mainstream CE; showcase CE impacts; creation of second-life market for products	
Waste to Energy	EU	WasM	Waste role in CE and EU; waste hierarchy; financial supports; recommended technologies	
Legislative proposal on Online Sales of Goods	EU	WasM; D	Protection of the customer in online sales; assurance of longevity of products; promotion of reuse	Complement with digital possibilities to the market creation
Legislative Proposal on Fertilizers	EU	F; WatM	Creation of second market for recovered nutrients	Focus on local markets
Directive on the Restriction of the Use of Certain Hazardous Substances in Electrical & Electronic Equipment	EU	WasM	Creation of second-life market for WEEE; substitution of hazardous components in electrical and electronic equipment	Focus on local markets
CE Package	EU	All	Set of several goals: 65% of municipal waste prepared for reuse/recycle and less than 10% to landfill, 75% of packaging prepared to reduce/recycle, reduce maritime litter by 30%, halve global food waste in retailers and consumers, by 2030; define priority sectors; discuss monitorization	Discuss city standardization, focus on indicators and available data
Programa Casa Eficiente	Portugal	B; D	Subsidies (100 M€ EIB + 100 M€ others) to retrofit buildings in efficiency; supported by digital platform	
Decreto-Lei No. 46/2008	Portugal	B; WasM	Discretization of waste in the building sector; barriers; role of municipalities; discretization of fiscal penalties	More legislation to allocate a bank of materials; heavier fiscal penalizations if non-registration of materials used in buildings
Lei No. 10/2014	Portugal	WatM	Showcase of Portuguese national water system; entities involved and how to regulate	Set of objectives and fiscal incentives/penalties
Decreto-Lei No. 141/2010	Portugal	R	Set of different goals to achieve by 2020: renewable share in energy consumption of 31% and increase by 10% in transports, reduce energy dependency by 74%; impact of this implementations; role of municipalities and autonomy in renewable energy	strategy for different technologies; incentives for decentralized production; citizen assessment
Decreto-Lei No. 82D/2014	Portugal	T	Fiscal benefits for ICE alternatives; bike-sharing implementation	
PNAEE 2017–2020	Portugal	All	Increase co-generation production; reduce energy consumption in buildings by 1.5%; increase fast-charging stations for EVs; interest in EVs, scooters, bike-sharing; interest in renewables sources	focus on decentralized production; discuss monitoring and upgrade of metering; discuss charging stations for EVs

Appendix B Table of Interviews

This paper was complemented by a set of interviews, listed in Table A4. These semi-structured interviews were always desired for the development of the framework.

Due to the holistic characteristic of CEs and circular cities, alongside the lack of indicators, data, and conceptualization, it was understood that insight from experts in different fields of CEs, sustainability, and sectors influencing a city would contribute positively to the framework.

The interviewed experts come from different sectors. However, each one has holistic understandings of the CE impact, certain opinions on the circular city, and expertise in their sector, as well as insight in other relevant sectors.

The interviews, independent of duration and type, started with particular questions regarding the position of the expert. They then followed the same procedure as per the literature review. After obtaining inputs of the different experts' conceptualization of CEs, their experience with it, how their sector and business contemplated circularity, and which policies were more influential, it was asked what indicators they could provide and recommend to monitor circularity—in a city, in their sector, and in their business.

The interview began with brainstorming. The future of each sector was discussed, different technologies and trends were analyzed, and the future circular city was considered. Insights were extremely useful due to the openness of these specific experts. They shared their knowledge and their opinions and expectations for a circular economy.

At one moment of the interview (but always during the brainstorming), the framework was presented. Here, it was tested: Questions were asked regarding its adaptability to different cities and how well the framework captured the circularity of a city. Finally, it was contextualized in the city of Porto. This part of the interview helped position the different fields in CCD and define their labels.

All the interviews contributed with powerful impacts. It is understood that, in such a holistic framework and analysis, a broader group of experts should be interviewed. The implementation of the Delphi method has been suggested in the future, with a broader group of interviewed experts [4].

Table A4. List of interviews.

Name	Position–Company	Interview Type/Duration
Elsa Rodrigues Monteiro	Head of Sustainability and Corporate Communication–Sonae Sierra	Face-to-Face/1 h
Diana Nicolau	Marketing, Education and Comunication Technician–Lipor	Distance/1 h
Joana Sousa Lara	Co-founder–Panana	Distance/1 h
Pedro Vieira e Moreira	Head of IT & Innovation–Águas do Porto	Distance/0:30 h
Vítor Martins	Head of Environmet–Modelo/Continente Supermarkets, S.A.	Distance/1 h
Nuno Ribeiro da Silva	Portugal Director & Invited Professor of Lisbon University–ENDESA	Face-to-Face/2 h
Pedro Pinto	Business Development & Franchise Director–Cooltra	Face-to-Face/1 h

References

1. Ellen MacArthur Foundation. Towards the Circular Economy. *J. Ind. Ecol.* **2013**, *2*, 23–44.
2. Petit-Boix, A.; Leipold, S. Circular economy in cities: Reviewing how environmental research aligns with local practices. *J. Clean. Prod.* **2018**, *195*, 1270–1281. [CrossRef]
3. Saidani, M.; Yannou, B.; Leroy, Y.; Cluzel, F.; Kendall, A.; Saidani, M.; Yannou, B.; Leroy, Y.; Cluzel, F.; Kendall, A. A taxonomy of circular economy indicators. *HAL* **2019**, *207*, 542–559. [CrossRef]
4. Azevedo, S.; Godina, R.; Matias, J. Proposal of a Sustainable Circular Index for Manufacturing Companies. *Resources* **2017**, *6*, 63. [CrossRef]
5. Fuso Nerini, F.; Tomei, J.; To, L.S.; Bisaga, I.; Parikh, P.; Black, M.; Borrion, A.; Spataru, C.; Castán Broto, V.; Anandarajah, G.; et al. Mapping synergies and trade-offs between energy and the Sustainable Development Goals. *Nat. Energy* **2018**, *3*, 10–15. [CrossRef]
6. Bourguignon, D. *Closing the Loop: NEW Circular Economy Package*; European Parliamentary Research Service: Brussels, Belgium, 2016.

7. European Commission. *Implementation of the Circular Economy Action Plan*; European Commission: Brussels, Belgium, 2017.
8. Ellen MacArthur Foundation; McKinsey & Company. *Towards the Circular Economy: Accelerating the Scale-Up across Global Supply Chains*; World Economic Forum: Geneva, Switzerland, 2014; pp. 1–64.
9. Vos, M.; Wullink, F.; de Lange, M.; Van Acoleyen, M.; van Staveren, D.; van Staveren, D. *The Circular Economy—What Is It and What Does It Mean for You?* Arcadis: London, UK, 2016.
10. Kalmykova, Y.; Sadagopan, M.; Rosado, L. Circular economy—From review of theories and practices to development of implementation tools. *Resour. Conserv. Recycl.* **2018**, *135*, 190–201. [CrossRef]
11. Williams, J. Circular cities: Challenges to implementing looping actions. *Sustainability* **2019**, *11*, 423. [CrossRef]
12. MacArthur, E.; Zumwinkel, K.; Stuchtey, M.R. *Growth within: A Circular Economy Vision for a Competitive Europe*; Ellen MacArthur Foundation: Cowes, UK, 2015.
13. Ellen MacArthur Foundation. *Towards a Circular Economy: Business Rationale for an Accelerated Transition*; Ellen MacArthur Foundation: Cowes, UK, 2015.
14. Circle Economy. *Circular Amsterdam: A Vision and Action Agenda for the City and Metropolitan Area*; City of Amsterdam: Amsterdam, The Netherlands, 2015.
15. Ellen MacArthur Foundation. *Circularity in the Built Environment: A Compilation of Case Studies from the CE100*; Ellen MacArthur Foundation: Cowes, UK, 2016.
16. Smol, M.; Kulczycka, J.; Avdiushchenko, A. Circular economy indicators in relation to eco-innovation in European regions. *Clean Technol. Environ. Policy* **2017**, *19*, 669–678. [CrossRef]
17. Sukhdev, A.; Vol, J.; Brandt, K.; Yeoman, R. *Cities in the Circular Economy: The Role of Digital Technology*; Ellen MacArthur Foundation: Cowes, UK, 2016.
18. World Economic Forum. *Circular Economy in Cities: Evolving the Model for a Sustainable Urban Future*; World Economic Forum: Cologny, Switzerland, 2018.
19. Circle Economy. *Circular Glasgow*; Glasgow Chamber of Commerce: Glasgow, UK, 2016.
20. Prendeville, S.; Cherim, E.; Bocken, N. Circular Cities: Mapping Six Cities in Transition. *Environ. Innov. Soc. Transit.* **2018**, *26*, 171–194. [CrossRef]
21. Santonen, T.; Creazzo, L.; Griffon, A.; Bódi, Z.; Aversano, P. *Cities as Living Labs—Increasing the Impact of Investment in the Circular Economy for Sustainable Cities*; European Commission: Brussels, Belgium, 2017.
22. Williams, J. Circular cities. *Urban Stud.* **2019**. [CrossRef]
23. Bonato, D.; Orsini, R. *Urban Circular Economy: The New Frontier for European Cities' Sustainable Development. The New Frontier for European Cities' Sustainable Development*; Elsevier Inc.: Amsterdam, The Netherlands, 2017.
24. República Portuguesa. *Apoiar a Transição para uma Economia Circular*; República Portuguesa Ambiente: Lisbon, Portugal, 2017.
25. Câmara Municipal do Porto. *Porto promove a Economia Circular*; Câmara Municipal do Porto: Porto, Portugal, 2018.
26. Câmara Municipal do Porto. *Roadmap para a Cidade do Porto Circular em 2030*; Câmara Municipal do Porto: Porto, Portugal, 2018.
27. AMP. *Área Metropolitana Do Porto*; AMP: Porto, Portugal, 2013.
28. Ellen MacArthur Foundation. *Cities in the Circular Economy: An Initial Exploration*; Ellen MacArthur Foundation: Cowes, UK, 2017.
29. República Portuguesa. *Liderar a Transição*; República Portuguesa Ambiente: Lisbon, Portugal, 2017.
30. Lönngren, Ö.; Nilson, L.A. *Stockholm 2050*; Stockholms stad: Stockholm, Sweden, 2014.
31. European Commission. *Tackling IUU Fishing*; European Commission: Brussels, Belgium, 2015.
32. European Commission. *Communication from the Commission to the European Parliament, the Council, the European Economic and Social Committee and the Committee of the Regions—The Role of Waste-to-Energy in the Circular Economy*; European Commission: Brussels, Belgium, 2017.
33. European Commission. *Annex—COM(2015)*; European Commission: Brussels, Belgium, 2015.
34. República Portuguesa. *Lei No. 10/2014*; Diário da República Electrónico: Lisbon, Portugal, 2018. Available online: https://data.dre.pt/eli/lei/10/2014/03/06/p/dre/pt/html (accessed on 01 March 2019).
35. República Portuguesa. *Decreto-Lei No. 46/2008*; Diário da República Electrónico: Lisbon, Portugal, 2018. Available online: https://data.dre.pt/eli/dec-lei/46/2008/03/12/p/dre/pt/html (accessed on 1 March 2019).

36. República Portuguesa. *Decreto-Lei No. 78/2004*; Diário da República Electrónico: Lisbon, Portugal, 2018. Available online: https://data.dre.pt/eli/dec-lei/78/2004/p/cons/20180611/pt/html (accessed on 1 March 2019).
37. República Portuguesa. *Lei No. 82-D/2014*; Diário da República Electrónico: Lisbon, Portugal, 2018. Available online: https://data.dre.pt/eli/lei/82-d/2014/p/cons/20171229/pt/html (accessed on 1 March 2019).
38. República Portuguesa. *Portaria No. 420-B/2015*; Diário da República Electrónico: Lisbon, Portugal, 2018. Available online: https://data.dre.pt/eli/port/420-b/2015/12/31/p/dre/pt/html (accessed on 1 March 2019).
39. República Portuguesa. *Portaria No. 10/2017*; Diário da República Electrónico: Lisbon, Portugal, 2018. Available online: https://data.dre.pt/eli/port/10/2017/p/cons/20171228/pt/html (accessed on 1 March 2019).
40. Assembleia da República. *Lei No. 85/2009*; Diário da República: Lisbon, Portugal, 2009; pp. 5635–5636.
41. Assembleia da República. *Decreto-Lei No. 141/2010*; Diário da República: Lisbon, Portugal, 2018; pp. 1–16.
42. República Portuguesa. *Governo Lança o Programa Casa Eficiente 2020 no Montante de 200 Milhões de Euros*; República Portuguesa Ambiente: Lisbon, Portugal, 2018.
43. *Plano de Ação para a Economia Circular em Portugal*; República Portuguesa Ambiente: Lisbon, Portugal, 2017. Available online: https://www.portugal.gov.pt/download-ficheiros/ficheiro.aspx?v=71fc795e-90a7-48ab-acd8-e49cbbb83d1f (accessed on 1 March 2019).
44. IAPMEI. *Sistemas de Incentivos à Economia Circular*; IAPMEI: Lisbon, Portugal, 2018.
45. República Portuguesa. *Cidades Circulares*; República Portuguesa Ambiente: Porto, Portugal, 2018.
46. European Academies' Science Advisory Council (EASAC). *Indicators for a Circular Economy*; EASAC: Halle, Germany, 2016.
47. Pauliuk, S. Critical appraisal of the circular economy standard BS 8001:2017 and a dashboard of quantitative system indicators for its implementation in organizations. *Resour. Conserv. Recycl.* **2018**, *129*, 81–92. [CrossRef]
48. Douma, A.; de Winter, J.; Ramkumar, S.; Raspail, N.; Dufourmont, J. *Circular Jobs*; Glodshmeding Foundation: Amsterdam, The Netherlands, 2015.
49. United Nations, Annex: Global Indicator Framework for the Sustainable Development Goals and Targets of the 2030 Agenda for Sustainable Development. 2018. Available online: https://unstats.un.org/sdgs/indicators/GlobalIndicatorFramework_A.RES.71.313Annex.pdf (accessed on 1 March 2019).
50. United Nations. *Indicators of Sustainable Development: Guidelines and Methodologies*; United Nations: New York, NY, USA, 2007.
51. Ford-Alexandraki, E. *EU Resource Efficiency Scoreboard 2015*; European Commission: Brussels, Belgium, 2016.
52. Sustainable Development Solutions Network. *The U.S. Cities Sustainable Development Index*; Sustainable Development Solutions Networl: New York, NY, USA, 2018.
53. World Bank Group. *The Litle Green Data Book*; World Development Indicators: Washington, DC, USA, 2017.
54. De Oliveira Neto, G.C.; de Jesus Cardoso Correia, A.; Schroeder, A.M. Economic and environmental assessment of recycling and reuse of electronic waste: Multiple case studies in Brazil and Switzerland. *Resour. Conserv. Recycl.* **2017**, *127*, 42–55. [CrossRef]
55. IRENA. DTU Global Wind Atlas 1 km Resolution. 2015. Available online: https://irena.masdar.ac.ae/gallery/#map/103 (accessed on 1 March 2019).
56. IRENA. Solar Irradiation accross Africa, Europe and Latin America. 2005. Available online: https://irena.masdar.ac.ae/gallery/#map/529 (accessed on 1 March 2019).
57. PORDATA. Imports and Exports. 2016. Available online: https://www.pordata.pt/Municipios/Valor+dos+bens+importados+e+exportados+pelas+empresas-393 (accessed on 1 March 2019).
58. APREN. Evolution of the Installed Capacity of the Different Sources of Electricity Generation in Portugal between 2000 and 2016. 2017. Available online: http://apren.pt/en/renewable-energies/power/ (accessed on 1 March 2019).
59. EDP. *Plano de Desenvolvimento e Investimento da Rede de Distribuição*; EDP: Lisbon, Portugal, 2016.
60. Associação Porto Digital. Relatório de Gestão. 2016. Available online: https://www.portodigital.pt/files/conteudos/Reports/RC2016-compressed.pdf (accessed on 1 March 2019).
61. Carvalho, P. Orçamento da Câmara do Porto Cresce Quase 18% e é o Maior da Última Década. 2016. Available online: https://www.publico.pt/2016/10/14/local/noticia/orcamento-da-camara-do-porto-cresce-quase-18-e-e-o-maior-da-ultima-decada-1747297 (accessed on 1 March 2019).

62. Sonae. *Sustentabilidade: O Impulso de Negócios Duradouros*; Sonae MC: Lisbon, Portugal, 2018.
63. Ferreira, M.P.; Reis, N.R.; Santos, J.C. *Mudança no Setor Alimentar:O Pingo Doce*; No. Caso de Estudo No. 8; globADVANTAGE: Leiria, Portugal, 2011; pp. 1–14.
64. INE. *Estatísticas da Construção e Habitação*; INE: Lisbon, Portugal, 2016.
65. IMT. *Mobilidade e Transportes*; IMT: Lisbon, Portugal, 2015.
66. Byrne, C.; Pedro, P. *Vencer o desafio da Mobilidade Elétrica em Portugal*; Plataforma para o Crescimento Sustentável: Lisbon, Portugal, 2016.
67. Amorim. *Relatório Anual Consolidado*; Corticeira Amorim: Mozelos, Portugal, 2017; Available online: https://www.amorim.com/xms/files/Investidores/5_Relatorio_e_Contas/2018CASGPSRelatorioAnualConsolidado.pdf (accessed on 1 March 2019).
68. Amorim. *Relatório de Sustentabilidade 2015*; Corticeira Amorim: Mozelos, Portugal, 2018; Available online: https://www.amorim.com/xms/files/Sustentabilidade/Relatorios/FINAL_Amorim_Rel_Sustentabilidade_2017_web_protect.pdf (accessed on 1 March 2019).
69. Águas do Porto. *Relatório & Contas*; Águas do Porto: Porto, Portugal, 2017.
70. Portugal, B. *Caso de Estudo: Economia Circular*; BCSD: Lisbon, Portugal, 2014; Available online: http://bcsdportugal.org/wp-content/uploads/2013/10/2014-CS-LIPOR-EconomiaCircular.pdf (accessed on 1 March 2019).
71. Lipor. Painel de Controlo. 2017. Available online: https://portal.lipor.pt/pls/apex/f?p=2020:1:0 (accessed on 1 March 2019).
72. PORDATA. Education Quitting Rate. 2017. Available online: https://www.pordata.pt/Municipios/Taxa+de+abandono+precoce+de+educaç~ao+e+formaç~ao+total+e+por+sexo-801 (accessed on 1 March 2019).
73. PORDATA. Education Level. 2011. Available online: https://www.pordata.pt/Municipios/Populaç~ao+residente+com+15+e+mais+anos+segundo+os+Censos+total+e+por+nível+de+escolaridade+completo+mais+elevado-69 (accessed on 1 March 2019).
74. SAPOTEK. 6.5 Milhões de Portugueses têm Smartphone. 2017. Available online: https://tek.sapo.pt/noticias/telecomunicacoes/artigos/65-milhoes-de-portugueses-tem-smartphone (accessed on 1 March 2019).
75. INE. Porto. 2017. Available online: https://www.citypopulation.de/php/portugal-admin.php?adm2id=1141312 (accessed on 1 March 2019).
76. Boseley, S. Rwanda: A revolution in rights for women. *The Guardian*, 28 May 2010.
77. Câmara Municipal do Porto. Executivo Câmara do Porto. 2018. Available online: http://www.cm-porto.pt/executivo (accessed on 1 March 2019).
78. Schanes, K.; Dobernig, K.; Gözet, B. Food waste matters—A systematic review of household food waste practices and their policy implications. *J. Clean. Prod.* **2018**, *182*, 978–991. [CrossRef]
79. República Portuguesa. *Decreto-Lei No. 379/93*; Diário da República Electrónico: Lisbon, Portugal, 2018. Available online: https://data.dre.pt/eli/dec-lei/379/1993/11/05/p/dre/pt/html (accessed on 1 March 2019).
80. Marques, N.; Felício, R. *A longa estrada para o carro elétrico*; Expresso: Lisbon, Portugal, 2017.
81. ARWU. Academic Ranking of World Universities 2018. 2018. Available online: http://www.shanghairanking.com/ARWU2018.html (accessed on 1 March 2019).
82. Público. Ranking das Escolas 2018. 2018. Available online: https://www.publico.pt/ranking-escolas-2018/em-que-lugar-ficou-a-sua-escola#- (accessed on 1 March 2019).
83. Kirchherr, J.; Reike, D.; Hekkert, M. Conceptualizing the circular economy: An analysis of 114 definitions. *Resour. Conserv. Recycl.* **2017**, *127*, 221–232. [CrossRef]
84. WBCSD; Climate-KIC. *Circular Metrics Landscape Analysis Executive Summary*; WBCSD: Geneva, Switzerland; Climate-KIC: Paris, France, 2018.
85. Fuso Nerini, F.; Hughes, N.; Cozzi, L.; Cosgrave, E.; Howells, M.; Sovacool, B.; Tavoni, M.; Tomei, J.; Zerriffi, H.; Milligan, B. Use SDGs to guide climate action. *Nature* **2018**, *557*, 31. [CrossRef] [PubMed]
86. Nerini, F.F.; Slob, A.; Segestrom, R.; Trutnevyte, E. A Research and Innovation Agenda for Zero-Emission European Cities. *Sustainability* **2019**, *11*, 1692. [CrossRef]

© 2019 by the authors. Licensee MDPI, Basel, Switzerland. This article is an open access article distributed under the terms and conditions of the Creative Commons Attribution (CC BY) license (http://creativecommons.org/licenses/by/4.0/).

Article

Smart Cities: The Main Drivers for Increasing the Intelligence of Cities

André Luis Azevedo Guedes [1,*], Jeferson Carvalho Alvarenga [1], Maurício dos Santos Sgarbi Goulart [1], Martius Vicente Rodriguez y Rodriguez [2] and Carlos Alberto Pereira Soares [1]

[1] Fluminense Federal University, Passo da Pátria Street 156 São Domingos, Niterói 24210-240, Brazil; jeferson.c.alvarenga@gmail.com (J.C.A.); msgarbi@id.uff.br (M.d.S.S.G.); capsoares@id.uff.br (C.A.P.S.)
[2] Fluminense Federal University, Mario Santos Braga Street 94 Centro, Niterói 24020-140, Brazil; martiusrodriguez@id.uff.br
* Correspondence: andre.guedes@gmail.com; Tel.: +55-21-98231-9173

Received: 31 July 2018; Accepted: 29 August 2018; Published: 31 August 2018

Abstract: Since the concept of smart cities was introduced, there has been a growing number of surveys aiming to identify the dimensions that characterize them. However, there is still no consensus on the main factors that should be considered to make a city more intelligent and sustainable. This report contributes to the topic by identifying the most important smart city drivers from the perspective of professionals from four broad areas of expertise: applied social sciences, engineering, exact and Earth sciences, and human sciences, which provide important insights for the understanding of smart and sustainable cities. In this study, we conducted a wide and detailed literature review, in which 20 potential smart city drivers were identified. The drivers were prioritized from the results of a survey conducted with 807 professionals that work in the concerned field. The results showed that the seven drivers identified as the most important to increase the intelligence of cities are related to the governance of cities.

Keywords: smart city; sustainable city; smart governance; drivers

1. Introduction

In a context of the accelerated growth of cities and the increasing demand for solutions that enable more appropriate responses to sustainability challenges, researchers have become more interested in issues related to smart cities. Because of this, recent debates on sustainable urban development have been intrinsically related to smart cities [1–3]. In fact, it is currently difficult to think of a smart city without associating it with aspects of sustainability and vice versa.

The concept of a smart city is not new and has evolved in recent decades [4], mainly as an answer to the challenges imposed by growing urbanization, digital revolution, and the demands of society for more efficient and sustainable urban services and the improvement of quality of life.

As a matter of fact, the concept of smart cities has been expanded over time, incorporating variables that reflect ways of dealing with challenges imposed by the transformations resulting from the way cities are owned and perceived by society. Thus, these variables, which could signify possible solutions to the growing challenges, have been assuming a much more reactive character than a proactive and strategic way of thinking of cities.

Recently, several studies have been developed to better understand smart cities from the dimensions that characterize them [5–9]. These studies began to intensify the multidisciplinary character of a variety of domains and disciplines [10], which emphasizes different aspects of the phenomenon depending on the context [11,12]. Although a smart city is still a diffuse concept that can have several interpretations, [13,14] it is possible to identify the convergence over time of the concepts of an intelligent city and a sustainable city [15]. The consensus is that it must be inclusive, secure, resilient, sustainable, and based on information technologies [12,16,17].

Other studies have also been developed focusing on the challenge of transforming today's cities into "smarter cities", searching for possible drivers that potentiate this transformation. The main research on this subject can be grouped in studies of technology and governance, with these two approaches being present in most articles consulted. Technology-related approaches, in short, aim to improve the efficiency of services and infrastructure (e.g., communication, transport, supply, etc.), mainly related to information and communication technologies (ICT). On the other hand, the approaches related to governance focus on management and the interactions between the various stakeholders in the city, connecting and developing socioeconomic and productive interactions among networks of urban actors.

Therefore, a more current and comprehensive way of understanding a smart city from the integration of existing knowledge and experiences is that of an innovative city, which combines aspects of intelligence and sustainability through a governance that integrates stakeholder interactions and that uses the technology to optimize services and infrastructure to improve quality of life. It is an orchestrated city in its actions and projects, interconnected and more intelligent, with the intensive use of technologies, such as the ones of sensing, information, and communication, in order to increase the efficiency of energy networks, transportation, and other logistical operations. The technology provides the means for the improvement and the connection of actors and services aiming to achieve a sustainable urban development, upgrading the socioeconomic, ecological, logistical, managerial, and competitive performance of the city and the quality of life of its population, thus ensuring that the needs of present and future generations are met [15,16,18–24].

Towards the aim of understanding the dimensions that characterize smart cities and the drivers that stimulate today's cities to become "smarter", studies have also been developed to classify how smart a city is. These studies focused mainly on the development of rankings from terms such as technology, economics, people, governance, mobility, health, environment, and quality of life, among others. However, the word smart was always attached to a set of indicators to explain the cities performance factors from certain contexts [25,26]. However, even today, there is no consensus on the main factors that should be considered to make cities smarter and sustainable. Studies on this topic are scarce.

This study addresses this gap and contributes to the literature regarding smart cities; in particular, it adds to the literature on what factors make cities smarter and sustainable. Therefore, the identification of the drivers was made more important by the researchers who published on the subject, from a broad and detailed bibliographic search.

Another contribution is the prioritization of these drivers based on the vision of 807 Brazilian experts, who have expertise in the priority areas pointed out by the literature on smart cities and work in four main areas of knowledge: applied social sciences, engineering, exact and Earth sciences, and human sciences.

The results showed that the twenty drivers identified as important in the literature were also considered important by experts, and from these, 15 drivers mainly focus on the governance of cities and the other five focus on technology. In addition, five drivers were rated as "extremely important" by all experts. The importance of identifying this smaller set of drivers considered as a priority is that leaders of these cities need to focus on those that are most important, considering the Brazilian scenario of scarcity of resources and that a great majority of Brazilian cities have the same main problems.

Consider that of the 20 drivers, 15 are mainly focused on the governance of cities. This, at first, suggests that governance is the main problem faced by cities. We also take this opportunity to reflect on possible solutions based on governance actions.

2. Materials and Methods

2.1. General Approach

The main research question of this study was "what are the main drivers for increasing the intelligence of cities?" To answer this question, we designed an approach in four steps: bibliographic research, identification of smart city drivers, survey with expert's opinions, and data analysis.

2.2. Bibliographic Research

Due to the multidisciplinary nature of the studies on smart cities, a wide and detailed bibliographical search was carried out. Several search engines and databases were used, especially those available at the "Portal Periódicos da Coordenação de Aperfeiçoamento de Pessoal de Nível Superior" of the Coordination for the Improvement of Higher Education Personnel (CAPES), Brazil. This tool provides access to the full texts available in more than 38,000 international and national periodicals, as well as to several databases (Web of Science, Scopus, Scielo, etc.). The "Portal Periódicos da Coordenação de Aperfeiçoamento de Pessoal de Nível Superior" includes references and abstracts of academic and scientific studies to technical standards, theses, and dissertations, among other types of materials, covering all areas of knowledge. The search was also carried out on the website of the main scientific periodicals and Google Scholar.

The literature search included papers published in the last 10 years, so that the drivers were more representative of current reality. The keywords searched were "smart cities", "smart city", "smarter cities", "smarter planet", "digital cities", "sustainable cities", and "ecological cities", which were combined with the terms "drivers", "dimensions", "rankings", and "components".

To accomplish the bibliographical search, we adopted the recommendations of Webster and Watson (2002) [27] and of the Preferred Reporting Items for Systematic Reviews and Meta Analyzes (PRISMA). The main strategy was to initially conduct an exploratory reading based on a brief study of titles and abstracts in order to exclude all articles that did not have some evidence or information on the issues addressed. After that, a selective reading was carried out. The articles whose abstracts were selected went through a full reading, excluding those who did not have relevant primary information to the research questions. The bibliographic research was finished when we concluded that we were not finding new papers with relevant information.

Thus, from the keywords, we identified 1827 articles from the last 10 years. After excluding the 418 duplicates, the number of papers was reduced to 1409. From the exploratory reading of titles and abstracts, we discarded 1150 articles. The exclusion criteria were abstracts that were not clear enough to identify relevance to our study or whose content did not express this relevance. Papers published in journals without a peer review system or that did not provide full text were also excluded, as were articles whose language was not English or Portuguese.

For the remaining 259 articles, we performed a selective reading to verify if our perception of the contribution to the research from the abstracts was proven. This step resulted in the exclusion of 116 papers. The exclusion criteria were non-original articles, those which insufficiently described investigation methods, results that did not contribute to the study, and results whose methodology did not support their validity.

The remaining 143 articles were analyzed in detail. For the present study, 110 articles were effectively used, of which 61 were the basis for the choice of drivers. From these articles a spreadsheet was created containing the most relevant sections to support and answer the research problem. Figure 1 summarizes the literature search using the PRISMA flowchart.

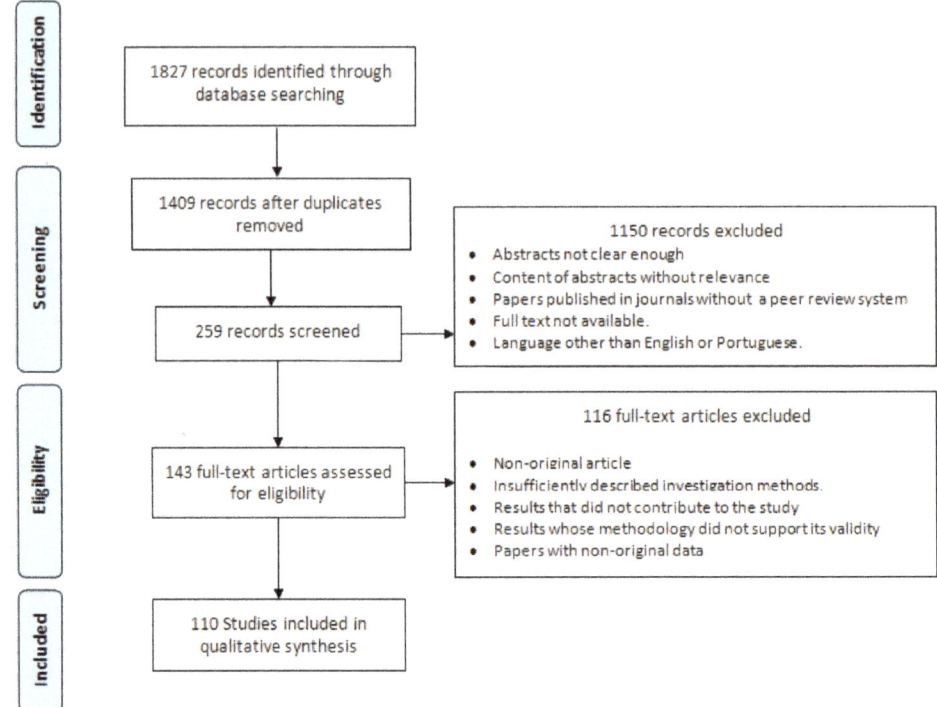

Figure 1. Literature search from the Preferred Reporting Items for Systematic Reviews and Meta Analyzes (PRISMA) flowchart.

2.3. Identification of Smart City Drivers

For the identification of the potential drivers for smart cities, a reflective and interpretive reading of the articles selected in the bibliographic search was carried out, and an examination of the perspectives, multiplicity, and plurality of approaches was performed. This aimed to understand what was already done concerning the proposed research and the latest developments in the field of smart cities. The strategy was to obtain a solid theoretical basis for our study, to order and summarize the information, to relate the main concepts and knowledge from already published papers to the scope of the research, and, finally, to identify a set of potential drivers.

In order for a potential driver to be considered of great relevance for increasing the intelligence of cities, its importance had to be portrayed in at least two works that did not refer to each other. As a result, we obtained a list containing twenty drivers.

The prioritization of the drivers based on their importance to the increase of the intelligence of the cities was carried out from the results of the survey of expert's opinions.

2.4. Survey of Expert Opinions

To carry out the survey, we used a questionnaire developed in an online platform (Google Forms), containing questions regarding demographic data and questions addressing the importance of the selected drivers in the bibliographic research. In the questionnaire, experts expressed their professional opinion about the importance of the contribution of each driver to make a city smarter, according to a five-point Likert scale, ranging from extremely important to minimally important. The drivers were randomly presented to avoid responses being influenced by the order in which they appeared.

We invited professionals that work in the main fields focused on in this report to answer the pre-test and the reviewed questionnaire. The inclusion criteria were to have expertise in the priority areas indicated in the literature for smart cities, to have training experience of more than five years, and to hold a degree in one of the following areas: applied social sciences, engineering, exact and Earth science, and human sciences.

These areas of knowledge are organized by Coordination for the CAPES from the clustering of several areas of formation, due to the affinity of their objects, cognitive methods, and instrumental resources reflecting specific sociopolitical contexts. The areas considered were Applied social sciences encompass the interdisciplinary areas of knowledge that deal with aspects related to public and private administration, accounting and tourism, architecture, urbanism and design, communication and information, law, economics, urban and regional planning/demography, and social services.

Human sciences have a human-centered approach and focus on the connections with history, beliefs, and the time/local space that can connect them. In this sense, human sciences involve themes related to anthropology, archeology, political science and international relations, religion and theology sciences, education, philosophy, geography, history, psychology, and sociology.

Engineering is characterized by the study and application of several branches of technology in order to materialize ideas in reality through techniques to solve problems and satisfy human needs—that is, applying methods and scientific vision for solving problems. It includes all engineering courses.

Exact and Earth sciences encompass disciplines based on physical-mathematical calculations, such as astronomy, physics, computer science, geosciences, mathematics, probability, statistics, and chemistry.

To get the experts opinions, we followed two strategies. The first was to use events in 2017 in which the authors participated in the organizing committee or as speakers in order to invite the speakers and participants with a potential to respond to the research. The events were "Corporate dilemmas—A critical view of the current scenario and practical solutions" held on 25 April 2017, "Corporate Dilemmas and Smart Cities" held on 25 May 2017, "Smart Cities Connecting with the Future" held on 24 October 2017, "Smart Cities and Creative Solutions" held on 6 November 2017, "Smart Cities and Creative Solutions—second meeting" on 16 November 2017, and "International Seminar on Policies, Incentives, Technology and Regulation of Smart Grids" held on 4 December 2017.

The second strategy was to request the coordinators of expert networks on issues related to smart cities working in Brazil to appoint specialists. Accordingly, several coordinators of various agencies cooperated with the questionnaire, such as the Innovation Agency of the Federal University of Fluminense (AGIR/UFF); the Laboratory of Innovation, Technology, and Sustainability of UFF (LITS/UFF); the Center for Smart Technologies (CTSMART); Rede Brasileira de Cidades Inteligentes e Humanas (RBCIH); the Smart City Business America (SCBA); and the Project Management Office (EGP/Niterói) of the Niterói City Hall.

The experts were invited in person, by e-mail, by Whatsapp, and by Linkedin. Nine hundred and ninety experts from various regions of Brazil were invited, of which 895 agreed to participate.

To receive the answers of a minimum number of respondents per knowledge area that were interested in participating in the survey, we used all the sources described above, and the survey took 10 months to complete.

With regards to qualification and professional experience, we decided that participants had to have at least one of the specialties that make up the four areas of knowledge researched in this paper. Also, they had to work in fields related to smart cities and have five years of professional experience or more since, in Brazil, this is usually the minimum amount of professional experience required to carry out specific activities that require deep knowledge.

The pre-test was executed in person with 10 specialists, using printed questionnaires, to identify possible doubts and eliminate inconsistencies. Thus, the respondents expressed their opinion about the overall design of the questionnaire, the clarity and pertinence of the questions, the preferred layout, and the order of the questions. The questionnaire was reviewed based on the comments received.

All questions of the survey were completed by the 895 respondents in 16 weeks (from 19 August 2017 to 8 December 2017). The professionals who did not have a complete higher education level and a minimum of five years of experience were excluded from the sample, as well as professionals from other areas, which resulted in a sample containing 807 respondents.

2.5. Data Analysis

After completing the data collection, we used Cronbach's alpha to evaluate the reliability of the data collection tool and the respondents. For that, the measurement of the variance of the responses of each item and the variance of the responses of each respondent were made [28]. Cronbach's alpha is one of the most important and widespread statistical tools in research involving the construction of tests and their application, because it accounts for the variance attributed to the subjects and the variance attributed to the interaction between subjects and items, resulting in an index used to evaluate the magnitude to which the items of an instrument are correlated. Thus, this makes it possible to evaluate the average of the correlations between the items that are part of an instrument and the extent to which the factor measured is present in each item [29].

To prioritize the data, we created the concept of relative median, which is represented by an indicator that allows for the hierarchization of the drivers in each semantic classification of the Likert scale. Taking the two lines of Figure 2 as an example, which presents a median equal to four, we can see that the median in the first line is much closer to the frequency represented by the number three. In the second line, when you add more cells to the frequency represented by the number five, the median moves farther to the right. When comparing the first with the second line, although both have medians equal to four, the driver of the second line can be interpreted as more important, since it received more classifications as five and kept the other frequencies.

1	2	2	3	3	3	3	3	**4**	4	4	4	4	5	5	5	5									
1	2	2	3	3	3	3	3	4	4	4	4	4	**5**	5	5	5	5	5	5	5	5	5	5	5	5

Figure 2. Example of the median position.

The formula used to calculate the relative medians was

$$RM = \begin{cases} 1 & x = 1 \\ m + \frac{Pmed - \sum_{i=1}^{m-1} j_i}{j_i} & 2 \leq x \leq N \\ N & x = N \end{cases}$$

where RM is the relative median, m is the median, $Pmed$ is the position of the median, N is the number of respondents, and j_i is the number of respondents who were assigned a semantic classification of "i".

3. Results and Discussion

Through the methodology described above, we obtained two main results. The first is the set of drivers identified from the papers selected in the bibliographic search. The second is the summary of the information obtained from the survey.

3.1. Selected Drivers

Twenty drivers were selected according to the criteria described in the materials and methods, as shown in Table 1.

Table 1. Selected drivers.

Driver	Source
Urban planning: Territorial management through the use of tools and indexes, including urban environmental quality, air quality, and well-being	[12,15,30–44]
City infrastructure: Management of the basic networks of rainwater, sanitation and water, and sewage services	[30,33,42,45–49]
Smart grids (energy): Intelligent management of energy sources and energy networks	[33,47,49–51]
Smart buildings: Use of sensors to minimize energy consumption without compromising comfort and safety (e.g., temperature, lighting, air quality, and natural ventilation)	[32,33,42,49,51–55]
Urban risks: Vulnerabilities, monitoring, prevention, and response to disasters in cities	[32,47,48,56–58]
Sustainability: Efficient management of natural resources to increase the quality of life of citizens for present and future generations	[12,15,33,34,38,48,59,60]
Mobility: Multimodal transport (individual and collective), intelligent urban mobility	[15,32,33,47,49,54,55,60–62]
Logistic solutions: Stocking, storage, transport, and distribution of products with optimization of the logistics chain	[33,62–67]
Logistic applications: Radio-frequency identification (RFID), geographic information systems (GIS), electronic routing of goods, drones	[47,51–53,62,68–70]
Public safety: Prevention and control of crime and violence by public entities	[32,33,36,47,50,51,71–74]
Health: Quality of public health and care (elective and emergency)	[33,40,46,55,75–80]
Innovation: Development of culture, intelligence, and collective co-creation for new products, services, businesses, or processes	[15,30,33,35,48,81–84]
Business networks management: Network of strategic partnerships (stakeholders) to boost innovation	[12,47,48,55,61,81,85,86]
Funding of new solutions: Public or private financial support or through public-private partnerships (PPP)	[12,32,33,47,48,55,72,85,87–91]
Relationship management: Analysis of the influence of the actors that compose the city as a social group	[12,30,33,41,42,81,85,92]
Technological applications for cities: Use of information and communication technologies (ICT) for smarter solutions	[12,30,32,33,42,46,48,52,54,55,62,73,76,82,85,93–97]
The sociotechnical impacts of digitization: Impact of technology on productive and labor tasks	[12,32,35,36,54,85,96,98,99]
Public policies: Planning and development of public policies for an intelligent city	[4,12,32,33,35,41,46,48,55,60,62,85,97,100–103]
Self-regulation: Elaboration and establishment by the community itself of the rules that discipline the market with the adoption of ethical standards	[32,85,97,101,104–107]
Regulation: Set of rules developed by state agencies to guide the economy and mechanisms of social control	[30,32,55,72,96,97,101,108–110]

Of the 20 selected drivers, 15 focus mainly on city governance and 5 focus on technology (Table 2). The drivers were considered for table composition after an interpretive process. In the "Source" column of Table 1 we cite authors who helped in the construction of the thoughts about smarter cities.

Table 2. Drivers grouped from their approaches.

Governance	Technology
Urban planning	Smart grids energy
Cities infrastructure	Smart buildings
Urban risks	Logistics applications
Sustainability	Technological applications for cities
Mobility	The sociotechnical impacts of digitalization
Logistic solutions	
Public safety	
Health	
Innovation	
Business network management	
Funding of new solutions	
Relationship management	
Public policies	
Self-regulation	
Regulation	

3.2. Survey Results

Initially, we calculated the Cronbach's Alpha, the value of which was 0.904 and confirmed the reliability of the questionnaire and the data. Next, we used the demographic data from the first section of the questionnaire to identify the profile of the respondents, considering their educational area and their professional experience (Figure 3). For all four areas, at least 70% of the respondents had more than 10 years' experience.

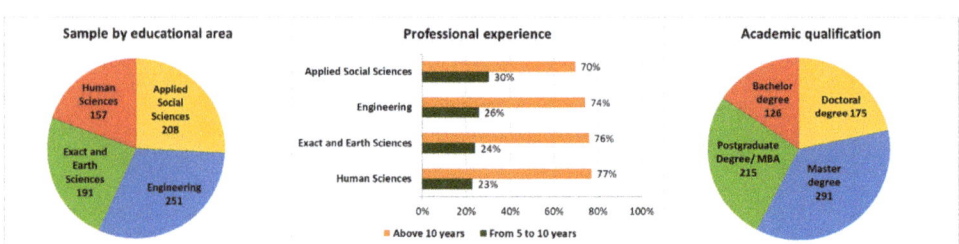

Figure 3. Demographic data.

Figure 4 shows the drivers ranked by the relative median. The drivers were classified from the judgment of the specialists of each training area. Figure 5 presents the same classification for all the respondents.

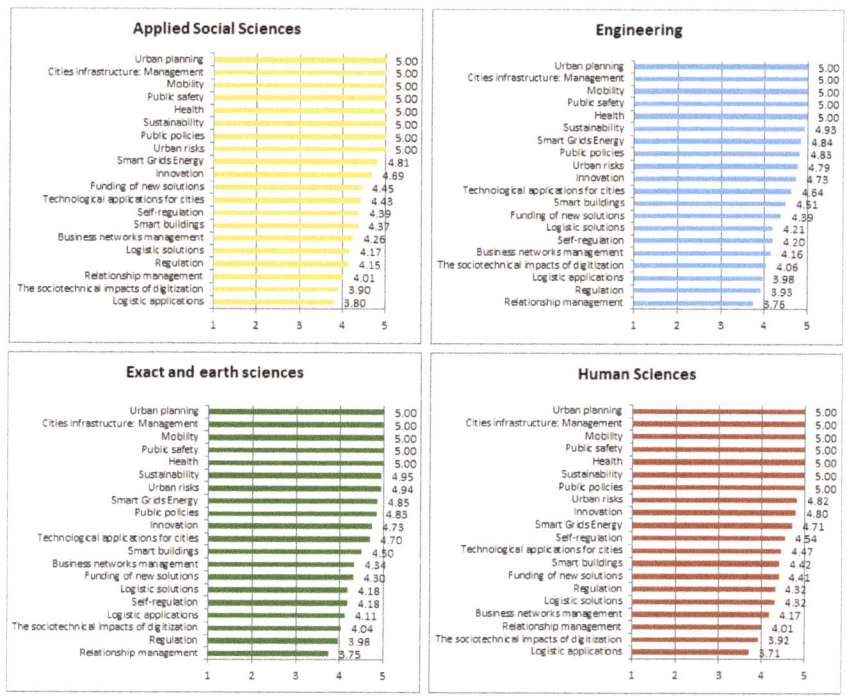

Figure 4. Drivers ranked by the relative median for the four areas of knowledge.

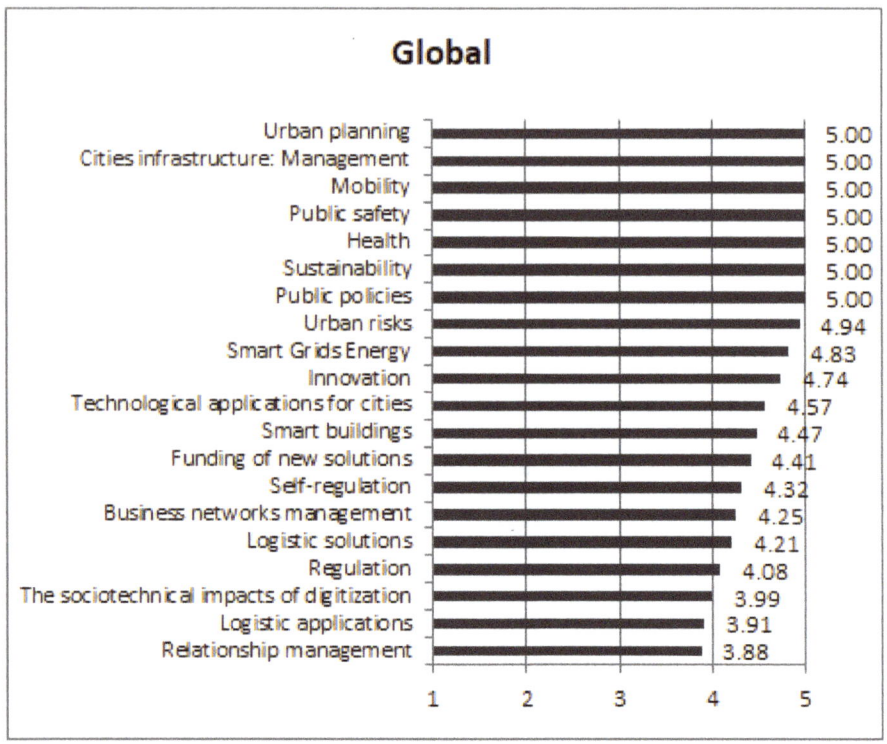

Figure 5. Drivers ranked by the relative median based on the total respondents.

Table 3 lists the drivers that were rated by experts as "extremely important" (equal to 5) from the relative median.

Table 3. Drivers ranked as "extremely important".

Drivers	Applied Social Sciences	Engineering	Exact and Earth Sciences	Human Sciences	Entire Sample
Urban planning	⊙	⊙	⊙	⊙	⊙
Cities infrastructure	⊙	⊙	⊙	⊙	⊙
Mobility	⊙	⊙	⊙	⊙	⊙
Public safety	⊙	⊙	⊙	⊙	⊙
Health	⊙	⊙	⊙	⊙	⊙
Sustainability	⊙			⊙	⊙
Public policies	⊙			⊙	⊙
Urban risks	⊙				

Figures 4 and 5 showed the drivers ranked from the relative median. It is possible to observe that all these drivers were considered important by the specialists (the relative medians were higher than 3.0), corroborating with the view of the researchers who published on the subject. From this result, the drivers considered as "extremely important" (equal to five) by training area were investigated (Table 3), and eight drivers met this requirement (urban planning, cities infrastructure, mobility, public safety and health, sustainability, public policies, and urban risks).

Taking into account this set of eight drivers, it was investigated which drivers were ranked as "extremely important" for the training areas, considering as the most relevant those that received the top rating evaluation for all the areas. Only five drivers met this requirement (urban planning, cities infrastructure, mobility, public safety, and health). From this evaluation, this set of five drivers was considered as the most important for stakeholders to prioritize their decisions, being denominated as the top five group.

By continuing the analysis of Table 3, it was observed that two other drivers (i.e., sustainability and public policies) received the top rating in at least two of the four training areas (human sciences and applied social sciences). This finding corroborated with the analysis by the relative median based on the total respondents in the sample as a whole, also shown in Table 3. Thus, by adding these two drivers to the top 5, a top 7 were composed. The driver "urban risks" has only been rated as "extremely important" by applied social sciences experts.

Figures 6 and 7 present the behavior of the drivers when the evaluations by training areas are compared with the evaluations carried out by the whole sample.

At the bottom of the scale of importance in Figure 6, three drivers that were evaluated as "important" stand out as being important but not a priority for all respondents: the sociotechnical impacts of digitalization, logistic applications, and relationship management, whose relative medians are between 3 and 4.

Eight drivers showed variations between the relative medians 4.01 and 4.99. These drivers are considered important but secondary in priority: smart grid energy, innovation, technological applications for cities, smart buildings, funding of new solutions, self-regulation, business networks management, and logistics solutions.

Two drivers were presented as borderline. The "urban risks" driver stands out as an "extremely important" driver for the applied social sciences group, and the "regulation" driver tends to be considered as a low priority, being very close to the three least priority ones, for three of the four professional groups researched.

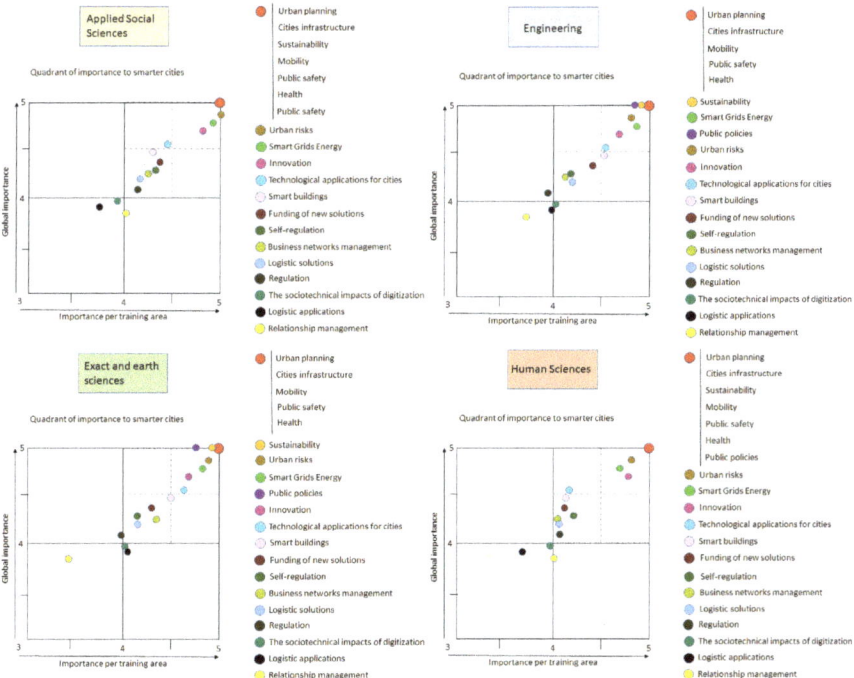

Figure 6. Drivers' behavior by training areas, related to the whole sample.

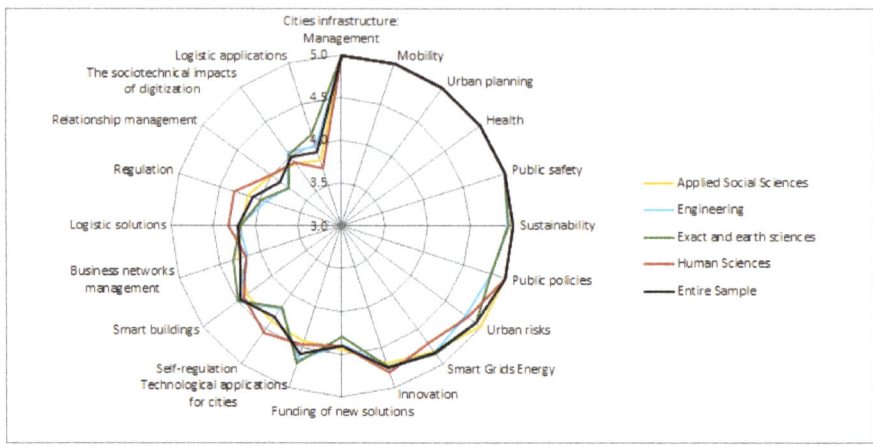

Figure 7. Drivers' behavior.

Considering the results obtained, and that all seven drivers are related to city governance, two fundamental questions arose. The first is, "why do drivers with a technological approach not appear among the top seven given that, in the existing literature, technology is widely addressed?" Using an analogy with the human body, we think of the heart and the brain as the most important for its functioning, and we hardly think of the circulatory system, although it is what maintains the life of these organs. We think of these organs as living parts and, consequently, imply the inclusion of what is necessary to keep them alive. We believe that something similar happened during the trial. The fact that, today, technological resources are massively present in our lives means that we do not think of most of the things we do with the technology being used. For example, when we use our smartphones, laptops, etc., we know that they are made possible by technological resources, but we do not think of these. This reasoning means that technological resources are not felt in isolation but are incorporated into something. Thereby, the technology layer appears in a transverse way, contributing to the improvement and efficiency of the services and infrastructure of the cities. The literature points to a strong correlation of these concepts with the governance [32,60,98,99,103] and orchestration of services [5,10,30,42,87] in cities.

The second question is, "can the top seven drivers be considered as the most important for cities of all countries?" We believe that they should be considered in relation to the reality of each country, because the way the city is perceived and owned by society is strongly influenced by the context in which the cities are inserted.

This understanding is because cities of each country have characteristics that differentiate them (e.g., government profile, socio-environmental culture, financing capacity, citizen participation, etc.). Thus, in several cities of other countries, the perception of the problems is different from those in Brazil, as Brazilian's cities problems are lack of planning, lack of infrastructure, and lack of adequate basic services, such as health care.

In recent years, Brazilian society has experienced a serious political and financial crisis, which has intensified the deterioration of services and urban infrastructure without most management bodies being able to propose solutions. In this sense, there is a perception of lack of planning, lack of infrastructure, and lack of adequate basic services, such as those related to health. Thus, concerning Brazilian cities, the results are fully justifiable. On the other hand, if we consider the studies consulted during the bibliographic search (Table 1), the results found can also be easily understood from the following understanding:

Urban planning: the management of territories through tools and indexes, including urban environmental quality, air quality, and well-being. This connects with all areas of the city because, to develop cities, planning is a fundamental tool for defining the priorities that operationalize the public policies, enabling cities to become more intelligent and sustainable.

City infrastructure: this includes the management of basic rainwater networks, sanitation, and water and sewage services. These must be managed as living systems, with efficient operation and management. For this driver, it is reasonable to note the need for large-scale management to provide a reasonable minimum sustainability of finite resources to citizens.

Mobility: multimodal transport (individual and collective) and intelligent urban mobility are the key sectors of smart cities. In the future we will have autonomous and electric vehicles providing an immediate impact on the transport systems [49]. This driver corroborates with what some authors point out: that there are more favorable conditions for smart city initiatives with these configurations aimed at public transport [33].

Public safety: the prevention and the control of crime and violence by public entities can use the potential of an intelligent city, where camera systems, motion detectors, electronic surveillance by control and command centers, real-time monitoring of security teams (patrolling), and monitoring of incidents can enhance the safety of smart cities. One well-known success story is New York's 911 that associated technologies with a political response to security.

Health: the quality of public health and elective and emergency services are being transformed in smarter cities. Through the adoption of advanced tools and technologies, the deficiencies found in municipalities can be supplied with health services that use concepts disseminated in private health, such as m-health, e-health, telemedicine [77], or the concept of smart health (s-health) that uses information and communication technologies for the good of individuals and of life in society [46].

Sustainability: the efficient management of natural resources contributes to raising the quality of life of the citizens for present and future generations. Social, economic, and environmental sustainability are strategic vectors for smart cities.

Public policies: the planning and development of public policies in favor of an intelligent city appears crucial for all the groups surveyed, since the municipal administrations are the entities that depend heavily on local policies to manage the projects, actions, and services. As these management groups involve various actors, they sometimes may seem to disagree. This view approaches the theoretical context pointed out by some authors such as Melo, Macedo, and Baptista [62].

Of the twenty drivers identified in the literature review, fifteen have as their main focus the governance of cities. This, at first, suggests that this is the main problem faced by cities. In Brazil this is true, since the eight drivers that received a maximum rating are also related to governance. Thus, possible solutions go through governance actions.

Developing a smarter city cannot be a top-down process. Drivers such as urban planning, city infrastructure, mobility, public safety, health, sustainability, and public policies demand a holistic and integrated vision focused on the priorities of society. The participation of the citizens in the initiatives of smart cities is fundamental for Brazil to avoid a utopia or a biased tendency towards the solution of having cities with an exclusively business vision.

Governance challenges can build on the helix quadruple, uniting the forces and intelligences of universities, the market, society, and governments for an integrated and combined solution to local priorities. A new way of governance for Brazilian cities should be based on intelligent collaboration and the use of information and communication technologies as a transverse and integrating resource.

In addition, policies aimed at the transformation of cities must be more comprehensive, effective, and have the integrated participation of all levels of government. In this context, municipal managers play a fundamental role and should incorporate better efficiency and effectiveness in the intelligent application of resources into city planning and strategic project execution to improve management.

We must overcome challenges with more innovative solutions. The intensification of public awareness and engagement programs in the monitoring of the application of resources can also help to avoid the deterioration of services and infrastructure in Brazilian cities, since few resources for new investments are available.

4. Conclusions

The concept of a smart city has incorporated evolutions and expansions over time. This is mainly from the convergence of the concepts of smart city and sustainable city. It is also due to the incorporation of variables that reflect ways of dealing with challenges imposed by the transformations caused by how cities are owned and perceived by society. However, even today, there is no consensus on the main factors that should be considered to make a city smarter and sustainable.

This paper sought to contribute to the subject by proposing the main question of this research: "What are the main drivers for increasing the intelligence of cities?" In order to answer this question, we identified and prioritized potential smart city drivers from a broad review of the literature and a survey conducted with specialists in the concerned fields who had undergraduate degrees in one of the following areas of expertise: applied social sciences, engineering, exact and Earth sciences, and human sciences.

The results demonstrated that of the 20 drivers identified in the literature, seven (i.e., urban planning, cities infrastructure, sustainability, mobility, public safety, health, and public policies) were considered as the highest priority for the development of more intelligent and sustainable cities. An eighth driver (i.e., urban risks), did not integrate into this group because it was evaluated with priority by only one of the areas of knowledge. In addition, three drivers (i.e., the sociotechnical impacts of digitization, logistics applications, and relationship management) were evaluated as not being a priority. We also observed that all seven priority drivers are related to city governance.

It was questioned why the drivers with technological approaches do not appear among the top seven since in the existing literature, technology is widely addressed. It was also asked if the top seven drivers could be considered as the most important for cities of all countries. Our conclusion is that the technology layer appears in a transverse way, contributing to the improvement and efficiency of the services and infrastructure of the cities. The literature consulted points to a strong correlation of these concepts to the governance and orchestration of services in the cities. However, it is important that the results are considered in the light of the reality of each country, since the way a city is perceived and owned by society is strongly influenced by the context in which they are inserted.

The present study has some limitations. First, even though we have done extensive and detailed bibliographic research, there is always the risk that some important contribution may not have been addressed in our analysis. The second is that for the prioritization of drivers, we relied only on the evaluations of Brazilian experts, who have certainly been influenced by the reality of the Brazilian cities. Thus, local realities should be considered. However, it is important to note that the reality experienced in most underdeveloped and developing countries are like those experienced by Brazilian cities.

Considering that citizens perceive the cities from their characteristics, this paper did not aim to compare these perceptions (and consequently the most important drivers from these perceptions) to the different cities of the world, which would be an interesting development for a further study.

Author Contributions: A.L.A.G., C.A.P.S., and M.V.R.y.R. conceptualized the study. A.L.A.G. and C.A.P.S. contributed to the methodology. J.C.A., C.A.P.S., and A.L.A.G. contributed the software. Formal analysis was performed by A.L.A.G., J.C.A., and C.A.P.S. Data curation was performed by A.L.A.G. and C.A.P.S. Validation was done by A.L.A.G., C.A.P.S., M.V.R.y.R., and M.d.S.S.G. The original draft was written by A.L.A.G. and C.A.P.S. Review and editing were done by C.A.P.S., M.V.R.y.R., and M.d.S.S.G.

Funding: This research was funded by the Coordination for the Improvement of Higher Education Personnel (CAPES) and authors funded the APC.

Acknowledgments: The study is an integral part of a PhD research project being carried out at Fluminense Federal University and funded by the Coordination for the CAPES. The authors would like to thank all the experts who answered the survey. The authors also express their gratitude to the editor and anonymous reviewers for comments and suggestions.

Conflicts of Interest: The authors declare no conflict of interest.

References

1. Garau, C.; Pavan, V.M. Evaluating urban quality: Indicators and assessment tools for smart sustainable cities. *Sustainability* **2018**, *10*, 575. [CrossRef]
2. Haarstad, H. Constructing the Sustainable City: Examining the Role of Sustainability in the 'Smart City' Discourse. *J. Environ. Policy Plan.* **2017**, *19*, 423–437. [CrossRef]
3. Komninos, N.; Pallot, M.; Schaffers, H. Special Issue on Smart Cities and the Future Internet in Europe. *J. Knowl. Econ.* **2013**, *4*, 119–134. [CrossRef]
4. Šiurytė, A.; Davidavičienė, V. An Analysis of Key Factors in Developing a Smart City/Pagrindinių Faktorių Kuriant Išmanų Miestą Analizė. *Sci. Future Lith.* **2016**, *8*, 254–262. [CrossRef]
5. Chourabi, H.; Nam, T.; Walker, S.; Gil-Garcia, J.R.; Mellouli, S.; Nahon, K.; Pardo, T.A.; Scholl, H.J. Understanding Smart Cities: An Integrative Framework. In Proceedings of the 2012 45th Hawaii International Conference on System Sciences, Maui, HI, USA, 4–7 January 2012; pp. 2289–2297.
6. Barrionuevo, J.M.; Berrone, P.; Ricart, J.E. Smart Cities, Sustainable Progress: Opportunities for Urban Development. *IESE Insight Rev.* **2012**, *14*, 50–57. [CrossRef]
7. Nam, T.; Pardo, T.A. *Conceptualizing Smart City with Dimensions of Technology, People, and Institutions*; ACM Press: New York, NY, USA, 2011; p. 282.
8. Giffinger, R.; Haindl, G. Smart Cities Ranking: An Effective Instrument for the Positioning of Cities? *ACE* **2010**, *4*, 7–26.
9. Mahizhnan, A. Smart Cities: The Singapore Case. *Cities* **1999**, *16*, 13–18. [CrossRef]
10. Lytras, D.M.; Visvizi, A. Who Uses Smart City Services and What to Make of It: Toward Interdisciplinary Smart Cities Research. *Sustainability* **2018**, *10*, 1998. [CrossRef]
11. Bibri, S.E.; Krogstie, J. Smart Sustainable Cities of the Future: An Extensive Interdisciplinary Literature Review. *Sustain. Cities Soc.* **2017**, *31*, 183–212. [CrossRef]
12. Hollands, R.G. Will the Real Smart City Please Stand Up? *City* **2008**, *12*, 303–320. [CrossRef]
13. Errichiello, L.; Micera, R. Leveraging Smart Open Innovation for Achieving Cultural Sustainability: Learning from a New City Museum Project. *Sustainability* **2018**, *10*, 1964. [CrossRef]
14. O'Grady, M.; O'Hare, G. How Smart Is Your City? *Science* **2012**, *335*, 1581–1582. [CrossRef] [PubMed]
15. Kobayashi, A.R.K.; Kniess, C.T.; Serra, F.A.R.; Ferraz, R.R.N.; Ruiz, M.S. Smart Sustainable Cities: Bibliometric Study and Patent Information. *Int. J. Innov.* **2017**, *5*. [CrossRef]
16. Pierce, P.; Ricciardi, F.; Zardini, A. Smart Cities as Organizational Fields: A Framework for Mapping Sustainability-Enabling Configurations. *Sustainability* **2017**, *9*, 1506. [CrossRef]
17. Vianna Crespo, A.R.; Puerta, J.M. *Evaluation of the IDB's Emerging and Sustainable Cities Initiative*; Thematic Evaluation; Inter-American Development Bank: Washington, DC, USA, 2016.
18. Harrison, C.; Eckman, B.; Hamilton, R.; Hartswick, P.; Kalagnanam, J.; Paraszczak, J.; Williams, P. Foundations for Smarter Cities. *IBM J. Res. Dev.* **2010**, *54*, 1–16. [CrossRef]
19. Hsieh, H.N.; Chou, C.Y.; Chen, C.C.; Chen, Y.Y. The Evaluating Indices and Promoting Strategies of Intelligent City in Taiwan. In Proceedings of the 2011 International Conference on Multimedia Technology, Hangzhou, China, 26–28 July 2011; pp. 6704–6709.
20. Bakıcı, T.; Almirall, E.; Wareham, J. A Smart City Initiative: The Case of Barcelona. *J. Knowl. Econ.* **2013**, *4*, 135–148. [CrossRef]
21. Cretu, L.-G. Smart Cities Design Using Event-Driven Paradigm and Semantic Web. *Inf. Econ.* **2012**, *16*, 57.
22. Lombardi, P.; Giordano, S.; Farouh, H.; Yousef, W. Modelling the Smart City Performance. *Innov. Eur. J. Soc. Sci. Res.* **2012**, *25*, 137–149. [CrossRef]
23. Kourtit, K.; Nijkamp, P. Smart Cities in the Innovation Age. *Innov. Eur. J. Soc. Sci. Res.* **2012**, *25*, 93–95. [CrossRef]
24. ITU. *The UNECE–ITU Smart Sustainable Cities Indicators*; Economic and Social Council: Geneva, Switzerland, 2015.
25. Deakin, M. Smart Cities: The state-of-the-art and governance challenge. *Triple Helix* **2014**, *1*, 7. [CrossRef]
26. Giffinger, R.; Haindlmaier, G.; Kramar, H. The Role of Rankings in Growing City Competition. *Urban Res. Pract.* **2008**, *3*, 299–312. [CrossRef]
27. Webster, J.; Watson, R.T. Analyzing the Past to Prepare for the Future: Writing a Literature Review. *MIS Q.* **2002**, *26*, xiii–xxiii.

28. Martins, L.; Pereira, L.; Maravilha de Almeida, L.; da Hora, H.; Costa, H. Estudo sobre escala mais adequada em questionários: Um experimento com o modelo de Kano. *Vértices* **2011**, *13*, 73–100. [CrossRef]
29. Almeida, D.; Santos, M.A.R.D.; Costa, A.F.B. Aplicação do coeficiente alfa de Crombach nos resultados de um questionário para avaliação de desempenho da saúde pública. *ENEGEP* **2010**, *15*, 1–12.
30. Allwinkle, S.; Cruickshank, P. Creating Smart-Er Cities: An Overview. *J. Urban Technol.* **2011**, *18*, 1–16. [CrossRef]
31. Anthopoulos, L.G.; Vakali, A. Urban Planning and Smart Cities: Interrelations and Reciprocities. In *The Future Internet*; Lecture Notes in Computer Science; Springer: Berlin/Heidelberg, Germany, 2012; pp. 178–189.
32. Kitchin, R. The Real-Time City? Big Data and Smart Urbanism. *GeoJournal* **2014**, *79*, 1–14. [CrossRef]
33. Neirotti, P.; De Marco, A.; Cagliano, A.C.; Mangano, G.; Scorrano, F. Current Trends in Smart City Initiatives: Some Stylised Facts. *Cities* **2014**, *38*, 25–36. [CrossRef]
34. Zali, N.; Tajiik, A.; Gholipour, M. An Application of AHP for Physical Sustainability Assessment on New Town of Andisheh, Tehran—Iran. *Raega* **2014**, *31*, 69–90. [CrossRef]
35. Datta, A. New Urban Utopias of Postcolonial India: 'Entrepreneurial Urbanization' in Dholera Smart City, Gujarat. *Dialogues Hum. Geogr.* **2015**, *5*, 3–22. [CrossRef]
36. Firmino, R.; Duarte, F. Private Video Monitoring of Public Spaces: The Construction of New Invisible Territories. *Urban Stud.* **2015**, *53*, 741–754. [CrossRef]
37. Greco, I.; Cresta, A. A Smart Planning for Smart City: The Concept of Smart City as an Opportunity to Re-Think the Planning Models of the Contemporary City. In *Computational Science and Its Applications—ICCSA 2015*; Gervasi, O., Murgante, B., Misra, S., Gavrilova, M.L., Rocha, A.M.A.C., Torre, C., Taniar, D., Apduhan, B.O., Eds.; Springer International Publishing: Basel, Switzerland, 2015; pp. 563–576.
38. Prado, A.L. Desenvolvimento urbano sustentável: De paradigma a mito. *Oculum Ensaios* **2015**, *12*, 83–97. [CrossRef]
39. Hajduk, S. The Concept of a Smart City in Urban Management. *Bus. Manag. Educ.* **2016**, *14*, 34–49. [CrossRef]
40. Sperandio, A.M.G.; Filho, F.; Luiz, L.; Mattos, T.P.; Sperandio, A.M.G.; Filho, F.; Luiz, L.; Mattos, T.P. Política de Promoção Da Saúde e Planejamento Urbano: Articulações Para o Desenvolvimento Da Cidade Saudável. *Ciência Saúde Coletiva* **2016**, *21*, 1931–1938. [CrossRef] [PubMed]
41. Antonello, I.T. Perspectivas dos instrumentos democráticos de planejamento e gestão do território urbano: As formas de participação da sociedade. *Raega—O Espaço Geográfico em Análise* **2017**, *42*, 133–148. [CrossRef]
42. Finger, M.; Razaghi, M. Conceptualizing "Smart Cities". *Informatik-Spektrum* **2017**, *40*, 6–13. [CrossRef]
43. Laurini, R. Towards Smart Urban Planning through Knowledge Infrastructure. *Diamond* **2017**, 75–80. Available online: http://www.academia.edu/32117774/Towards_Smart_Urban_Planning_through_Knowledge_Infrastructure (accessed on 1 March 2018).
44. Trindade, E.P.; Hinnig, M.P.F.; da Costa, E.M.; Marques, J.S.; Bastos, R.C.; Yigitcanlar, T. Sustainable Development of Smart Cities: A Systematic Review of the Literature. *J. Open Innov. Technol. Mark. Complex.* **2017**, *3*, 11. [CrossRef]
45. van Leeuwen, C.J. City Blueprints: Baseline Assessments of Sustainable Water Management in 11 Cities of the Future. *Water Resour. Manag.* **2013**, *27*, 5191–5206. [CrossRef]
46. Solanas, A.; Patsakis, C.; Conti, M.; Vlachos, I.S.; Ramos, V.; Falcone, F.; Postolache, O.; Pérez-Martínez, P.A.; Di Pietro, R.; Perrea, D.N.; et al. Smart Health: A Context-Aware Health Paradigm within Smart Cities. *IEEE Commun. Mag.* **2014**, *52*, 74–81. [CrossRef]
47. Hayat, P. Smart Cities: A Global Perspective. *India Q.* **2016**, *72*, 177–191. [CrossRef]
48. Koop, S.H.A.; van Leeuwen, C.J. The Challenges of Water, Waste and Climate Change in Cities. *Environ. Dev. Sustain.* **2017**, *19*, 385–418. [CrossRef]
49. Morello, R.; Mukhopadhyay, S.C.; Liu, Z.; Slomovitz, D.; Samantaray, S.R. Advances on Sensing Technologies for Smart Cities and Power Grids: A Review. *IEEE Sens. J.* **2017**, *17*, 7596–7610. [CrossRef]
50. Newcombe, T. Santander: The Smartest Smart City. Available online: http://www.governing.com/topics/urban/gov-santander-spain-smart-city.html (accessed on 24 July 2018).
51. Baig, Z.A.; Szewczyk, P.; Valli, C.; Rabadia, P.; Hannay, P.; Chernyshev, M.; Johnstone, M.; Kerai, P.; Ibrahim, A.; Sansurooah, K.; et al. Future Challenges for Smart Cities: Cyber-Security and Digital Forensics. *Digit. Investig.* **2017**, *22*, 3–13. [CrossRef]
52. Duarte, K.; Cecilio, M.J. Information and Assisted Navigation System for Blind People. In Proceedings of the 8th International Conference on Sensing Technology, Liverpool, UK, 2–4 September 2014; Volume 4.

53. Lubell, S. Virtual Singapore Looks Just like Singapore IRL—But with More Data. Available online: https://www.wired.com/2017/02/virtual-singapore-looks-just-like-singapore-irl-data/ (accessed on 24 July 2018).
54. Mariani, E.; Giacaglia, M.E. Guidelines for Electronic Systems Designed for Aiding the Visually Impaired People in Metro Networks. In *Advances in Human Aspects of Transportation*; Stanton, N.A., Ed.; Springer International Publishing: Basel, Switzerland, 2018; pp. 1010–1021.
55. New, J.; Castro, D.; Beckwith, M. *How National Governments Can Help Smart Cities Succeed*; Center for Data Innovation: Washington, DC, USA, 2017.
56. Yarime, M. Facilitating Data-Intensive Approaches to Innovation for Sustainability: Opportunities and Challenges in Building Smart Cities. *Sustain. Sci.* **2017**, *12*, 881–885. [CrossRef]
57. Ueyama, J.; Faiçal, B.S.; Mano, L.Y.; Bayer, G.; Pessin, G.; Gomes, P.H. Enhancing Reliability in Wireless Sensor Networks for Adaptive River Monitoring Systems: Reflections on Their Long-Term Deployment in Brazil. *Comput. Environ. Urban Syst.* **2017**, *65*, 41–52. [CrossRef]
58. Zhang, X.; Bayulken, B.; Skitmore, M.; Lu, W.; Huisingh, D. Sustainable Urban Transformations towards Smarter, Healthier Cities: Theories, Agendas and Pathways. *J. Clean. Prod.* **2018**, *173*, 1–10. [CrossRef]
59. Bugliarello, G. Critical New Bio-Socio-Technological Challenges in Urban Sustainability. *J. Urban Technol.* **2011**, *18*, 3–23. [CrossRef]
60. Caragliu, A.; Del Bo, C.; Nijkamp, P. Smart Cities in Europe. *J. Urban Technol.* **2011**, *18*, 65–82. [CrossRef]
61. Chen, Y.; Ardila-Gomez, A.; Frame, G. Achieving Energy Savings by Intelligent Transportation Systems Investments in the Context of Smart Cities. *Transp. Res. Part D Transp. Environ.* **2017**, *54*, 381–396. [CrossRef]
62. Melo, S.; Macedo, J.; Baptista, P. Guiding Cities to Pursue a Smart Mobility Paradigm: An Example from Vehicle Routing Guidance and Its Traffic and Operational Effects. *Res. Transp. Econ.* **2017**, *65*, 24–33. [CrossRef]
63. Botti, A.; Monda, A.; Pellicano, M.; Torre, C. The Re-Conceptualization of the Port Supply Chain as a Smart Port Service System: The Case of the Port of Salerno. *Systems* **2017**, *5*, 35. [CrossRef]
64. Melo, S.; Baptista, P. Evaluating the Impacts of Using Cargo Cycles on Urban Logistics: Integrating Traffic, Environmental and Operational Boundaries. *Eur. Transp. Res. Rev.* **2017**, *9*, 30. [CrossRef]
65. Öberg, C.; Graham, G. How Smart Cities Will Change Supply Chain Management: A Technical Viewpoint. *Prod. Plan. Control* **2016**, *27*, 529–538. [CrossRef]
66. Montes-Sancho, M.J.; Alvarez-Gil, M.J.; Tachizawa, E.M. How "Smart Cities" Will Change Supply Chain Management. *Supply Chain Manag.* **2015**, *20*, 237–248. [CrossRef]
67. Zhang, Q.; Huang, T.; Zhu, Y.; Qiu, M. A Case Study of Sensor Data Collection and Analysis in Smart City: Provenance in Smart Food Supply Chain. *Int. J. Distrib. Sens. Netw.* **2013**, *9*, 382132. [CrossRef]
68. Erkollar, A.; Oberer, B. FLEXTRANS 4.0—Smart Logistics for Smart Cities. *Sigma* **2017**, *8*, 269–277.
69. Bilal, M.; Oyedele, L.O.; Qadir, J.; Munir, K.; Ajayi, S.O.; Akinade, O.O.; Owolabi, H.A.; Alaka, H.A.; Pasha, M. Big Data in the Construction Industry: A Review of Present Status, Opportunities, and Future Trends. *Adv. Eng. Inform.* **2016**, *30*, 500–521. [CrossRef]
70. Lom, M.; Pribyl, O.; Svitek, M. Industry 4.0 as a Part of Smart Cities. In Proceedings of the 2016 Smart Cities Symposium Prague (SCSP), Prague, Czech Republic, 26–27 May 2016; pp. 1–6. [CrossRef]
71. Calavia, L.; Baladrón, C.; Aguiar, J.M.; Carro, B.; Sánchez-Esguevillas, A. A Semantic Autonomous Video Surveillance System for Dense Camera Networks in Smart Cities. *Sensors* **2012**, *12*, 407. [CrossRef] [PubMed]
72. New, J.; Castro, D. *Why Countries Need National Strategies for the Internet of Things*; Center for Data Innovation: Washington, DC, USA, 2015; Volume 25.
73. Jin, D.; Hannon, C.; Li, Z.; Cortes, P.; Ramaraju, S.; Burgess, P.; Buch, N.; Shahidehpour, M. Smart Street Lighting System: A Platform for Innovative Smart City Applications and a New Frontier for Cyber-Security. *Electr. J.* **2016**, *29*, 28–35. [CrossRef]
74. Li, Z.; Shahidehpour, M. Deployment of Cybersecurity for Managing Traffic Efficiency and Safety in Smart Cities. *Electr. J.* **2017**, *30*, 52–61. [CrossRef]
75. Hossain, M.S. Patient State Recognition System for Healthcare Using Speech and Facial Expressions. *J. Med. Syst.* **2016**, *40*, 272. [CrossRef] [PubMed]
76. Muhammad, G.; Alsulaiman, M.; Amin, S.U.; Ghoneim, A.; Alhamid, M.F. A Facial-Expression Monitoring System for Improved Healthcare in Smart Cities. *IEEE Access* **2017**, *5*, 10871–10881. [CrossRef]

77. Pramanik, M.I.; Lau, R.Y.K.; Demirkan, H.; Azad, M.A.K. Smart Health: Big Data Enabled Health Paradigm within Smart Cities. *Expert Syst. Appl.* **2017**, *87*, 370–383. [CrossRef]
78. Ahmed, F.; Ahmed, N.; Heitmueller, A.; Gray, M.; Atun, R. Smart Cities: Health and Safety for All. *Lancet Public Health* **2017**, *2*, e398. [CrossRef]
79. Batagan, L. Smart Cities and Sustainability Models. *Inf. Econ.* **2011**, *15*, 80–87.
80. Hussain, A.; Wenbi, R.; da Silva, A.L.; Nadher, M.; Mudhish, M. Health and Emergency-Care Platform for the Elderly and Disabled People in the Smart City. *J. Syst. Softw.* **2015**, *110*, 253–263. [CrossRef]
81. Trivellato, B. How Can 'Smart' Also Be Socially Sustainable? Insights from the Case of Milan. *Eur. Urban Reg. Stud.* **2016**, *24*, 337–351. [CrossRef]
82. Robert, J.; Kubler, S.; Kolbe, N.; Cerioni, A.; Gastaud, E.; Främling, K. Open IoT Ecosystem for Enhanced Interoperability in Smart Cities—Example of Métropole De Lyon. *Sensors* **2017**, *17*, 2849. [CrossRef] [PubMed]
83. Bifulco, F.; Tregua, M.; Amitrano, C.C. Co-Governing Smart Cities through Living Labs. Top Evidences from EU. *Transylv. Rev. Adm. Sci.* **2017**, *13*, 21–37. [CrossRef]
84. Anttiroiko, A.-V. City-as-a-Platform: The Rise of Participatory Innovation Platforms in Finnish Cities. *Sustainability* **2016**, *8*, 922. [CrossRef]
85. Vanolo, A. Smartmentality: The Smart City as Disciplinary Strategy. *Urban Stud.* **2013**, *51*, 883–898. [CrossRef]
86. Kraus, S.; Richter, C.; Papagiannidis, S.; Durst, S. Innovating and Exploiting Entrepreneurial Opportunities in Smart Cities: Evidence from Germany. *Creat. Innov. Manag.* **2015**, *24*, 601–616. [CrossRef]
87. Au-Yong, R. Vision of a Smart Nation Is to Make Life Better: PM Lee. Available online: https://www.straitstimes.com/singapore/vision-of-a-smart-nation-is-to-make-life-better-pm-lee (accessed on 23 July 2018).
88. Arbes, R.; Bethea, C. Songdo, South Korea: City of the Future? Available online: https://www.theatlantic.com/international/archive/2014/09/songdo-south-korea-the-city-of-the-future/380849/ (accessed on 23 July 2018).
89. Tham, I. Smart Nation Push to See $2.8b Worth of Tenders This Year. Available online: https://www.straitstimes.com/singapore/smart-nation-push-to-see-28b-worth-of-tenders-this-year (accessed on 24 July 2018).
90. Vadgama, C.V.; Khutwad, A.; Damle, M.; Patil, S. Smart Funding Options for Developing Smart Cities: A Proposal for India. *Indian J. Sci. Technol.* **2015**, *8*. [CrossRef]
91. Fishman, T.D.; Flynn, M. *Using Public-Private Partnerships to Advance Smart Cities*; Deloitte Center for Government Insights: London, UK, 2018.
92. Dyer, J. Chicago's High-Tech Surveillance Experiment Brings Privacy Fears. Available online: https://news.vice.com/article/chicagos-high-tech-surveillance-experiment-brings-privacy-fears (accessed on 23 July 2018).
93. O'Connell, P.L. Korea's High-Tech Utopia, Where Everything Is Observed. Available online: https://www.nytimes.com/2005/10/05/technology/techspecial/koreas-hightech-utopia-where-everything-is-observed.html (accessed on 24 July 2018).
94. Intelligent Cities—Routes to a Sustainable, Efficient and Livable City. 2013. Available online: https://www.sibyllerock.com/app/download/10563888823/intelligent+city+report+-+management+summary+English.pdf?t=1507572167 (accessed on 23 July 2018).
95. Nagel, D. Oregon State Brings Smart Sensors to Accessible Parking. Available online: https://campustechnology.com/articles/2013/01/14/oregon-state-brings-smart-sensors-to-accessible-parking.aspx (accessed on 24 July 2018).
96. Decker, M. The next Generation of Robots for the next Generation of Humans? *Robot. Auton. Syst.* **2017**, *88*, 154–156. [CrossRef]
97. Frecè, J.T.; Selzam, T. Tokenized Ecosystem of Personal Data—Exemplified on the Context of the Smart City. *JeDEM-eJournal eDemocracy Open Gov.* **2017**, *9*, 110–133. [CrossRef]
98. Tompson, T. Understanding the Contextual Development of Smart City Initiatives: A Pragmatist Methodology. *She Ji J. Des. Econ. Innov.* **2017**, *3*, 210–228. [CrossRef]
99. Meijer, A.; Bolívar, M.P.R. Governing the Smart City: A Review of the Literature on Smart Urban Governance. *Int. Rev. Adm. Sci.* **2015**, *82*, 392–408. [CrossRef]
100. Fernández, E.G. Novos instrumentos de participação: Entre a participação e a deliberação. In *Experiências Internacionais de Participação*; Pensando a Democracia Participativa; Cortez Editora: São Paulo, Brazil, 2010; pp. 19–40.
101. Yurie Dias, P. Regulação da Internet como Administração da Privacidade/Internet Regulation as Governance of Privacy. *Law State Telecommun. Rev./Revista de Direito, Estado e Telecomunicações* **2017**, *9*, 167–182.

102. Moraci, F.; Errigo, F.M.; Fazia, C.; Burgio, G.; Foresta, S. Making less Vulnerable Cities: Resilience as a New Paradigm of Smart Planning. *Sustainability* **2018**, *10*, 755. [CrossRef]
103. Allam, Z.; Newman, P. Redefining the Smart City: Culture, Metabolism and Governance. *Smart Cities* **2018**, *1*. [CrossRef]
104. Yang, H.; Chen, F.; Aliyu, S. Modern Software Cybernetics: New Trends. *J. Syst. Softw.* **2017**, *124*, 169–186. [CrossRef]
105. do Monte Silva, L.; Guimarães, P.B.V. Autorregulação jurídica no urbanismo contemporâneo: Smart cities e mobilidade urbana/Self regulation in the contemporary urbanism: Smart cities and urban mobility. *Revista de Direito da Cidade* **2016**, *8*, 1231–1253. [CrossRef]
106. Bandura, A. Social Cognitive Theory of Self-Regulation. *Organ. Behav. Hum. Decis. Process.* **1991**, *50*, 248–287. [CrossRef]
107. Ianuale, N.; Schiavon, D.; Capobianco, E. Smart Cities, Big Data, and Communities: Reasoning from the Viewpoint of Attractors. *IEEE Access* **2016**, *4*, 41–47. [CrossRef]
108. Daws, R. Britain Wants HyperCat to Reign the Internet of Things. Available online: https://www.telecomstechnews.com/news/2014/aug/21/britain-wants-hypercat-reign-internet-things/ (accessed on 23 July 2018).
109. Gori, P.; Parcu, P.L.; Stasi, M. *Smart Cities and Sharing Economy*; SSRN Scholarly Paper ID 2706603; Social Science Research Network: Rochester, NY, USA, 2015.
110. *Competitive Cities and Climate Change*; OECD Regional Development Working Papers 2009/02; OECD: Paris, France, 2009.

© 2018 by the authors. Licensee MDPI, Basel, Switzerland. This article is an open access article distributed under the terms and conditions of the Creative Commons Attribution (CC BY) license (http://creativecommons.org/licenses/by/4.0/).

Article

Evaluating the Environmental Performance of Solar Energy Systems Through a Combined Life Cycle Assessment and Cost Analysis

Maria Milousi [1], Manolis Souliotis [2], George Arampatzis [1] and Spiros Papaefthimiou [1],*

[1] Industrial, Energy and Environmental Systems Laboratory, School of Production Engineering and Management, Technical University of Crete, 73100 Chania, Greece; mmilousi@isc.tuc.gr (M.M.); garampatzis@pem.tuc.gr (G.A.)
[2] Department of Environmental Engineering, University of Western Macedonia, 50132 Kozani, Greece; msouliotis@uowm.gr
* Correspondence: spiros@pem.tuc.gr

Received: 25 March 2019; Accepted: 23 April 2019; Published: 1 May 2019

Abstract: The paper presents a holistic evaluation of the energy and environmental profile of two renewable energy technologies: Photovoltaics (thin-film and crystalline) and solar thermal collectors (flat plate and vacuum tube). The selected renewable systems exhibit size scalability (i.e., photovoltaics can vary from small to large scale applications) and can easily fit to residential applications (i.e., solar thermal systems). Various technical variations were considered for each of the studied technologies. The environmental implications were assessed through detailed life cycle assessment (LCA), implemented from raw material extraction through manufacture, use, and end of life of the selected energy systems. The methodological order followed comprises two steps: i. LCA and uncertainty analysis (conducted via SimaPro), and ii. techno-economic assessment (conducted via RETScreen). All studied technologies exhibit environmental impacts during their production phase and through their operation they manage to mitigate significant amounts of emitted greenhouse gases due to the avoided use of fossil fuels. The life cycle carbon footprint was calculated for the studied solar systems and was compared to other energy production technologies (either renewables or fossil-fuel based) and the results fall within the range defined by the global literature. The study showed that the implementation of photovoltaics and solar thermal projects in areas with high average insolation (i.e., Crete, Southern Greece) can be financially viable even in the case of low feed-in-tariffs. The results of the combined evaluation provide insight on choosing the most appropriate technologies from multiple perspectives, including financial and environmental.

Keywords: Life cycle assessment (LCA); carbon footprint; renewable energy systems; photovoltaics; solar thermal collectors

1. Introduction

Between 1973 and 2016, world electricity generation increased from 6131 to 24,973 TWh, i.e., 4.07 times, while today almost 81.1% of the world total primary energy supply originates from fossil fuels (i.e., coal, natural gas, and oil). Emissions of greenhouse gases (GHG), such as CO_2 and CH_4, from energy generation have been assessed in numerous studies, which often play a key role in developing GHG mitigation strategies for the energy sector [1,2].

The renewable power generating capacity exhibited the largest annual increase ever in 2017, with an estimated 178 GW installed world-wide, thus increasing the global total by almost 9% over 2016. Photovoltaics (PV) led the way, accounting for nearly 55% of the newly installed renewable power capacity and practically more PV capacity was added in 2017 than the net additions of fossil fuels and nuclear power combined. The total renewable power capacity more than doubled in the decade of 2007 to 2017, while non-hydropower renewables increased more than six-fold. In addition, investments in the new renewable power capacity (including all hydropower) was three times the investments in the fossil fuel generating capacity, and more than double the investments in fossil fuel and nuclear power generation combined [2–4].

Cost for electricity from solar PV and wind is rapidly falling. Record-breaking tenders for solar PV occurred in Argentina, Chile, India, Jordan, Saudi Arabia, and the United Arab Emirates, with bids in some markets below 0.03 $/kWh. Parallel developments in the wind power sector saw record low bids in several countries, including Chile, India, Mexico, and Morocco. Record low bids in offshore wind power tenders in Denmark and the Netherlands brought Europe's industry closer to its goal to produce offshore wind power cheaper than coal by 2025 [1,2,5,6].

Global voices for the decarbonization of the energy sector continuously increase, and thus renewables are expected to become the backbone of future power systems [3,4,6]. Typically, in such analyses, the GHG emissions are estimated without accounting for the impacts of the complete life cycle of the studied energy production systems. Life cycle assessment (LCA), carbon footprinting, and other GHG accounting approaches are commonly used for decision support. In LCA, potential environmental impacts associated with the life cycle of a product and/or service are assessed based on a life cycle inventory (LCI), which includes relevant input/output data and emissions compiled for the system associated with the product/service in question. The comprehensive scope of LCA is useful in avoiding problem shifting from one life cycle phase to another, from one region to another, or from one environmental problem to another [7,8]. Although the carbon footprint may have more appeal than LCA due to the simplicity of the approach, carbon footprints involve only a single indicator and thus this may result in oversimplification. By optimizing the system performance based only on GHG emissions, new environmental burdens may be introduced from other environmental emissions (e.g., NO_x and SO_2). A holistic or system-level perspective is therefore essential in the assessment, and the range of emission types included in a study may critically affect the outcome [9–11].

LCA is the methodology to be used when comparing the environmental performance (strengths and weaknesses) of different energy technologies, among them renewable systems. The idea behind a life cycle perspective in the context of power generation is that the environmental impacts of electricity are not only due to the power production process itself, but also originate from the production chains of installed components, materials used, energy carriers, and necessary services. Through an LCA analysis, a product is investigated throughout the entire life cycle ("Cradle-to-Grave") [12–14].

The main scope and motivation of the paper is to utilize detailed LCA and techno-economic results and present a holistic evaluation of the energy, environmental, and economic profile of two renewable energy technologies, photovoltaics and solar thermal collectors, both installed in a non-interconnected island with high average insolation. The former technology has been chosen as it can be employed from small scale applications to large power plants, while solar thermal systems are mainly focused to residential applications, but can play an important role in energy saving schemes as they practically deal with domestic hot water production and can cover significant thermal needs. Various technical variations will be presented for each of the studied technologies. For the evaluation of each of the renewable energy systems studied in the paper, the methodological approach followed comprises two steps: i. LCA and uncertainty analysis (conducted via SimaPro [15]) and ii. techno-economic assessment (conducted via RETScreen [16]). The paper employs the most recent LCA data and techno-economic parameters, thus presenting a complete, credible, and updated evaluation of the studied solar energy based technologies.

2. Materials and Methods

Renewable energy sources (i.e., biomass, hydropower, shallow and deep geothermal, solar, wind, and marine energies) are considered to be those that are primary, clean, low risk, and inexhaustible [5,6]. Sustainable development requires methods and tools to measure and compare the environmental impacts of human activities for various products. In order to understand where net savings in GHG emissions can be accomplished and the magnitude of the relevant opportunities, renewable energy systems should be analyzed and compared with the energy systems they would replace. The LCA methodology has been widely used to study the environmental burdens of energy produced from various renewable and non-renewable sources [17,18]. An LCA study is generally carried out by iterating four distinct phases [13]:

Step 1. Goal and scope definition. During the first step, the goal and scope of the study are defined as well as the selection of the functional unit (FU) and the system's boundaries. The meaningful selection and definition of system boundaries and the system's analysis are important tasks within every LCA. The functional unit relates to the product function rather than a particular physical quantity and is typically time-bound.

Step 2. Inventory analysis (LCI). In the second step, a life cycle inventory analysis, of relevant energy and material inputs and environmental releases, is made up identifying and quantifying inputs and outputs at every stage of the life cycle. In addition, the characteristics of data collection and calculation procedures are defined.

Step 3. Life Cycle Impact assessment (LCIA). This is the phase of LCA, with particular respect to sustainability assessment. During the impact assessment step, the elaboration of which has deliberately been left open by ISO (International Organization for Standardization) guidelines, the potential environmental impacts associated with identified inputs and releases are categorized in different midpoint and endpoint impact categories. LCIA translates emissions and resource extractions into a limited number of environmental impact scores by means of so-called characterization factors. There are two mainstream ways to derive these factors, i.e., at the midpoint level and at the endpoint level. Midpoint indicators focus on single environmental problems, for example, climate change or acidification. Endpoint indicators show the environmental impact on three higher aggregation levels, being the (1) effect on human health, (2) biodiversity, and (3) resource scarcity.

Step 4. Interpretation of results. In the last step, the results of the inventory analysis and the impact assessment should be interpreted and combined, to help decision makers make a more informative and sound decision. Furthermore, a sensitivity analysis is performed to validate the consistency of the results.

Depending on the scope of the LCA study, the life stages of energy production systems may include all or part of: **i.** Fuel consumption (i.e., to also account for the non-consumable portion of the produced fuel) and transportation to the plant, **ii.** facility construction, **iii.** facility operation and maintenance, and **iv.** dismantling. In this section, we present the technical details for the two studied renewable energy systems: **i.** Photovoltaics and **ii.** solar thermal collectors.

2.1. Photovoltaics

Photovoltaics based power generation employs solar panels to produce power on both a standalone basis using batteries or on a grid-connected basis using an inverter and electrical utility lines. Currently, commercially available PV modules are considered as not highly efficient (with typical efficiencies of ~16%), and thus there are intense research and development efforts for the development of new technological solutions to the challenge of producing commercial PV with increased efficiencies [10,19]. The rapid decline in installed costs (prices per installed MW have fallen by about 60% since 2008) has significantly improved the economic viability of PV around the world, with the global installed capacity escalated at 402 GW in 2017 compared to 8 GW back in 2007 [2,20]. Most of this growth has come from grid connected systems, though the off-grid market has also continued to expand [21]. Governmental subsidies and other supporting schemes were the initial driving force that allowed the market penetration of PV systems, but nowadays PV grid parity is a fast approaching reality in many countries [20,22].

For this research paper, four PV technologies will be evaluated: **i.** Single crystalline silicon (sc-Si), **ii.** multi-crystalline silicon (mc-Si), **iii.** Copper-Indium-diSelenide (CIS), and **iv.** amorphous silicon (a-Si). A limited number of comprehensive life cycle analyses based on industrial data for PV systems are available in the literature [23–26] and refer primarily to sc-Si and a-Si cells. Detailed technical information on PV module efficiencies are provided in Table 1.

Table 1. Technical characteristics of PV cell technologies used in this paper.

	Photovoltaic Technology	Technical Characteristics
Crystalline technologies	Single crystalline silicon cells (sc-Si)	The active material is made from a single crystal without grain boundaries. The sc-Si cells have the highest efficiencies (for commercial cells: 13%–18%).
	Multi-crystalline silicon cells (mc-Si)	The cell material consists of different crystals. The cells have a lower efficiency, but it is cheaper in production. Commercial mc-Si cells have efficiencies in the range of 11% to 16%.
Thin-Film Technologies	Copper-Indium-diSelenide (CIS)	CIS modules are constructed by depositing extremely thin layers of photovoltaic materials on a low cost layer (such as glass, stainless steel, or plastic). Material costs are lower because less semi-conductor material is required; secondly, labor costs are reduced because the thin films are produced as large, complete modules and not as individual cells that have to be mounted in frames and wired together. The efficiency is about 8% to 11%.
	Amorphous cells (a-Si)	The efficiency of amorphous cells is about 6% to 9% and decreases during the first 100 operation hours. A recently developed thin-film technology is hydrogenated amorphous silicon.

Thin-film technologies are less expensive overall in the production stages versus crystalline silicon because the materials and processes to manufacture the wafer-based silicon are far more expensive than producing thin-film based technologies. The main advantages of thin films are not their conversion efficiency, but their capital cost and their relatively low consumption of raw materials, high automation, and production efficiency. Thin films are also easier from integration on residential and commercial infrastructure. The current drawbacks are that the lower conversion efficiencies require more modules, which require more roof top space, which is limited on residential and commercial properties.

2.2. Solar Thermal Collectors

There have been a limited number of life cycle analyses looking specifically at solar thermal technologies. Emissions of GHGs (g CO_2-eq/kWh) have been estimated for central receiver systems between 36.2 and 43, while emissions from parabolic trough technologies have been estimated to 196 g CO_2-eq/kWh [27–29].

The most commonly used types of solar thermal collectors are the flat plate and the evacuated (or vacuum) tubes systems. Flat plate collectors consist of airtight boxes fitted with a glass (or other transparent material) cover, all installed on a suitable frame. They typically operate via the thermosyphonic effect and thus they need no electricity for circulation of the heat transfer fluid. The typical absorber area for residential applications ranges between 3 and 4 m^2, while a storage tank with a capacity between 150 and 180 L is capable of meeting the hot water demands for a family. An auxiliary electric immersion heater and/or a heat exchanger, for central heating assisted hot water production, are used in winter during periods of low solar insolation. Vacuum tube collectors are more advanced systems employing evacuated sealed glass tubes containing the solar radiation absorbers in order to minimize heat losses. These collectors exhibit a significantly higher performance compared to their flat plate counterparts, but at higher cost and they typically fit in more demanding applications (i.e., northern climates and lower ambient temperatures).

3. Results and Discussion

In order to validate the environmental impacts a Cradle-to-Grave LCA was implemented for each of the studied renewable technology. For this purpose, SimaPro 8.5 with ecoinvent version 3.4 was employed, while ReCiPe 2016 Midpoint Hierarchist (H) was chosen as the LCIA method in this study, as it provides the most extensive set of midpoint impact categories [15,30].

ReCiPe 2016 is the successor of the Eco-indicator and CML-IA. The purpose at the beginning of its development was to integrate the "problem oriented approach" of CML-IA and the "damage oriented approach" of Eco-indicator. The "problem oriented approach" defines the impact categories at a midpoint level. The uncertainty of the results at this point is relatively low. The drawback of this solution is that it leads to many different impact categories which makes the drawing of conclusions with the obtained results complex. On the other hand, the damage oriented approach of Eco-indicator results in only three impact categories, which makes the interpretation of the results easier. However, the uncertainty in the results is higher. ReCiPe implements both strategies and has both midpoint (problem oriented) and endpoint (damage oriented) impact categories.

Midpoint level indicators are direct measurements of the impacts arising from the considered phenomena. A total of 18 physical quantities were computed from the LCI results, providing a quantitative description of the single drivers of the environmental impact associated with the study. These include soil acidification (measured in kg SO_2-eq), the emission of GHGs (measured in kg CO_2-eq), ozone depletion (measured in kg CFC11-eq), and so forth.

The hierarchist perspective was chosen as it is the most balanced model based on common policy principles over a common time frame, compared to the individualistic and egalitarian perspectives, which consider a short and a long time frame, respectively [31].

Uncertainty analysis focuses on the extent of uncertainties produced in model outputs due to the existed uncertainties in input values. One of the several methods that propagate uncertainties is Monte Carlo simulation. This method makes use of an algorithm capable of producing a series of random numbers, within the uncertainty value of every input and output taken into account in the scenarios created, for which it assumes a lognormal distribution, with a certain confidence interval. For the studied systems in this paper, a Monte Carlo analysis was performed using SimaPro 8.5 software for each scenario and impact category.

3.1. Photovoltaics

3.1.1. LCA Analysis of PV Systems

The LCA results were used for the evaluation of the environmental impacts of various types of PV technologies. Four different PV systems using crystalline and thin-film technologies (as described in Table 1) were evaluated in this paper, all having the same nominal capacity of 3 kW. In this section, the detailed results from the LCA of the studied PV systems are presented in order to determine which technologies are more hazardous to human health and ecosystem quality in a comparative assessment, distinguish which lifecycle stage of the PV energy production represents the majority of these impacts, and finally evaluate their overall energy performance.

The LCA of a PV system starts with the extraction of raw materials and follows along the product to the end of its life and the disposal of the PV components. The first stage of the process entails the mining of raw materials, for example, quartz sand for silicon based PVs, followed by further processing and purification stages, to achieve the required high purities, which typically entails a large amount of energy consumption and related emissions. Other raw materials included are those for balance of system (BoS) components, for example, silica for glass, copper ore for cables, and iron and zinc ores for mounting structures. At the end of their lifetime, PV systems are decommissioned and the valuable parts and materials are disposed.

Although PV power systems do not require finite energy sources (fossil, nuclear) during their operation, a considerable amount of energy and emissions are released for their production.

The environmental issues associated with this energy use for PV manufacturing will also affect the environmental profile of PV power systems. The environmental themes that are strongly related to the PV energy system are: Exhaustion of finite resources, human health implications, and climate change [25,32,33].

The goal and scope of this LCA study was to evaluate over the lifecycle the impacts of the electricity produced by four different grid-tied 3 kW PV installations and the functional unit was the production of 1 kWh of produced electricity. The LCIA method used for the characterization of PV technologies was ReCiPe Midpoint, aiming to highlight the global warming potential and GHG emissions, fossil fuels, and climate change impacts related to each technology. The results were ranked from worst to best environmental performance and used to validate the environmental impacts of each PV system. The objective of conducting the LCA study was to make a comparative environmental analysis of different PV systems with a focus on comparing crystalline with thin film technologies.

The system boundaries account for all the impacts related to production, transportation, and system disposal of PV systems. The main parts of the studied systems are: **i.** The PV-panels, **ii.** the inverter, **iii.** the electric installation, and **iv.** the roof mounting structure. The process data for a 3 kW PV installation includes quartz reduction, silicon purification, wafer, panel and laminate production, and manufacturing of inverter, mounting, cabling, and infrastructure, assuming a 30 years operational lifetime. The following items were studied for each production stage as far as data were available:

- Energy consumption;
- Air and waterborne process-specific pollutants at all production stages (materials, chemicals, etc.);
- Transport of materials, energy carriers, semi-finished products, and the complete power plant;
- Waste treatment processes for production wastes;
- Dismantling of all components;
- Infrastructure for all production facilities with its land use.

The PV systems have the same nominal installed capacity (i.e., 3 kW) and differ according to the cell type (single- and multi-crystalline silicon, thin film cells with amorphous silicon, and CIS). All systems were assumed to be installed on existing buildings (slanted roof installation).

Life cycle inventory analysis involves creating an inventory of flows from and to nature for a product system. The Ecoinvent v3.4 database was employed for the inventories of PV systems, which can be assumed to be representative for typical PV installations. The Ecoinvent database provides detailed and transparent background data for a range of materials and services used in the production chain of photovoltaics. The delivery of the different PV parts to the final construction place was assumed as 100 km by a delivery van. This includes the transport of the construction workers. It was assumed that 20% of the panels are produced overseas and thus must be imported to Europe by ship. The lifetime of the inverter was assumed to be 15 years.

In Figure 1, the process network for the studied mc-Si PV system is depicted for the cut-off threshold of 10% (similar figures represent the data for the other three PV types). The thick red line in the network trees is known as the elementary flow and indicates the environmental bottleneck or burden in each process.

For the CIS system, 64.2% of all total inflows and outflows are due to the production of the photovoltaic panel. The installation phase and the inverter require 23.3% and 9.5%, respectively, of the energy and materials inflow. The main environmental impacts include the panel and cell production, inverter, and installation/construction phases. There are also impacts associated with the electricity, transportation, and system disposal, which are taken into consideration. Similar values stand for the case of a-Si panel: 56.9% for the production phase, and 32.5% and 8% for the installation phase and the inverter, respectively. For the sc-Si and mc-Si panels, 77.6% and 72.5%, respectively, of all total inflows and outflows are due to the production of the photovoltaic panel, installation is 13.1% and 16.5%, respectively, while the inverter accounts for 7% and 8.3%, respectively.

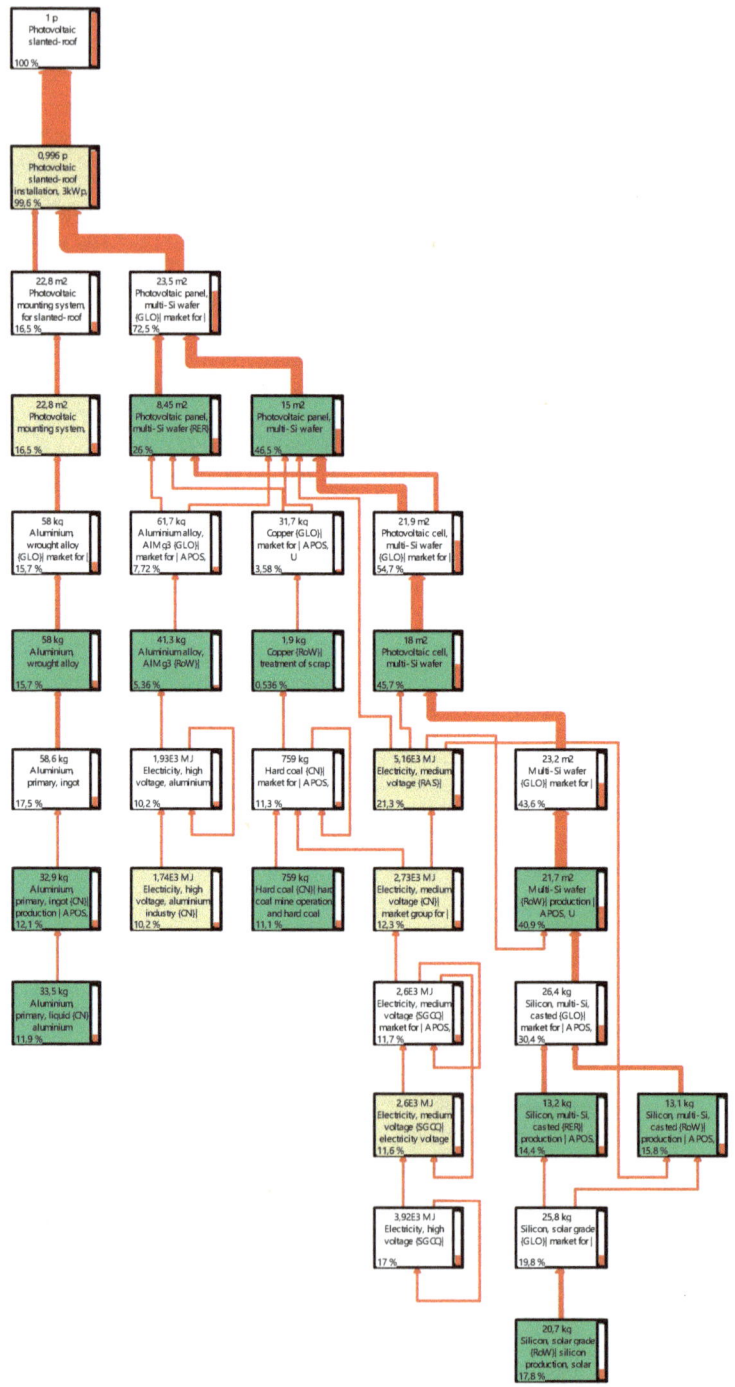

Figure 1. Process network for mc-Si PV system. Cut-off threshold: 10%, total nodes: 11,607.

From the process networks, it is evident that the production stage contributes the most important part of the environmental impacts in the life cycle of all studied PV technologies. The elementary flows indicate that most inflows of materials and energy for both thin-film and crystalline technologies occur during the cell and panel production phase. Subsequently, large emissions and impacts to the environment and human health follow this stage of the PV systems' lifecycle. Based on the above, we can conclude that the cell and panel production phase are the most important inputs to the development of a 3 kW PV system, followed by the inverter and construction of the mounting systems.

The environmental impacts of PV systems were calculated through the conducted LCA. The typical operation of PV systems was taken under consideration. In Table 2 and Figure 2, the aggregated LCA inventory results for the studied PV systems are presented. These are harmonized data representing the LCA results (for each impact category) per total electricity exported to the grid (in kWh) by each 3 kW PV system, thus providing a holistic evaluation indicator (i.e., environmental burden per total energy produced).

Table 2. Aggregated LCA inventory results for the studied PV systems.

Impact Category	Unit	a-Si	CIS	mc-Si	sc-Si
Global warming	kg CO_2-eq /kWh	4.35×10^{-2}	3.95×10^{-2}	4.43×10^{-2}	5.24×10^{-2}
Stratospheric ozone depletion	kg CFC11-eq/kWh	1.70×10^{-8}	1.75×10^{-8}	2.06×10^{-8}	2.45×10^{-8}
Ionizing radiation	kBq Co-60-eq/kWh	3.95×10^{-3}	3.96×10^{-3}	4.08×10^{-3}	4.45×10^{-3}
Ozone formation, human health	kg NO_x-eq/kWh	9.83×10^{-5}	9.09×10^{-5}	1.05×10^{-4}	1.20×10^{-4}
Fine particulate matter formation	kg $PM_{2.5}$-eq/kWh	1.09×10^{-4}	9.39×10^{-5}	1.04×10^{-4}	1.23×10^{-4}
Ozone formation, terrestrial ecosystems	kg NO_x-eq/kWh	1.01×10^{-4}	9.26×10^{-5}	1.10×10^{-4}	1.25×10^{-4}
Terrestrial acidification	kg SO_2-eq/kWh	2.25×10^{-4}	2.07×10^{-4}	2.21×10^{-4}	2.47×10^{-4}
Freshwater eutrophication	kg P-eq/kWh	3.55×10^{-5}	4.62×10^{-5}	3.78×10^{-5}	4.07×10^{-5}
Terrestrial ecotoxicity	kg1,4-DCB-eq/kWh	4.69×10^{-1}	4.62×10^{-1}	1.17	1.13
Freshwater ecotoxicity	kg1,4-DCB-eq/kWh	1.11×10^{-2}	1.30×10^{-2}	1.16×10^{-2}	1.17×10^{-2}
Marine ecotoxicity	kg1,4-DBC-eq/kWh	1.43×10^{-2}	1.69×10^{-2}	1.53×10^{-2}	1.54×10^{-2}
Human carcinogenic toxicity	kg1,4-DBC-eq/kWh	6.50×10^{-3}	4.19×10^{-3}	4.17×10^{-3}	4.33×10^{-3}
Human non-carcinogenic toxicity	kg1,4-DBC-eq/kWh	1.46×10^{-1}	2.00×10^{-1}	1.63×10^{-1}	1.64×10^{-1}
Land use	m^2a crop-eq/kWh	1.13×10^{-3}	9.60×10^{-4}	1.23×10^{-3}	1.23×10^{-3}
Mineral resource scarcity	kg Cu-eq/kWh	6.60×10^{-4}	8.21×10^{-4}	5.54×10^{-4}	5.42×10^{-4}
Fossil resource scarcity	kg oil-eq/kWh	1.04×10^{-2}	9.40×10^{-3}	1.08×10^{-2}	1.27×10^{-2}
Water consumption	m^3/kWh	4.51×10^{-4}	3.22×10^{-4}	1.35×10^{-3}	1.17×10^{-3}

In Figure 2 the relative contributions to the impact categories (based on the ReCiPe 2016 midpoint evaluation) for the studied PV systems are shown. The cumulative CO_2-eq emissions per kWh over the whole life cycle of the PV systems vary between approximately by 3.9×10^{-2} and 5.2×10^{-2} kg CO_2-eq/kWh.

During the lifecycle of a PV system, initially, the extraction of resources leads to emissions that affect human health, including carcinogens and respiratory inorganics, while at a second level, the use of fossil fuel during the production and manufacturing processes releases large amounts of greenhouse gases in the atmosphere, causing climate change. Processes occurring during the panel production phase can significantly affect air quality as hazardous substances are emitted into the atmosphere and biosphere.

According to this analysis, the most severe burdens seem to be gathered to the following categories: Global warming, fossil fuel resource scarcity, carcinogens, ecotoxicity, and land use. The crystalline technologies (mc-Si and sc-Si) have increased values in almost all impact categories. Thin-film CIS exhibits lower impacts in most categories and seems to be an optimum selection from an environmental perspective compared to its other counterparts. Results indicate that there are impacts in all indicators, especially those affecting human health from the substances released into the air and water. The manufacturing of a-Si PV cells and panels requires silicon and typically the energy intensive "Siemens process" [34]. On the other hand, thin film PV systems have lower efficiencies and thus a 3 kW installation will require a larger number of cells and panels and more materials for the mounting systems. According to this analysis, thin-film technologies require less materials' inflows for their construction and installation phases compared to crystalline systems and this coincides with reduced airborne pollutants, emissions, and energy (also connected with transportation, distribution, and mounting of the systems).

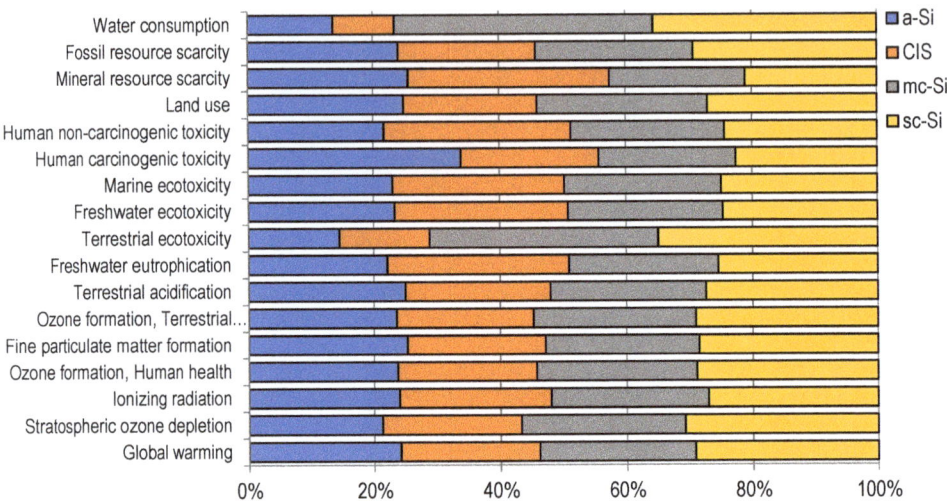

Figure 2. LCA results for the studied PV systems: relative contributions to the impact categories.

For the purposes of this study, two Monte Carlo analyses of the LCA results (repeated for 5000 iterations) were implemented for a comparison between the PV systems in each studied technology (i.e., crystalline and thin film). The aim of these analyses was to provide an additional validation (based on a statistical evaluation) for the credibility of the presented results. The first analysis was conducted between A: a-Si and B: CIS PV systems. During the Monte Carlo analysis, a stochastic variation of the parameters in the initial inventory database for each of the studied two cases (i.e., A and B) was performed, altering the LCA results and thus affecting the A−B outcome. A random variable was selected for each parameter within the specified uncertainty range and the impact assessment results were recalculated. The same process was repeated by taking different samples (within the uncertainty range) and all results were stored. After repeating the procedure for a set number of times (e.g., 5000), 5000 different results were obtained, thus forming the uncertainty distribution of the impacts (LCIA), with a confidence interval of 95%.

The results in a bar chart form are depicted in Figure 3 showing the percentage of times when system A has a greater impact than system B (A−B ≥ 0, in orange) and vice versa (A−B < 0, in blue). This is a balanced graph and, in general, we can conclude that A has increased impacts compared to B in most of the studied midpoint categories. This is quite evident for the human carcinogenic toxicity category, in which A has distinctively increased impacts compared to B for 96.6% of the completed iterations. Respectively, human non-carcinogenic toxicity and freshwater eutrophication are the two cases that A has a lower impact than B, for almost 80% of the completed iterations.

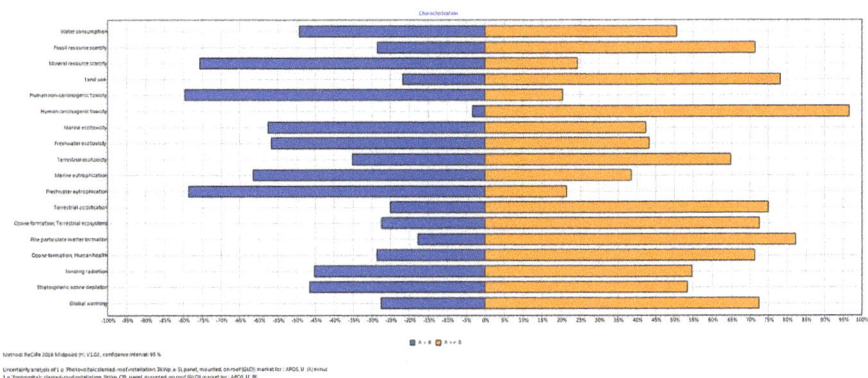

Figure 3. Monte-Carlo simulation results of LCIA uncertainties between a-Si (A) and CIS (B) PV systems.

The second Monte Carlo analysis was conducted between A: mc-Si and B: sc-Si PV systems. Figure 4 presents the results in a bar chart form, showing the percentage of times when system A has a greater impact than system B (A−B ≥ 0, in orange) and vice versa (A−B < 0, in blue). In this case, it is evident that case A has lower impacts compared to B in most of the studied midpoint categories. The impact categories that a balanced result is observed are water consumption, land use, human non-carcinogenic toxicity, marine, freshwater, and terrestrial ecotoxicity.

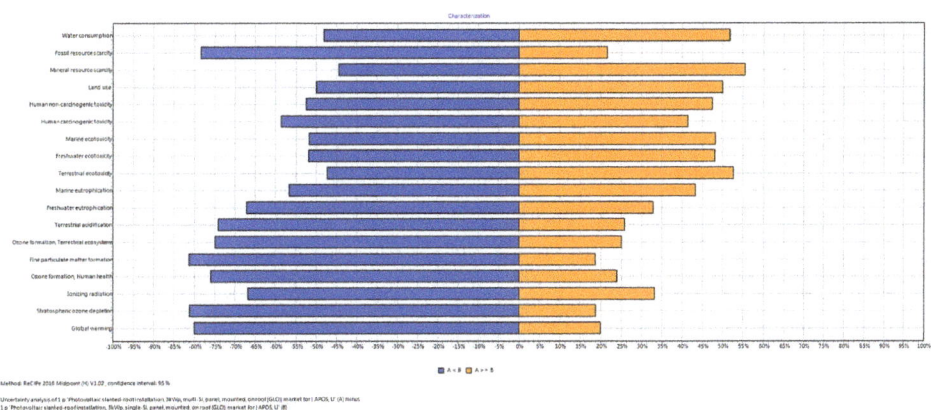

Figure 4. Monte-Carlo simulation results of LCIA uncertainties between mc-Si (A) and sc-Si (B) PV systems.

It is very important to stress the fact that the results depicted in Figures 3 and 4 refer to the comparison of the raw LCA data and not the harmonized results as mentioned in Table 2 and Figure 2 (i.e., LCA results for each impact category per total electricity exported to the grid for each PV system). Thus, these data do not include the provision for varying energy production for each of the studied systems.

Various additional technical components, the so-called balance of system (BoS) elements, can also play an increasingly important role for the comparison of different types of PV technologies with different efficiencies and thus different sizes of mounting systems for the same electric output. These BoS elements can have a significant share of 30% to 50%. On the one hand, this is due to the improvements, which could be observed for the production chain until the output of the final photovoltaic cell. On the other hand, now a more detailed investigation of these additional elements is available, which, for example, also includes the electronic components of the inverter. The low efficiency systems need larger amounts of the mounting structure and cabling, which partly outweighs the better performance per kWp of the module alone [26]. Overall, in the entire life cycle of both types of PV technologies, it was observed that the magnitude of environmental impacts of crystalline was greater than that of the thin film.

3.1.2. Energy and Economic Assessment of PV Systems

The first step in a pre-feasibility study of a solar (i.e., PV) project is to define the solar energy potential of the region in which the PV systems will be installed. This serves as a planning tool to quantify the anticipated electricity production and plant costs. The evaluation of these PV technology costs require in-depth analysis of site-specific solar energy potential; costs of solar technologies; customer types; meter types; utility types; physiographic conditions; local, regional, and national laws and regulations; feed-in-tariffs and financial mechanisms; etc. The techno-economic analysis carried out in this part of the paper quantifies the energy output and the economic income associated with each of the studied 3 kW PV power plants. The proposed area for installation of the PV systems is the island of Crete located in the southern part of Greece, which was selected as a typical representation of regions with a mild climate and high average insolation that lasts almost throughout the year (with greater intensity from April to October). These climatic conditions render Crete as one of the best available locations in Greece for installation of solar systems. The island is not interconnected to the mainland distribution grid and the necessary electricity is produced via diesel burning conventional thermal stations, thus increasing the cost (environmental and economic) per produced energy unit. In addition, Crete presents extreme variations in energy demand throughout the year, with significant peaks during the summer due to the tremendous increase of the population due to visiting tourists and increased air-conditioning needs. Thus, the need for decentralized production of electricity is more than obligatory as the solar grid parity in non-interconnected islands can already be considered as a fact [22]. On the other hand, the deficiencies in the existing electricity grid and local supporting schemes/governmental rules for renewables have created a vague scenery for potential investors. The economic and energy assessment of PV systems was carried out using the RETScreen software. The completed study involves quantifiable results for energy—economic impacts and savings for the chosen PV system. The site location for the installation of the PV systems was chosen to be the Acrotiri area in Chania, while all meteorological data (in the form of the annual time series of average climate conditions) were extracted from RETScreen referring to a weather station of Souda Bay, Chania.

The results of the RETScreen economic analysis provide a reliable and comprehensive evaluation of the anticipated technology, the energy production, potential emissions reduction, necessary investment cost, financial viability, and risks associated with the specific project. The accuracy of RETScreen is considered to be more than sufficient for preliminary feasibility studies and a small reduction in accuracy due to the use of monthly rather than hourly solar radiation data is more than compensated for due to the ease-of-use of the software.

After selecting the location area, the complete RETScreen analysis for each one of the studied PV systems was conducted. This analysis comprised four discrete steps: **i.** Selection of the technology (i.e., sc-Si, mc-Si, CIS, a-Si) and specification of the technical parameters, **ii.** energy analysis (see results in Table 3), **iii.** emissions analysis), and **iv.** financial analysis.

Table 3. Results of the techno-economic assessment for the studied PV systems.

	PV Technology	Cell Efficiency [%]	Frame Area [m²]	Capacity Per Unit [W]	Total Area [m²]	Cost [€/kW]	Capacity Factor [%]	Total Electricity Exported to Grid [MWh]	Annual Revenue [€/yr]	IRR [%]	Payback Time [years]
Crystalline	sc-Si	17	1.18	200	17.7	1600	20.6	162.6	542	11.5	10
	mc-Si	12.3	1.02	125	24.5	1500	20.6	162.6	542	12.3	9.3
Thin-film	CIS	10.6	0.94	100	28.2	1600	20.2	159.3	531	12	9.6
	a-Si	6.1	0.82	50	49.2	1500	21.8	171.6	572	13.1	8.8

For all financial calculations, the electricity price was set to 0.10 €/kWh and we considered that the installation was funded by own means (no bank loan). For Greece, the employed feed-in-tariff for roof top PV will decline to 0.8 €/kWh by the end of 2019, but residential installations up to 10 kWp can benefit from a net-metering scheme, which can allow for compensation at prices up to 0.15 €/kWh [35–37]. In Table 3, the main results of the RETScreen analysis for all studied PV systems are presented. The cell efficiencies of the PV systems vary (from 6.1% to 17%), but this parameter does not play an important role as the nominal capacity of all systems is set to 3 kW. On the other hand, the larger the efficiency of the panel, the less the area needed for the installation (from 17.7 m^2 to 49.2 m^2). The simple payback period is 8.8 to 10 years (for regions with same insolation, i.e., Andalucía in Spain, the corresponding values for residential PV are 7.6 to 12.1 [38]) and IRR values vary from 11.5 to 13.1. The a-Si system seems to have a higher annual energy yield, and this is practically due to the ability of these systems to produce more electricity under hazy or cloudy conditions and thus their capacity factor is increased (21.8%) compared to their counterparts. The electricity produced allows for the mitigation of ~4 tons of CO_2-eq annually for all PV systems.

According to the comparison of the different PV technologies, the anticipated energy production, emissions reduction, investment cost, financial viability, and risks associated with the four technologies are approximately the same. All technologies portray relatively equal cost benefit ratios and financial parameters. This is mainly due to the fact that our selection of comparing 3 kW systems harmonizes the influence of all technical advantages amongst technologies. On the other hand, the sc-Si system is the most efficient per cell, thus needing less area per installation compared to the other cases.

3.2. Solar Thermal Systems

3.2.1. LCA Analysis of Solar Thermal Systems

In this section, the detailed results from the LCA of solar thermal collectors will be presented. The two studied systems are: **i.** Flat plate collector with copper absorber and **ii.** vacuum (or evacuated) tube collector. In order to validate the environmental impacts, a detailed LCA was implemented for both studied systems.

The goal and scope of this LCA study is to evaluate over the lifecycle, the impacts of the thermal energy converted to hot water needs and consequently to the equivalent avoided electricity (thus the functional unit was the saving of 1 kWh electricity for hot water production), for the two types of solar collectors for use in a typical single house family. For this purpose, SimaPro 8.5 was employed, while ReCiPe 2016 Midpoint Hierarchist (H) was chosen as the LCIA method as it provides the most extensive set of midpoint impact categories, aiming to highlight the global warming potential and GHG emissions, fossil fuels, and climate change impacts related to each technology. The results are ranked from worst to best environmental performance. These results will be used to distinguish the impacts of each solar system and can be used during the combined environmental and technical assessment of installing such solar energy harvesting technologies.

The system boundaries account for all the impacts related to production, transportation, and disposal for both complete solar systems (excluding auxiliary heating), including various technical components, heat exchange fluid, installation of copper pipes, transportation of parts, delivery with a van, and montage on the roof. The main parts of the studied systems are: **i.** The solar collectors and absorbers (with an aperture area of 12.3 m^2 and 10.5 m^2 for the flat plate and the vacuum tube collectors, respectively), **ii.** the 200 L heat storage tank, and **iii.** the roof mounting structure. Both systems are aimed for installation on existing buildings (slanted roof installation) and their operational lifetime was assumed to be 20 years. Life cycle inventory analysis involves creating an inventory of flows from and to nature for a product system. The database, Ecoinvent 3.4, was employed for the inventories of solar collectors, as it provides detailed and transparent background data for a range of materials and services used in the production chain of solar collectors.

In Figure 5, the process network for the studied vacuum tube solar collector is depicted for a cut-off threshold of 10%. For the flat plate system, 57% and 27.1% of all total inflows and outflows are due to the production of the collector and the tank, respectively, while for the vacuum tube system, the corresponding values are 45.3% and 34.8%. Thus, as the networks clearly show, the production stage of the collector component contributes the most important part of the environmental impacts in the life cycle for both studied systems.

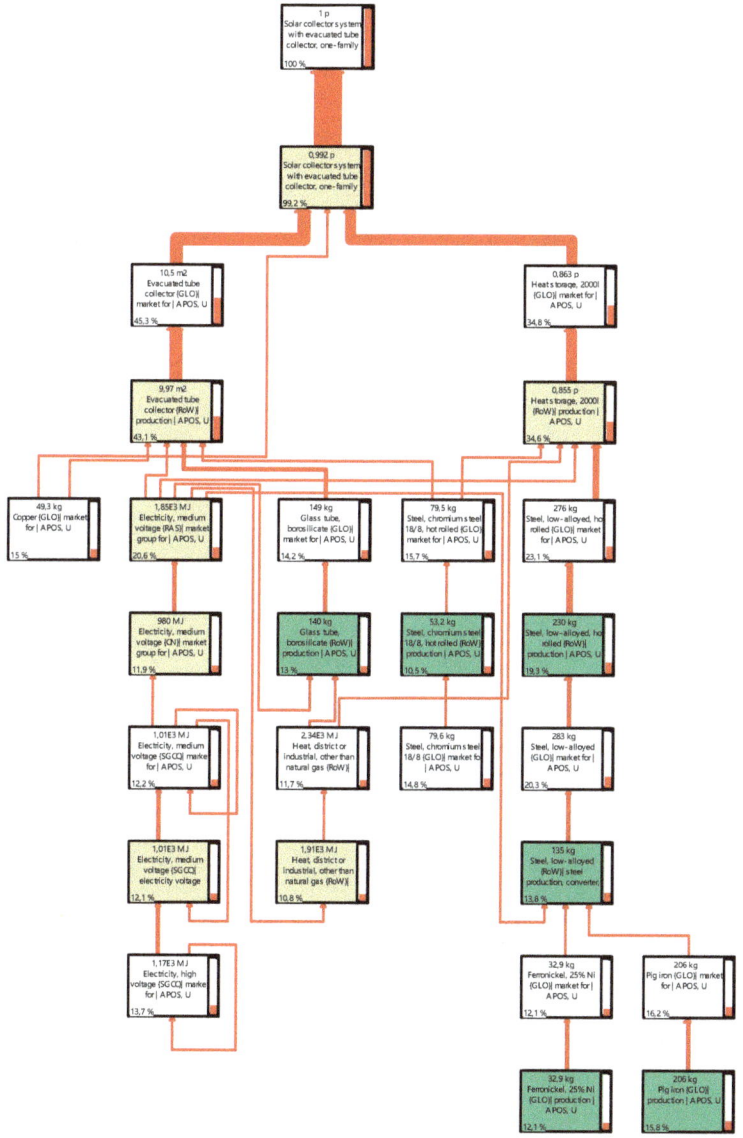

Figure 5. Process network for the vacuum tube solar collector. Cut-off threshold: 10%, total nodes: 11,607.

In Table 4 and Figure 6, the aggregated LCA inventory results for the studied solar thermal systems are depicted. These are harmonized data representing the LCA results (for each impact category) per total energy produced per aperture area (in kWh/m^2) by each solar collector, thus providing a holistic evaluation indicator (i.e., environmental burden per total energy produced). It is important to stress the fact that the electricity mentioned above in kWh corresponds to the necessary energy for heating water, which is substituted by the operation of the solar collectors, which convert solar radiation to heat transferred to a stored hot water in their tank. As depicted in Table 4, the cumulative CO$_2$-eq emissions over the whole life cycle of the solar systems are quite close, varying between 2.22×10^{-2} and 2.38×10^{-2} kg CO$_2$-eq/kWh·m^2, and the lowest value corresponds to the vacuum tube collector.

Table 4. Aggregated LCA inventory results for the studied solar thermal systems.

Impact Category	Unit (per m^2)	Flat Plate Collector	Vacuum Tube Collector
Global warming	kg CO$_2$-eq/kWh	2.38×10^{-2}	2.22×10^{-2}
Stratospheric ozone depletion	kg CFC11-eq/kWh	1.29×10^{-8}	1.36×10^{-8}
Ionizing radiation	kBq Co-60-eq/kWh	1.61×10^{-3}	1.88×10^{-3}
Ozone formation, human health	kg NO$_x$-eq/kWh	6.50×10^{-5}	6.89×10^{-5}
Fine particulate matter formation	kg PM$_{2.5}$-eq/kWh	8.78×10^{-5}	8.61×10^{-5}
Ozone formation, terrestrial ecosystems	kg NO$_x$-eq/kWh	6.66×10^{-5}	7.07×10^{-5}
Terrestrial acidification	kg SO$_2$ eq/kWh	2.07×10^{-4}	2.01×10^{-4}
Freshwater eutrophication	kg P-eq/kWh	3.89×10^{-5}	4.16×10^{-5}
Terrestrial ecotoxicity	kg1,4-DCB-eq/kWh	8.55×10^{-1}	9.31×10^{-1}
Freshwater ecotoxicity	kg1,4-DCB-eq/kWh	6.42×10^{-3}	6.94×10^{-3}
Marine ecotoxicity	kg1,4-DBC-eq/kWh	9.27×10^{-3}	1.00×10^{-2}
Human carcinogenic toxicity	kg1,4-DBC-eq/kWh	6.56×10^{-3}	6.53×10^{-3}
Human non-carcinogenic toxicity	kg1,4-DBC-eq/kWh	2.24×10^{-1}	2.44×10^{-1}
Land use	m^{2}a crop-eq/kWh	1.25×10^{-3}	1.52×10^{-3}
Mineral resource scarcity	kg Cu-eq/kWh	1.02×10^{-3}	1.03×10^{-3}
Fossil resource scarcity	kg oil-eq/kWh	5.45×10^{-3}	5.38×10^{-3}
Water consumption	m^3/kWh	2.39×10^{-4}	2.33×10^{-4}

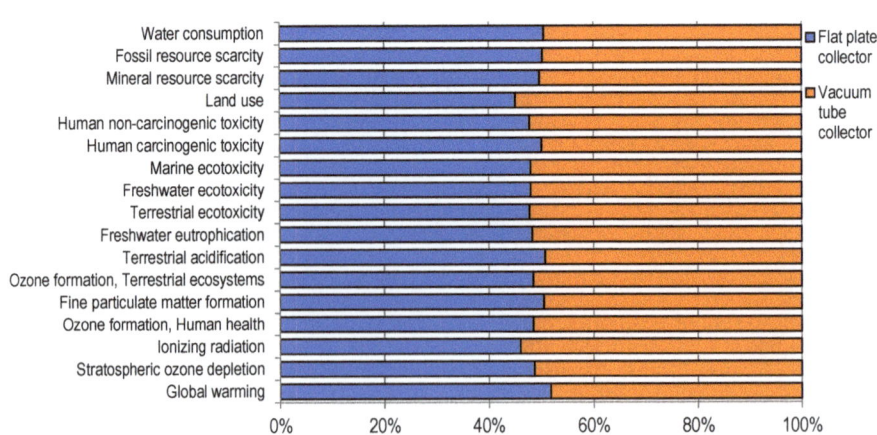

Figure 6. LCA results for the studied solar thermal systems: relative contributions to the impact categories.

In Figure 6, the relative contributions to the impact categories (based on the ReCiPe 2016 midpoint evaluation) for the solar systems are depicted. The results are mixed, with the two systems exhibiting similar environmental impacts in most categories, but the vacuum tube collector has the highest values in most cases.

For the purposes of this study, a Monte Carlo analysis of the LCA results was implemented through a comparison between the two studied solar collectors (A: Flat plate and B: Vacuum tube collector), which was repeated for 5000 iterations. In Figure 7, the results of the uncertainty analysis are depicted in a bar chart form, showing the percentage of times when collector A has a greater impact than collector B (A−B ≥ 0, in orange) and vice versa (A−B < 0, in blue). It is clear that for the studied solar collectors, A has increased impacts compared to B in most of the studied midpoint categories. Land use is the only case that A has a lower impact than B, for 53.4% of the completed iterations. It is important to keep in mind that these outcomes refer to the direct LCA results, which are non-harmonized (i.e., they do not take into account the environmental impacts per energy production and per aperture area for each system).

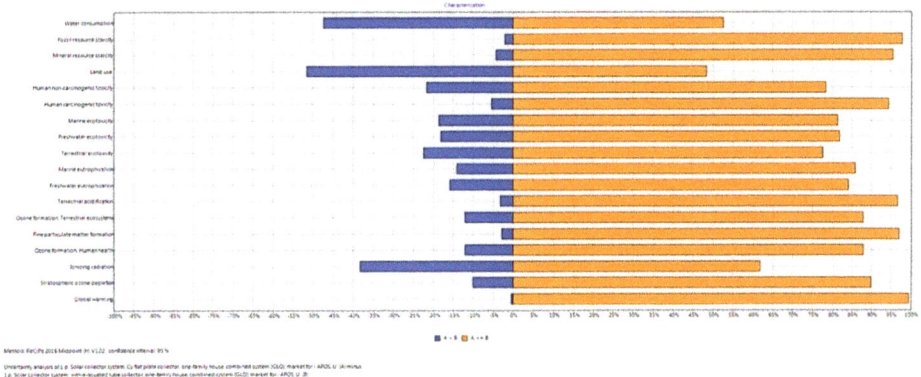

Figure 7. Monte-Carlo simulation results of LCIA uncertainties between flat plate (A) and vacuum tube (B) collectors.

3.2.2. Energy and Economic Assessment of Solar Thermal Systems

The comparative techno-economic assessment of the installation of the two solar thermal collectors was carried out through RETScreen. The installation location site was chosen to be the Acrotiri area in Chania, while all meteorological data (in the form of annual time series of average climate conditions) were extracted from RETScreen referring to a weather station of Souda Bay, Chania. After selecting the location area, the complete RETScreen analysis for each solar collector was conducted. This analysis comprised the following discrete steps: **i.** Determination of the annual hot water needs for the studied single family house, **ii.** selection of the auxiliary hot water heating system (i.e., diesel based heating equipment), **iii.** selection of the solar collector technology (i.e., flat plate and vacuum tube) and specification of the technical parameters, **iv.** energy analysis (see aggregated results in Table 5), and **v.** financial analysis.

For all financial calculations, the electricity price was set to 0.15 €/kWh and we considered that the installation was funded by own means (no bank loan). The hot water needs for a typical family house with four occupants (taking as granted a 100% occupancy rate and 24 operating hours per day) were estimated to be 2817 kWh per year. A typical auxiliary hot water heating system burning diesel was considered for backup. In Table 5, the main results of the RETScreen analysis for the studied solar thermal collectors are presented. Both selected systems are typical flat plate and vacuum solar collectors installed in Greek houses and they can be considered as top-class products, while the purchase cost of the vacuum tube collector is significantly higher, i.e., 1300 € vs. 900 € [39].

Table 5. Results of the techno-economic assessment for the studied solar thermal collectors.

Solar Collector Type	Aperture Area [m^2]	F$_r$UL [(W/m^2)/°C]	Cost [€]	Total Energy Saved [kWh]	Total Energy Saved Per Aperture Area [kWh/m^2]	Solar Fraction [%]	Annual Savings [€/yr]	IRR [%]	Payback Time [years]
Flat plate	2.32	4.6	900	27,260	11,750	55.3	352	41.8	2.6
Vacuum tube	2.61	1.7	1300	29,980	11,487	62.7	341	28.5	3.8

The thermal losses coefficient, FrUL, is increased for the flat plate collector compared to the vacuum tube system, i.e., 4.6 vs. 1.7 (W/m^2)/°C, respectively. This is due to the completely different thermal losses suppression design followed in each system, which practically makes the vacuum tube collector unaffected by variations in the ambient temperature. In addition, the solar fraction value (practically denoting the percentage of hot water needs covered by the system annually) for the vacuum tube system is higher that the flat plate collector (i.e., 62.7% vs. 55.3%, respectively). On the other hand, it is evident that overall, this parameter does not play an important role in the energy outcome of the systems, as finally the flat plate collector provides slightly more energy per aperture area throughout the year. This is mainly due to two reasons: i. The weather conditions in Crete (high intensity solar radiation for extended time periods and with increased ambient temperatures throughout the year) are favorable for solar systems and thus the advantageous thermal insulation and the ability to reach high temperatures of the vacuum system is not necessary, ii. the pump in the vacuum system requires more electricity due to increased friction in the collector (more complex circulation system).

The comparison of the annual energy-fuel consumption and the economic savings between the base case (auxiliary hot water heating system) and the solar collectors was performed for both the studied systems. Annual savings of 352 € (flat plate system) and 341 € (vacuum tube system) are anticipated, and their economic viability is obvious. The simple payback period is 2.6 and 3.8 years and IRR values of 41.8 and 28.5 for the flat plate and the vacuum tube system, respectively. The above mentioned results prove that the selection of a flat plate system is rather mandatory for typical installations in Crete (southern part of Greece) while vacuum tube systems could be selected for energy demanding applications or northern climates.

3.3. Life Cycle Carbon Footprint

As indicated in the previous analysis, the studied renewable energy systems have environmental impacts during their production phase, but through their operation (i.e., production of clean energy) they manage to mitigate significant amounts of emitted greenhouse gases due to the avoided use of fossil fuels. In the following section, we will comment on the overall environmental profile of various energy production technologies through the concept of a carbon footprint (thus focusing on global warming impacts). The measurement of life-cycle greenhouse gas emissions involves calculating the global-warming potential of electricity production through life-cycle assessment of each energy source. The findings are presented in units of global warming potential per unit of electrical energy generated by that source, i.e., gCO$_2$-eq/kWh. The goal of such evaluations is to analyze the complete life cycle of the energy generating technology, from material and fuel mining through construction to operation and waste management [40,41].

In Table 6, the values of the emitted, avoided, and the lifetime balance for the greenhouse gases and the total energy produced from photovoltaics and solar thermal systems are presented. Both technologies avoid the emission of significant amounts of GHG through their operation and energy production. It is evident that the magnitude of the total avoided emissions is higher for photovoltaics compared to solar thermal systems and this has to do with the difference in the concept and the installed capacity of the two technologies.

Table 6. Comparative life cycle carbon footprint results for the studied renewable energy systems.

	Carbon Footprint [g CO_2-eq/kWh] *	References	Carbon Footprint [g CO_2-eq/kWh] **	Total Emitted GHG [g CO_2-eq]	Total Avoided GHG [g CO_2-eq]	Lifetime GHG Balance [g CO_2-eq]	Total Energy Produced [kWh]
a-Si PV			43.5	7.47×10^6	1.24×10^8	1.17×10^8	1.72×10^5
CIS PV	26–60	[40,42–44]	39.5	6.29×10^6	1.15×10^8	1.09×10^8	1.59×10^5
mc-Si PV			44.3	7.20×10^6	1.17×10^8	1.10×10^8	1.63×10^5
sc-Si PV			52.4	8.52×10^6	1.17×10^8	1.08×10^8	1.63×10^5
Flat plate collector	20–45	[42,44]	23.8	3.44×10^6	2.60×10^7	2.26×10^7	3.12×10^4
Vacuum tube collector			22.2	2.68×10^6	2.50×10^7	2.23×10^7	3.54×10^4
Wind	9–35	[40,42–44]					
Geothermal plant	6–79	[40,43,44]					
Hydroelectric	1–24	[42–44]					
Nuclear	4–110	[40,44,45]					
Natural gas	410–650	[40,41,44]					
Oil	778	[40,41,44]					
Coal	740–1050	[40,41,44]					

* Results based on bibliographic references, ** Results from this study.

The carbon footprint for the studied renewable systems was calculated, and in addition, typical values for other energy production technologies (either renewables or fossil-fuel based) are also depicted in Table 6 [40–45]. The carbon footprint for solar thermal collectors is lower compared to photovoltaics, while both technologies alongside with wind, hydroelectric, and nuclear are quite far from fossil fuel based power plants (which exhibit carbon footprint values ranging from 400 to 1050). This is an expected result as the environmental advantage of renewable over conventional energy sources is unambiguous.

4. Concluding Remarks

The energy and environmental profile for photovoltaics and solar thermal collectors were presented in the previous sections of the paper. For each technology, various technical variations were presented, i.e., thin film-crystalline silicon photovoltaics and flat plate-vacuum tube solar collectors. In the following paragraphs, a synopsis of the results for each renewable technology is presented alongside the detailed discussion and conclusions.

Regarding the photovoltaics, all studied systems were selected to have the same nominal installed capacity of 3 kW, representing a typical choice for residential applications. The production stage contributes the most important part of the environmental impacts in the life cycle of all studied PV technologies (followed by the inverter and construction of the mounting systems), as 60% to 70% (depending on the system) of inflows of materials and energy for both thin-film and crystalline PV systems occur during the cell and panel production phase.

The crystalline technologies (mc-Si and sc-Si) have increased values in almost all environmental impact categories. Thin-film CIS exhibits lower impacts in most categories and seems to be an optimum selection from an environmental perspective compared to its other counterparts. On the other hand, a-Si PV cells require an energy intensive manufacturing process, which affects their environmental profile. The cumulative CO_2-eq emissions per kWh over the whole life cycle of the studied PV systems vary between approximately 3.9×10^{-2} and 5.2×10^{-2} kg CO_2-eq/kWh.

The efficiencies vary from 6.1% to 17%, with thin-films based PV systems exhibiting the lowest values, but this parameter does not play an important role as the nominal capacity of all systems is identical (i.e., 3 kW). On the other hand, the larger the efficiency of the panel, the less the area needed for the installation (from 17.7 m^2 to 49.2 m^2) and less materials will be required for the mounting systems. The simple payback period of the systems is 8.8 to 10.0 years and IRR values vary from 11.5 to 13.1. The a-Si based systems seems to have higher annual energy yields due to their ability to produce more electricity under hazy or cloudy conditions and thus their capacity factor is increased (21.8%) compared to their counterparts (values ~20.5). The electricity produced allows for the mitigation of ~4 tons of CO_2-eq annually for all PV systems. In general, the anticipated values for energy production, emissions reduction, investment cost, financial viability, and risks associated with the four 3 kW PV technologies are quite similar. For real case installations, parameters, like total cost and necessary area for installation, might play a decisive role for the final selection amongst the proposed technologies.

In terms of the studied solar thermal collectors, the comparison of flat plate and vacuum tube systems aimed at stressing the advantages and disadvantages of both technologies. The production stage of the collector component contributes the most important part of the environmental impacts in the life cycle for both studied systems. Thus, for the flat plate system, 57% and 27.1% of all total inflows and outflows are due to the production of the collector and the tank, respectively, while for the vacuum tube system, the corresponding values are 45.3% and 34.8%. The two systems exhibited similar environmental impacts in most categories, but the vacuum tube collector has the highest values in most cases. The cumulative CO_2-eq emissions over the whole life cycle of the solar systems are quite close, varying between 2.22×10^{-2} and 2.38×10^{-2} kg CO_2-eq/kWh·m^2, and the lowest value corresponds to the vacuum tube collector.

Both collectors can cover more than half of the annual hot water needs (equal to spending 2817 kWh in a typical auxiliary hot water heating system) for a family house with four occupants, as the solar fraction values are 62.7% and 55.3% for the vacuum tube and the flat plate collector, respectively. The vacuum tube collector is practically unaffected by variations in the ambient temperature due to its significantly lower thermal losses coefficient, but this technical advantage is not reflected in its final energy outcome mainly due to the favorable weather conditions (i.e., extended time periods with high intensity solar radiation and increased ambient temperatures) in the selected installation location, which make the flat plate collector equally efficient, and to the increased electricity consumption of its pump. In addition, the purchase cost of the vacuum collector is almost 45% higher, thus stressing the fact that for typical installations in southern climates (i.e., Greece), the flat plate system should be the principal option. The economic viability of both systems is proven as the simple payback period is 2.6 and 3.8 years for the flat plate and the vacuum tube system, respectively.

Author Contributions: Conceptualization, M.M., M.S., and S.P.; methodology, M.M., M.S., G.A., and S.P.; software, M.M.; validation, M.M. and S.P.; data curation, M.M. and G.A.; writing—original draft preparation, M.M. and S.P.; writing-review and editing, M.S. and S.P.; supervision, S.P.; funding acquisition, M.S. and S.P.

Funding: The research has been co-financed by the European Union and Greek national funds through the Operational Program Competitiveness, Entrepreneurship and Innovation, under the call RESEARCH—CREATE - INNOVATE (project code: T1EDK-01740).

Conflicts of Interest: The authors declare no conflict of interest

References

1. IEA. Key World Energy Statistics 2018. International Energy Agency 2018. Available online: www.iea.org/statistics (accessed on 22 March 2019).
2. REN21. Renewables 2018, Global Status Report. 2018. Available online: www.ren21.net (accessed on 22 March 2019).
3. IRENA 2018. *Renewable Energy Prospects for the European Union*; International Renewable Energy Agency: Abu Dhabi, United Arab Emirates, 2018.
4. IRENA 2017. *Planning for the Renewable Future: Long-Term Modelling and Tools to Expand Variable Renewable Power in Emerging Economies*; International Renewable Energy Agency: Abu Dhabi, United Arab Emirates, 2017.
5. IEA. Market Report Series: Renewables 2017, Analysis and Forecasts to 2022. International Energy Agency 2017. Available online: www.iea.org/publications/renewables2017 (accessed on 22 March 2019).
6. REN21. *Advancing the Global Renewable Energy Transition: Highlights of the REN21 Renewables 2017 Global Status Report in Perspective*; REN21 Secretariat: Paris, France, 2018.
7. Finnveden, G.; Hauschild, M.; Ekvall, T.; Guinee, J.; Heijungs, R.; Hellweg, S.; Koehler, A.; Pennington, D.; Suh, S. Recent developments in Life Cycle Assessment. *J. Environ. Manag.* **2009**, *91*, 1–21. [CrossRef] [PubMed]
8. Turconi, R.; Boldrin, A.; Astrup, T. Life cycle assessment (LCA) of electricity generation technologies: Overview, comparability and limitations. *Renew. Sustain. Energy Rev.* **2013**, *28*, 555–565. [CrossRef]
9. Weidema, B.P.; Thrane, M.; Christensen, P.; Schmidt, J.; Løkke, S. Carbon Footprint. *J. Ind. Ecol.* **2008**, *12*, 3–6. [CrossRef]
10. Mahmud, M.A.; Huda, N.; Farjana, S.H.; Lang, C. Environmental impacts of solar-photovoltaic and solar-thermal systems with life-cycle assessment. *Energies* **2018**, *11*, 2346. [CrossRef]
11. Constantino, G.; Freitas, M.; Fidelis, N.; Pereira, M.G. Adoption of photovoltaic systems along a sure path: A life-cycle assessment (LCA) study applied to the analysis of GHG emission impacts. *Energies* **2018**, *11*, 2806. [CrossRef]
12. Weisser, D. A guide to life-cycle greenhouse gas (GHG) emissions from electric supply technologies. *Energy* **2007**, *32*, 1543–1559. [CrossRef]
13. ISO. EN ISO 14040, Environmental Management—Life Cycle Assessment—Principles and Framework. 2006. Available online: https://www.iso.org/standard/37456.html (accessed on 22 March 2019).
14. EC-JRC. *International Reference Life Cycle Data System (ILCD), Handbook—General Guide for Life Cycle Assessment—Detailed Guidance*; Publications Office of the European Union: Luxembourg, 2010.
15. Sima Pro. 2018. Available online: simapro.com (accessed on 22 March 2019).

16. RETScreen Expert. 2018. Available online: www.nrcan.gc.ca/energy/software-tools/7465 (accessed on 22 March 2019).
17. Ghafghazi, S.; Sowlati, T.; Sokhansanj, S.; Bi, X.; Melin, S. Life cycle assessment of base-load heat sources for district heating system options. *Int. J. Life Cycle Assess.* **2011**, *16*, 212–223. [CrossRef]
18. Amponsah, N.Y.; Troldborg, M.; Kington, B.; Aalders, I.; Hough, R.L. Greenhouse gas emissions from renewable energy sources: A review of lifecycle considerations. *Renew. Sustain. Energy Rev.* **2014**, *39*, 461–475. [CrossRef]
19. Itten, R.; Stucki, M. Highly efficient 3rd generation multi-junction solar cells using silicon heterojunction and perovskite tandem: Prospective life cycle environmental impacts. *Energies* **2017**, *10*, 841. [CrossRef]
20. IRENA. *Renewable Power Generation Costs in 2017*; International Renewable Energy Agency: Abu Dhabi, United Arab Emirates, 2018.
21. Varun, R.P.; Bhat, I.K. Energy, economics and environmental impacts of renewable energy systems. *Renew. Sustain. Energy Rev.* **2009**, *13*, 2716–2721. [CrossRef]
22. Papaefthimiou, S.; Souliotis, M.; Andriosopoulos, K. Grid parity of solar energy: Imminent fact or future's fiction? *Energy J.* **2016**, *37*, 263–276. [CrossRef]
23. Jungbluth, N.; Stucki, M. *Life Cycle Inventories of Photovoltaics*; ESU-Services Ltd.: Uster, Switzerland, 2012; p. 250.
24. Frischknecht, R.; Itten, R.; Sinha, P.; de Wild-Scholten, M.; Zhang, J.; Fthenakis, V.; Kim, H.C.; Raugei, M.; Stucki, M. *Life Cycle Inventories and Life Cycle Assessment of Photovoltaic Systems*; PVPS 12, Report T12-04; International Energy Agency (IEA): Paris, France, 2015.
25. Fthenakis, V.M.; Kim, H.C. Photovoltaics: Life-cycle analyses. *Sol. Energy* **2011**, *85*, 1609–1628. [CrossRef]
26. Jungbluth, N.; Tuchschmid, M.; de Wild-Scholten, M. *Life Cycle Assessment of Photovoltaics: Update of Ecoinvent Data v2.0*; ESU-Services Ltd.: Uster, Switzerland, 2008.
27. Lenzen, M. Greenhouse gas analysis of solar-thermal electricity generation. *Sol. Energy* **1999**, *65*, 353–368. [CrossRef]
28. Kreith, F.; Norton, P.; Brown, D. A comparison of CO_2 emissions from fossil and solar power plants in the United States. *Energy* **1990**, *15*, 1181–1198. [CrossRef]
29. Lechón, Y.; de la Rúa, C.; Sáez, R. Life Cycle Environmental Impacts of Electricity Production by Solarthermal Power Plants in Spain. *J. Sol. Energy Eng.* **2008**, *130*, 021012. [CrossRef]
30. Weidema, B.P.; Bauer, C.; Hischier, R.; Mutel, C.; Nemecek, T.; Reinhard, J.; Vadenbo, C.O.; Wernet, G. *Overview and Methodology. Data Quality Guideline for the Ecoinvent Database Version 3*; The ecoinvent Centre: St. Gallen, Switzerland, 2013.
31. Goedkoop, M.; Heijungs, R.; de Schryver, A.; Struijs, J.; Van Zelm, R.; ReCiPe 2008. A life Cycle Impact Assessment Method Which Comprises Harmonised Category Indicators at the Midpoint and the Endpoint Level. Report I: Characterisation. 2013. Available online: https://www.pre-sustainability.com/download/ReCiPe_main_report_MAY_2013.pdf (accessed on 22 March 2019).
32. Fthenakis, V.M. End-of-life management and recycling of PV modules. *Energy Policy* **2000**, *28*, 1051–1058. [CrossRef]
33. Fthenakis, V.; Wang, W.; Kim, H.C. Life cycle inventory analysis of the production of metals used in photovoltaics. *Renew. Sustain. Energy Rev.* **2009**, *13*, 493–517. [CrossRef]
34. Pauls, K.; Mitchell, K.W.; Chesarak, W.; King, R.R. Silicon Concentrator Solar Cells Using Mass-Produced, Flat-Plate Cell Fabrication Technology. In Proceedings of the Conference Record of the Twenty-Third IEEE Photovoltaic Specialists Conference, Louisville, KY, USA, 10–14 May 1993; Volume 2, pp. 824–827.
35. RAE. Regulatory Authority for Energy (RAE). Available online: http://www.rae.gr/site/en_US/portal.csp (accessed on 11 April 2019).
36. Maroulis, G. Legal Sources on Renewable Energy, Feed-in Tariff (Rooftop PV). 2019. Available online: http://www.res-legal.eu/search-by-country/greece/single/s/res-e/t/promotion/aid/feed-in-tariff-ii-pv-on-rooftops/lastp/139/ (accessed on 11 April 2019).
37. HELAPCO. Hellenic Association of Photovoltaic Companies. 2014. Available online: http://helapco.gr/en/ (accessed on 11 April 2019).
38. Van der Vlies, D.; van Breevoort, P.; Winkel, T. *The Value of Distributed Solar PV in Spain*; GreenPeace: Amsterdam, The Netherlands, 2018.
39. Calpak. 2018. Available online: www.calpak.gr (accessed on 22 March 2019).

40. Schlömer, S.; Bruckner, T.; Fulton, L.; Hertwich, E.; McKinnon, A.; Perczyk, D.; Roy, J.; Schaeffer, R.; Sims, R.; Smith, P.; et al. ANNEX III: Technology-specific Cost and Performance Parameters. In *Climate Change 2014: Mitigation of Climate Change. Contribution of Working Group III to the Fifth Assessment Report of the Intergovernmental Panel on Climate Change*; Cambridge University Press: Cambridge, UK; New York, NY, USA, 2014.
41. Edenhofer, O.; Pichs-Madruga, R.; Sokona, Y.; Seyboth, K.; Matschoss, P.; Kadner, S.; Zwickel, T.; Eickemeier, P.; Hansen, G.; Schlömer, S.; von Stechow, C. Special Report on Renewable Energy Sources and Climate Change Mitigation. In *Summary for Policymakers and Technical Summary. A Report of Working Group III of the Intergovernmental Panel on Climate Change*; Cambridge University Press: Cambridge, UK; New York, NY, USA, 2011.
42. Moomaw, W.; Yamba, F.; Kamimoto, M.; Maurice, L.; Nyboer, J.; Urama, K.; Weir, T.; Bruckner, T.; Jäger-Waldau, A.; Krey, V.; et al. Renewable Energy and Climate Change. In *Renewable Energy Sources and Climate Change Mitigation*; Edenhofer, O., Pichs-Madruga, R., Sokona, Y., Seyboth, K., Matschoss, P., Kadner, S., Zwickel, T., Eickemeier, P., Hansen, G., Schlomer, S., et al., Eds.; Cambridge University Press: Cambridge, UK, 2011; pp. 161–208.
43. Sullivan, J.L.; Clark, C.E.; Han, J.; Wang, M. Life-Cycle Analysis Results of Geothermal Systems in Comparison to Other Power Systems. 2010. Available online: https://www.energy.gov/eere/geothermal/downloads/life-cycle-analysis-results-geothermal-systems-comparison-other-power (accessed on 22 March 2019).
44. Krey, V.; Masera, O.; Blanford, G.; Bruckner, T.; Cooke, R.; Fisher-Vanden, K.; Haberl, H.; Hertwich, E.; Kriegler, E.; Mueller, D.; et al. Annex II: Metrics & Methodology. In *Climate Change 2014: Mitigation of Climate Change. Contribution of Working Group III to the Fifth Assessment Report of the Intergovernmental Panel on Climate Change*; Edenhofer, Cambridge University Press: Cambridge, UK; New York, NY, USA, 2014.
45. Sovacool, B.K. Valuing the greenhouse gas emissions from nuclear power: A critical survey. *Energy Policy* **2008**, *36*, 2950–2963. [CrossRef]

© 2019 by the authors. Licensee MDPI, Basel, Switzerland. This article is an open access article distributed under the terms and conditions of the Creative Commons Attribution (CC BY) license (http://creativecommons.org/licenses/by/4.0/).

Article

An Evaluation of Input–Output Value for Sustainability in a Chinese Steel Production System Based on Emergy Analysis

Fengjiao Ma [1,2], A. Egrinya Eneji [3] and Yanbin Wu [1,2,*]

1. School of Management Science and Engineering, Hebei University of Economics and Business, Shijiazhuang 050061, Hebei, China; mafengjiao@heuet.edu.cn
2. GIS Big Data Platform for Socio-Economy in Hebei, Shijiazhuang 050061, Hebei, China
3. Department of Soil Science, Faculty of Agriculture, University of Calabar, Calabar PMB 1115, Nigeria; aeeneji@yahoo.co.uk
* Correspondence: wuyanbin080@126.com; Tel./Fax: +86-311-87656207

Received: 22 October 2018; Accepted: 8 December 2018; Published: 12 December 2018

Abstract: The social investment, natural resource consumption, and pollutant emissions involved in steel production can be evaluated comprehensively using the emergy analysis. We explored the sustainability of the steel production system from four aspects: input index, output index, input–output index, and sustainability index. The results showed that the maximum inputs were the intermediate product/recyclable materials produced within the production line; energy sources were mainly non-renewable and the emergy value of pollutants discharged was rather low. The environmental load rate of the pelletizing and sintering processes were the highest and the proportion of recycled materials for puddling and steel-making were the highest. The emergy investment rate of rolling was the highest; the emergy value of the pollutants discharged in each process was very small, and the emergy yield ratio was highest in the rolling process. Pelletizing, sintering, and steel-making were input consuming processes, but the sustainability index of puddling and rolling processes was sound. The whole process line can be sustainable, considering the useful intermediate and recyclable products.

Keywords: emergy analysis; pollution impact; resource consumption; steel production; sustainable development

1. Introduction

Steel is widely used in construction, transportation, packaging, renewable energy, and other industries and the world's crude steel output exceeded 1.6 billion tons in 2016 [1]. However, it is also an energy-intensive industry, whose carbon dioxide emissions account for 6% to 7% of global anthropogenic carbon dioxide emissions due to large amounts of fossil fuel consumption [2]. The treatment of solid waste such as steel slag, iron dust, and coal ash generated during production has caused a series of environmental problems [3]. Steel production relies on the natural ecosystem and human economic system feedback resources and the resulting waste flows into the natural ecosystem and could affect human health. A research framework that considers the human economic system, natural ecosystem, and the steel production system is required to evaluate the sustainable development of the steel industry. The ecological economics evaluation method that comprehensively considers economic development, resource consumption and environmental protection is an important tool for evaluating sustainable development. Its application to the steel industry is an important research topic for the sustainable management of the industry.

Among the existing eco-economic evaluation methods, the material flow analysis does not consider the contribution of the ecosystem to production [4,5]; the evaluation using life cycle assessment is based on human preferences [6]; economic analysis mainly depends on market and shadow prices, and its outcome is not objective enough; energy analysis usually does not consider the different effects provided by energy from different sources [7,8].

In contrast to other analytical methods, H.T. Odum considered the natural energy hierarchy of the universe in which many joules of one kind must be degraded to generate a few joules of another and propose the concept of "emergy" [9]. Odum measures, values, and aggregates energy of different types by their transformities. Transformities, defined as the emergy per unit energy, are calculated as the amount of one type of energy required to produce a heat equivalent of another type of energy. To account for the difference in quality of thermal equivalents among different energies, all energy costs are measured in solar emjoules (sej), the quantity of solar energy used to produce another type of energy. Fuels and materials with higher transformities require larger amounts of sunlight to produce and therefore are considered more economically useful [10]. The emergy analysis is an energy ecological method based on the principle of physical thermodynamics. The indicators of economic system and ecosystem can be uniformly converted into emergy values. By incorporating aspects of energy quality and ecological hierarchy to evaluate the contribution of the natural environment to the human-economic system, this methodology allows for balancing of the needs of both human and natural systems, expressing the socio-economic-environmental effects in common terms [11]. Emergy with corresponding indices and ratios has been proved to be an effective and robust tool to understand the resource flows supporting both the natural ecosystem and macro-economic system, and can be used to measure their overall performances and sometimes sustainability [12]. This method has been widely accepted as an effective ecological evaluation tool to assess comprehensive performances of all kinds of systems with different scales and functions [13–16].

In the field of industrial production, Brown and Ulgiati added ecological service indicators to the emergy production system to evaluate the power production system [17]. Geng et al. used emergy analysis to evaluate the environmental performance and sustainability of industrial parks [18] and Yuan et al. analyzed the recycling effects of different methods for construction waste through the emergy theory [19]. In the field of renewable energies industry, a comprehensive energy and economic assessment of biofuels was conducted by Ulgiati, based on economy, energy, and emergy and a proposal to integrate ethanol production with industrial activities with a "zero emission framework" was suggested [20]. Takahashi and Ortega made an emergy assessment of oleaginous crops cultivated in Brazil, available to produce biodiesel, to determine which crop is the most sustainable [21]. Zhou et al. analyzed a farm biogas based on emergy analysis and found that the farm biogas project has more reliant on the local renewable resources input, less environmental pressure and higher sustainability compared with other typical agricultural systems [22]. In the field of steel production, Zhang et al. used emergy analysis to assess the sustainability of Chinese steel production from 1998 to 2008, showing that its sustainability was very low and continued to decline [23]. Pan et al. evaluated the sustainability of Chinese steel eco-industrial parks based on the emergy theory and found that after the implementation of material recycling and energy cascade utilization, all indicators were superior to the traditional production chain [24].

In order to understand the energy efficiency, environmental impact, and sustainable development of steel industry, a systematic method to measure the comprehensive performances of steel enterprise is urgent. The emergy analysis can be an effective method for evaluating sustainable development, considering the social investment, natural resource consumption, and impacts of pollutant emission from the steel industry. However, the current application of emergy analysis to the steel industry has only focused on the sustainable development from a fixed resource type. A detailed inquiry into the various material resources for the steel production process is needed to analyze the productivity and sustainability of the steel industry. Therefore, we explored the detailed inputs of renewable and non-renewable resources from three aspects: natural ecosystem, human economic system, and steel production system. In addition, we analyzed each sub-link of the steel production line to explore the status and potential of energy consumption. Finally, the efficiency and sustainable development of steel production were examined in detail from the input-, output-, input–output- and comprehensive sustainability indexes of steel production. This will allow for the examination of the dependence of steel production on different systems as well as the role of recycling in the production process and identification of the sustainable development index that considers the environmental impacts and waste discharge.

2. Materials and Methodology

2.1. Evaluation Framework

Our research target was a steel production system (Figure 1), whose boundary is the area of steel production enterprise. Steel production consumes a lot of resources, and generates various wastes. Three categories of system are defined for emergy accounting and for the understanding of the system interactions. (1) The natural system represents the natural environment, which has not been substantially altered by human intervention. (2) The human economics system is dominated by human beings and deals with the production, distribution, and consumption of goods and services in a particular society. (3) The steel production system in this paper refers to an industrial system that contains pelletizing, sintering, puddling, steel-making, rolling, and other related auxiliary process. The resources are derived from natural and human economic systems; the products are sold to the human economic systems, the pollutants are returned to the natural ecosystem while affecting human health, and some wastes that could be reused (here defined as recyclable materials) are returned to the production line. Based on the emergy algorithm, we abbreviate the renewable resources from natural system as R, the non-renewable resources from natural system as N, the renewable resources from human economics system as F_R, the non-renewable resources from human economics system as F_N and the product for human economic system as Y. In addition, some products (such as sinter, pellet, etc.) are defined as intermediate products, because they can be sold on the market, but they are also used in other parts of the steel production system. However, the effect of pollutant emissions from the steel production plant on other systems is useless or even harmful. Here, we used dotted lines to describe their pathway (Figure 1). The production process could refer to the entire steel production line, but also to a sub-process, such as sintering process.

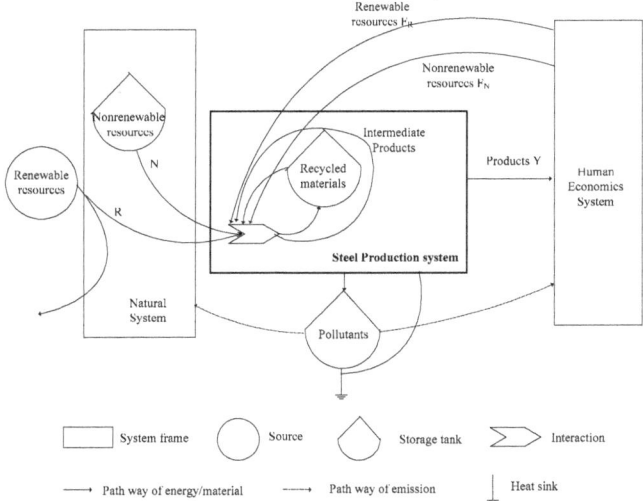

Figure 1. Material and energy flow diagram of the steel production process.

The inputs of different systems include renewable and non-renewable resources. The resource input from the human economic system is also needed, to be supported by the natural ecosystem. For example, the electricity supplied by the human economic system depends on coal or water resources supplied by the natural system. However, considering that the power system needs a large number of other production equipment, the proportion of natural resources input is relatively low, so the electricity is classified into the resource input of the human economic system. The natural resources, such as coal and lime, are direct supplies of the natural ecosystem.

2.2. Data Collection and Calculation of Emergy

2.2.1. Data Collection

This study explored the sustainable development of steel production system. The steel factory studied was a combined factory, consisting of sintering, pelletizing, puddling, steel making, steel rolling and power generation. It had an annual production capacity of 1.2 million tons of pellets, 9.15 million tons of sinter, 4.65 million tons of pig iron, 4.5 million tons of billets and 3.2 million tons of coils. Its production pathway is shown in Figure 2. The products of the factory can enter the market or the next production link of the factory directly.

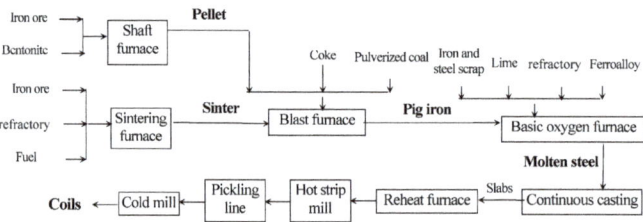

Figure 2. Flow diagram for the steel production process.

Data Source: Considering the different conditions of production across years and the imperfection of some material flow monitoring, we used data from the environmental impact assessment report of a standard steel factory—the Yuhua Steel Co. Ltd. in Wuan City, China. We collected the report directly from the authors, who conducted field investigation and technical demonstration on the entire steel enterprise. Data were collected for natural renewable and non-renewable resources, human economic renewable and non-renewable resources, intermediate products and recycled materials as the input raw data; pollutants, products, intermediate products, and recycled materials were the output raw data.

Data processing methods: Various input–output indexes must be comparable to evaluate the efficiency and sustainability of the whole system. In this study, the emergy analysis method was used. The specific algorithm was firstly to convert different input and output elements into energy or mass data, followed by calculation of the emergy conversion rate. Finally, the original data were multiplied by the emergy conversion rate to obtain the solar emergy value (sej) of each index (Table 1).

2.2.2. Impact Evaluation of Emissions

For the steel production process, although most input–output indexes could be calculated as the product of original data and emergy conversion rate, the pollutants produced in the process could not be simply multiplied by their emergy conversion rates. Because pollutants are harmful to people and environment rather than a useful resource, their emergy value should be calculated from their negative effects.

Even if the pollutants from the production process are within the national permissible limits after remediation, there are still some gaps between the emission concentration and the environmental quality standards suitable for human survival. These pollutants need a lot of air within the environment to dilute to the acceptable concentration. This environmental service is defined as services for diluting pollutants. However, these pollutants may cause ecological and economic losses (biodiversity loss, ecosystem degradation, damage to human health) before reaching the acceptable concentration. These losses contain certain emergy values. Therefore, the ecological impacts of effluent pollutants are in two parts: dilution of ecosystem services and emergy loss of emission.

The calculation of emergy for diluting ecological services was done according to Ulgiati and Brown [25]. However, the pollutants are regarded as by-products in the literature [25] and the service used for dilution is regarded as renewable resources provided by the environment. We considered the pollutants as harmful waste rather than by-products since they cannot be utilized under the current technical level of the research enterprise. As an effluent waste, pollutants can only be a harmful substance, whose disadvantages are expressed by the damage to natural and human resources. The value of this damage is essentially negative emergy.

① Calculation of Emergy for Diluting Ecological Services

Firstly, the environmental quality of diluted pollutants was calculated. The pollutant studied in this paper was only exhaust gas. Therefore, the air quality of diluted exhaust gas was calculated as

$$M_{d,air} = d \times \frac{W}{c} - M_{e,air} \tag{1}$$

where $M_{d,air}$ is the quality of the air used to dilute pollutants; $M_{e,air}$ is the quality of the air emitted from the steel production process; d is the air density (1.23 kg/m^3); W is the amount of pollutants discharged annually; c is the acceptable concentration of pollutants.

The mass of diluted air was converted to emergy by diluting the kinetic energy of air. By multiplying by the conversion rate of emergy, the emergy value of diluted air can be obtained.

$$Em_{d,air} = E_{d,air} \times Tr_{air} = \frac{1}{2} \times M_{d,air} \times v^2 \times Tr_{air} \tag{2}$$

where $Em_{d,air}$ is the emergy value of dilute air; $E_{d,air}$ is the kinetic energy of dilute air; Tr_{air} is the transformity of wind, here it is 1.50×10^3 sej/J; $M_{d,air}$ is the quality of the air used to dilute pollutants; v is average annual wind speed, here it was chosen as 1.5 m/s.

② Calculation of emergy loss of emission

Human resources were considered as a slow renewable resource, and the generation and use of pollutants would lead to irreversible losses. In this report, the DALY method proposed by WHO was used to quantitatively assess the damage of pollutants to human beings [22]. Emergy loss was calculated as

$$Em_{manpower} = \sum M_i \times DALY_i \times \tau_H \tag{3}$$

where $Em_{manpower}$ is the emergy of human resource loss, sej; i is the ith pollutant; M_i is the quality of the ith pollutant; $DALY$ is the impact factor of the ith pollutant; τ_H is the emergy unit of human resources per year, which is equal to the annual total emergy use of a country or region divided by its population, and here τ_H equaled 1.32×10^{16} sej/person [24,26].

The specific indicators and results of the calculation are shown in Table 1.

2.2.3. Emergy Evaluation of the Steel Industry

From the material and energy flow diagram of the production process, the different sub-processes of the steel factory were systematically sorted out. The input and output indexes were converted into heat or mass data. The original data were multiplied by the corresponding emergy conversion rate to obtain the solar emjoule value (sej) of each index. Because of the large amount of data, the summary results are shown in Table A1 at the end of this paper. The main body of this paper only gave the emergy input–output statistics of the steel production system (Table 2).

Table 1. Yearly emergy estimate of pollutants in the steel production process.

Production Process	Pollutant	Pollutant Discharge Mass/t	Discharge Volume of Waste Gas/m³	Acceptable Concentration/μg/m³ [27]	Mass of the Air Used to Dilute Pollutant/t	Emergy of the Air Used to Dilute Pollutant/sej	DALY/Kg of Emission	Emergy Loss of Emission/sej	Emergy of Total Impacts of Pollutants/sej
Pelletizing	SO_2	295.26	3.11×10^9	20	1.82×10^{10}	3.06×10^{16}	5.46×10^{-5} [28]	2.13×10^{17}	2.43×10^{17}
	Dust	85.51	5.10×10^9	80	1.31×10^9	2.21×10^{15}	3.75×10^{-4} [28]	4.23×10^{17}	4.25×10^{17}
	NO_x	530.91	3.12×10^9	50	1.31×10^{10}	2.20×10^{16}	8.87×10^{-5} [29]	6.22×10^{17}	6.44×10^{17}
Sintering	SO_2	2315.38	1.99×10^{10}	20	1.42×10^{11}	2.40×10^{17}	5.46×10^{-5} [28]	1.67×10^{18}	1.91×10^{18}
	Dust	967.46	4.23×10^{10}	80	1.48×10^{10}	2.50×10^{16}	3.75×10^{-4} [28]	4.79×10^{18}	4.81×10^{18}
	NO_x	4503.19	2.02×10^{10}	50	1.11×10^{11}	1.87×10^{17}	8.87×10^{-5} [29]	5.27×10^{18}	5.46×10^{18}
Puddling	SO_2	178.81	3.85×10^9	20	1.10×10^{10}	1.85×10^{16}	5.46×10^{-5} [28]	1.29×10^{17}	1.47×10^{17}
	Dust	713.47	3.54×10^{10}	80	1.09×10^{10}	1.84×10^{16}	3.75×10^{-4} [28]	3.53×10^{18}	3.55×10^{18}
	NO_x	449.03	3.85×10^9	50	1.00×10^{10}	1.86×10^{16}	8.87×10^{-5} [29]	5.26×10^{17}	5.44×10^{17}
Steel-making	Dust	528.30	2.24×10^{10}	80	8.11×10^9	1.37×10^{16}	3.75×10^{-4} [28]	2.62×10^{18}	2.63×10^{18}
Rolling	SO_2	84.33	2.00×10^9	20	5.18×10^9	8.75×10^{15}	5.46×10^{-5} [28]	6.08×10^{16}	6.95×10^{16}
	Dust	37.85	2.00×10^9	80	5.80×10^8	9.78×10^{14}	3.75×10^{-4} [28]	1.87×10^{17}	1.88×10^{17}
	NO_x	170.97	2.00×10^9	50	4.20×10^9	7.09×10^{15}	8.87×10^{-5} [29]	2.00×10^{17}	2.07×10^{17}
Power Plant	SO_2	535.52	1.08×10^{10}	20	3.29×10^{10}	5.56×10^{16}	5.46×10^{-5} [28]	3.86×10^{17}	4.42×10^{17}
	Dust	87.18	1.08×10^{10}	80	1.33×10^9	2.24×10^{15}	3.75×10^{-4} [28]	4.32×10^{17}	4.34×10^{17}
	NO_x	424.64	1.08×10^{10}	50	1.04×10^{10}	1.76×10^{16}	8.87×10^{-5} [29]	4.97×10^{17}	5.15×10^{17}

Table 2. Emergy input and output in the steel production system.

Items	Resource Type	Indexes	Emergy sej	Items	Resource Type	Indexes	Emergy sej
Input				**Output**			
Natural system	Renewable resources (R)	Fresh water	2.35×10^{18}	Natural system	Pollutants	SO_2	2.81×10^{18}
		Air	2.30×10^{20}			Dust	1.20×10^{19}
	Non-renewable resources (N)	Bentonite	1.92×10^{19}			NO_x	7.37×10^{18}
		Powdered iron	7.58×10^{21}	Human economics system	Products (Y)	Sinter	2.75×10^{21}
		Limestone	1.46×10^{21}			Pig iron	2.51×10^{21}
		High magnesium powder	8.59×10^{19}			Billet steel	3.88×10^{21}
		Iron ore	8.70×10^{19}			Rolled steel	9.89×10^{21}
		Coal	2.02×10^{20}	Steel production system	Intermediate products	Pellet	1.31×10^{21}
		Pulverized coal	2.62×10^{19}			Sinter	7.23×10^{21}
		Iron block	9.41×10^{19}			Pig iron	1.01×10^{22}
		Ferroalloy	1.14×10^{19}			Billet steel	1.00×10^{22}
		Doomite	6.55×10^{19}		Recycled materials	Desulphurizing Slag	6.20×10^{18}
		Flour	6.00×10^{17}			Dust and ash	1.57×10^{20}
		Soil loss	7.53×10^{20}			Desulphurized gypsum	2.36×10^{19}
Human economics system	Renewable resources (F_R)	Labor	2.64×10^{20}			Sinter reentry	8.26×10^{20}
		Investment in fixed assets	2.32×10^{20}			Blast furnace slag	1.21×10^{21}
	Non-renewable resources (F_N)	Thermal power electricity	6.78×10^{20}			Blast furnace gas	2.12×10^{21}
		Coke powder	4.15×10^{9}			Hot blast stove flue gas	2.29×10^{22}
		Coke	4.71×10^{20}			Steel slag	1.32×10^{21}
		Nut coke	7.76×10^{18}			Dust mud	5.38×10^{19}
		White ash	1.01×10^{21}			Iron oxide skin	2.38×10^{19}
Steel production system	Intermediate products	Sinter	7.23×10^{21}			Remainder residue	1.62×10^{19}
		Pellet	1.31×10^{21}			Steam production	5.58×10^{19}
		Pig iron	1.01×10^{22}			Converter gas production	2.87×10^{20}
		Billet steel	1.001×10^{22}			Iron oxide sludge	2.19×10^{19}
	Recycled materials	Blast furnace gas	2.121×10^{21}			Steel scrap	8.75×10^{19}
		Convertor gas	2.87×10^{20}			Electricity	3.03×10^{20}
		Dust and ash	4.00×10^{19}			Nitrogen	9.03×10^{20}
		Water treatment sludge	5.39×10^{19}			Oxygen	1.38×10^{20}
		Sinter reentry	2.01×10^{20}				
		Pellet return	1.09×10^{20}				
		Steam consumption	5.74×10^{19}				
		Steel scrap	9.89×10^{19}				
		Electricity	3.03×10^{20}				
		Nitrogen	9.03×10^{20}				
		Oxygen	1.38×10^{20}				

2.2.4. Emergy Indexes Used in This Study

The emergy evaluation indexes were compiled according to the input–output system and resource utilization of the steel industry (Table 3).

Table 3. Emergy indexes used in steel production system.

Items	Indexes	Formulation
Input index	Environment loading ratio (ELR)	Non-renewable resources $(N + F_N)$/Renewable resources $(R + F_R)$
	Proportion of recycled materials used (PRM)	Recycled materials used/Total input
	Emergy investment ratio (EIR)	Human economics system input/Natural system input
Output index	Environmental impact rate (EnIR)	Pollutants/Products
	Product rate (PR)	Products/Total output
Input–output index	Emergy yield ratio (EYR)	Products/Human economics system input
	Total emergy yield ratio (TEYR)	(Products + Intermediate products + Recycled materials)/(Human economics system input + Natural system input)
	Net emergy yield ratio (NEYR)	(Products + Intermediate products + Recycled materials-Pollutants)/(Human economics system input + Natural system input)
	Emergy input–output rate (EIOR)	Products/Total input
	Total emergy input–output ratio (TEIOR)	(Products + Intermediate products + Recycled materials)/Total input
	Net emergy input–output ratio (NEIOR)	(Products + Intermediate products + Recycled materials − Pollutants)/Total input
Sustainability index	Emergy sustainable development index (ESDI)	Emergy yield ratio (EYR)/Environment loading ratio (ELR)
	Total emergy sustainable development index (TESDI)	Total emergy yield ratio (TEYR)/Environment loading ratio (ELR)
	Net emergy sustainable development index (NESDI)	Net emergy yield ratio (NEYR)/Environment loading ratio (ELR)

3. Results and Discussion

3.1. Structure of Inputs and Outputs in the Steel Industry

The whole process of steel production was analyzed in terms of input and output. The beneficial effect of steel production on people was positive, and both its harmful effect on people and the use of human beneficial emergy were negative, as shown in Figure 3. There was little difference between input and output of the system; the emergy loss was minimal, and the emergy output rate was high.

Among the three systems, the steel production system had the highest overall input and output, which included intermediate and recyclable products. Apart from the intermediate products that could be sold and used directly, the proportion of recyclable materials in various input–output indicators was also the largest. In addition to gas and other resources, these materials were mostly solid wastes such as steel slag, dust particles, etc. After being treated and collected, they accounted for 43% of the emergy value of inputs. The harmless treatment of steel production process played an important role. If these materials were not properly recycled, more resources would need to be invested from the natural ecosystem and human economic system, and the impact of the associated direct emissions would be close to the beneficial emergy value derived from the product itself. Regardless of the input of steel production system, the input resources were mainly non-renewable resources, which was consistent with previous findings [23,24].

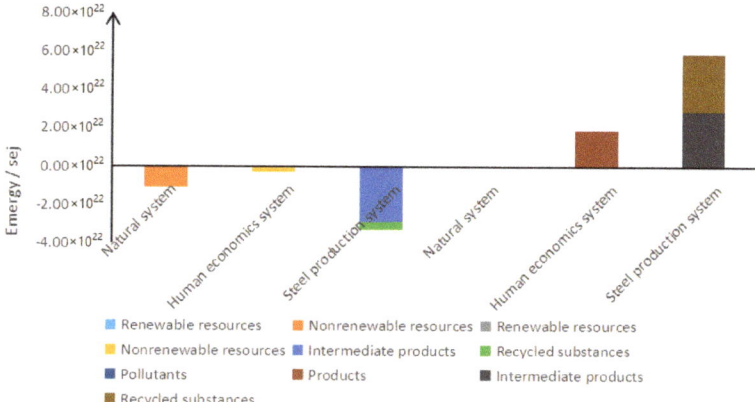

Figure 3. Structure of the emergy inputs and outputs in the steel production process.

In the input and output of each system, both renewable resources of natural ecosystem and human economic system accounted for a very low proportion, suggesting that the environmental load of steel production was high. In addition, the emergy value of pollutants discharged into the natural ecosystem was very low. Although the steel industry is a typical high pollution enterprise, the proportion of its pollution impact on overall input and output was not particularly serious based on the emergy analysis. Under the pressure of environmental protection, the steel factories have fared better. The impact of pollutants was relatively small when the pollutants were treated and reusable resources/wastes recovered as much as possible. In addition, compared with other steel enterprises in China [23,24], this research enterprise does not discharge waste water and the amount of waste gas pollutants was also significantly less. This indicated that the environmental protection technology of this enterprise was at the forefront in China.

3.2. Variation in Emergy of Different Production Links

Input–output indexes were analyzed from four aspects: input indexes, output indexes, the relationship between input–output indexes, and comprehensive sustainability indexes, considering different types of superimposed effects. The emergy production efficiency of each sub-production process and the whole production process was also analyzed (Table 4).

Table 4. Emergy indexes used in steel production system.

Items	Indexes	Pelletizing	Sintering	Puddling	Steel-Making	Rolling	Whole Process
Input index	Environment loading ratio (ELR)	31,191.381	145,338.975	14.418	21.239	39.481	17.242
	Proportion of recycled materials used (PRM)	0.014	0.061	0.090	0.098	0.024	0.093
	Emergy investment ratio (EIR)	0.177	0.038	1.588	4.876	39.481	0.251
Output index	Environmental impact rate (EnIR)	0.001	0.001	0.000	0.000	0.000	0.001
	Product rate (PR)	0.994	0.914	0.324	0.891	0.989	0.243
Input-output index	Emergy yield ratio (EYR)	6.987	32.669	20.920	12.234	42.917	8.777
	Total emergy yield ratio (TEYR)	7.020	35.708	64.647	13.735	43.397	21.992
	Net emergy yield ratio (NEYR)	7.013	35.668	64.640	13.732	43.395	21.982
	Emergy input–output rate (EIOR)	1.038	1.116	1.205	1.095	0.940	0.411
	Total emergy input–output ratio (TEIOR)	1.043	1.219	3.723	1.229	0.951	1.031
	Net emergy input–output ratio (NEIOR)	1.041	1.218	3.723	1.229	0.951	1.030
Sustainability index	Emergy sustainable development index (ESDI)	0.000	0.000	1.451	0.576	1.087	0.509
	Total emergy sustainable development index (TESDI)	0.000	587.384	4.484	0.647	1.099	1.276
	Net emergy sustainable development index (NESDI)	0.000	945.265	4.483	0.647	1.099	1.275

3.2.1. Input Indexes

By analyzing the relationship among different input indexes, the dependence of the production process on different systems and resource types could be understood. Environmental load rate (ELR) reflects the proportion of input of non-renewable and renewable resources. The ELR for pelletizing and sintering processes at the front end of production chain were much higher than other processes. Thus, the whole production process needed to invest a large amount of non-renewable resources to start production, then the demand for non-renewable resources was greatly reduced. The ELR of whole process was 17.242, which was lower than the values for other steel enterprises in China, but there was still some great environmental stress (ELR > 10) [23,24].

The proportion of recycled materials used (PRM) reflects the extent of waste disposal during steel production. If the waste materials cannot be recycled for further production, they can easily become an additional environmental burden. Therefore, the PRM index, acting much like the decomposer in the ecosystem, plays an important role in a sustainable industrial production system. Both puddling and steel-making processes have a high PRM, so that these processes can better absorb and digest the waste in the whole steel production process (Table 4).

Emergy investment ratio (EIR) is different from ELR and is used to explore the relationship between inputs from the human economic system and natural ecosystem. The EIR of the rolling process was higher because it had no input from the human economic system, except some electricity; so the proportion of natural resources input was significantly increased. In pelletizing, and sintering processes as well as the entire line, the input from the human economic system was much greater than that from the natural ecosystem, indicating that the dependence of steel production on human economic system was greater than on natural ecosystem. The EIR of the entire line was similar to other steel enterprises, though it varied with years [23].

3.2.2. Output Indexes

Environmental impact rate (EnIR) reflects the emergy of discharged pollutants per unit product, which can measure the negative environmental impact of products. The EnIR of each process was low, and the EnIR of puddling, steelmaking, and rolling, were less than 0.001 (Table 4). Thus, the environmental costs of production consistent with emission standards was relatively small, based on emergy analysis. In particular, this paper considered pollutants as emissions and were analyzed among the output emergy; however, some studies regarded pollution as the loss of input emergy [24,25].

Product rate (PR) refers to the proportion of target products relative to all other outputs. The very low PR for the whole process was due to the fact that only the final steel was used as the product in this analysis, excluding the huge intermediate products and recycling materials. The PR of the puddling process was much lower than that of other sub-processes. Many of the emergy invested in the puddling process was converted into recycled materials which are subject to further processing. The emergy efficiency of the production cycle could be greatly increased by increasing the PR of the puddling process.

3.2.3. Input–Output Indexes

The traditional Emergy yield ratio (EYR) reflects the emergy of the output (product) under a certain amount of purchased emergy. It can be seen from the Table 4 that the EYR of rolling processes was much higher than that of other processes. The main reason was that compared with pelletizing and sintering processes, the most important resource inputs for rolling were the intermediate products produced in the previous process, which needed not be purchased. The EYR of steelmaking was low due to the high input of natural resources such as oxygen and nitrogen. The EYR of whole process was 8.777, which was larger than other steel enterprises, meaning the research enterprise was more competitive [23–25]. Total emergy yield ratio (TEYR) represents the output of all products (including the sum of final products and recyclable materials) under a certain purchased emergy. The net emergy yield ratio (NEYR) represents the output of all products minus pollutants under a certain purchased emergy. Because the emergy value of pollutants was much lower than other outputs, there was little difference between the TEYR and the NEYR. The difference of TEYR and NEYR from each process was similar to the difference of the EYR. The TEYR and NEYR of puddling process were high, because it used less purchasable resources. Also, not only the pig iron products, but the recyclable materials were produced with a great deal of emergy value. So, the investment rate of puddling process was much higher than that of other processes.

The emergy input–output rate (EIOR) is developed to reflect the amount of emergy products produced by the system, considering all the input resources at the same time. The total emergy input–output ratio (TEIOR) explores the total output of products, intermediate products and recyclables that are produced after inputting resources. The net emergy input–output ratio (NEIOR) represents the total product mentioned above minus the pollutant emergy value after inputting resources. Compared with the EYR, the EIOR not only considers the input and output efficiency of purchased resources, but also comprehensively analyzes the conversion efficiency of all input resources in the process. As shown in Table 4, the EIOR differed little among the various processes and the input–output ratio of each process was almost slightly greater than 1, and the emergy efficiency of each process was high. The TEIOR and the NEIOR of the puddling process were greater than 3. The emergy production and conversion efficiency was rather high under the comprehensive consideration of various inputs and outputs.

3.2.4. Sustainability Indexes

The emergy sustainable development index (ESDI) reflects the sustainable development of the system. The relationship between the ESDI and sustainable development can be summed up as follows: when the ESDI is greater than 1 but less than 10, it indicates that the system is developing and relatively dynamic, and the emergy of sustainable development is in good condition; an ESDI greater than 10 indicates that the economy is underdeveloped; ESDI less than 1 indicates a consumption-oriented system and the development is unsustainable [30]. The ESDI of pelletizing, sintering, puddling, steel-making, and the whole process line were all less than 1, which meant that the production consumed a large amount of non-renewable resources, and the ELR was high. However, the ESDI of other steel enterprise were lower, being less than 0.1 [23,24]. The ESDI of the puddling and rolling processes were greater than 1 but less than 10, suggesting that the emergy sustainable development of puddling and rolling processes were in good condition. Considering all the useful outputs, such as intermediate and reusable products, emergy sustainability indexes (TESDI and NESDI) have been greatly improved. The TESDI and NESDI of puddling, rolling, and the whole process were within a reasonable range of 1 to 10. It could be seen that if the steel production line recycled the intermediate products of each process, it could achieve sustainable development; if not, the system has a high environmental load rate and cannot develop sustainably.

4. Conclusions and Recommendations

4.1. Conclusions

Based on the emergy of various input–output indicators, the total input and output emergy of the steel production line was not very different; the largest input was the intermediate products and recyclable materials produced in the production process; the recyclable materials accounted for 43% of the total input. The input emergy was mainly non-renewable resources, and the ELR was high; the emergy of pollutants discharged was very low, indicating that the environmental impact of steel production was small if the pollutants were discharged after treatment.

The ELR of pelletizing and sintering processes that occurs in the front-end production line was the highest; the proportions of recycled materials used for steel-making and puddling were the highest, and played the greatest role in 'waste' absorption. The EIR in rolling were the highest since its dependence on natural system was the greatest. The emergy value of pollutants from each process was very small, and the EnIR was close to or below 0.001. The PR was only 0.324 in the puddling process, and the emergy efficiency of production could greatly increase if the product rate of puddling was improved. The EYR of sintering and rolling processes were the highest. Both the TEYR and NEYR of puddling were the highest. There was little difference between the procedures in the EIOR, TEIOR, and NEIOR after considering all resource inputs simultaneously.

The ESDI of pelletizing, sintering and steel-making were less than 1, indicating an unsustainable production process but puddling and rolling processes were reasonable. Considering the intermediate products and recyclable materials, the TESDI and NESDI of puddling, rolling and the whole process were between 1 and 10, and the development was acceptable. Therefore, the steel production process could achieve sustainable development if various intermediate products could be recycled considerably.

4.2. Recommendations

This paper systematically analyzed the input and output of the steel production line, but the research process still needs to be improved and further explored. Pollutants discharged from the steel production process will have adverse effects on human and other biological health in the ecological environment. Due to absence of corresponding methods and data for assessing biological hazards, this part of the study was omitted for the time being. The pollutant could be evaluated more accurately once the biological hazards are considered in future studies. The type of pollutants from the steel production process were much more varied than the particulate matter, sulfur dioxide, and nitrogen oxides studied here. After determining the influence of other pollutants for inclusion in future evaluations, the results would be more comprehensive.

In addition to emergy analysis, other eco-economic assessments have also been tried to evaluate the sustainability of steel production. For example, the life cycle assessment method, which mainly concerns the environmental impact of goods and services, has been used at different scales [31–34]. Although each method has its own advantages and disadvantages, it may be more scientific and informative to combine several eco-economic assessments with emergy analysis.

Author Contributions: Y.W. and F.M. designed this research; F.M. performed calculation and analyzed the data; and A.E.E. and F.M. wrote the paper. All authors have read and approved the final manuscript.

Funding: This research was funded by the Program of the Humanities and Social Sciences Research of Ministry of Education of China (17YJCZH192).

Acknowledgments: The authors wish to acknowledge the anonymous reviewers for their suggestions that have greatly improved our study.

Conflicts of Interest: The authors declare no conflict of interest.

Appendix A

Table A1. Emergy input and output in the steel production sub-process.

Production Process	System	Resource Type	Indicator	Unit	Raw Data	Emergy Conversion Rate sej/unit	Emergy sej
Pelletizing	**Input**						
	Natural system	Renewable resources (R)	Fresh water	t	60,000	6.64×10^{11} a	3.98×10^{16}
		Non-renewable resources (N)	Bentonite	t	19,200	1.00×10^{15} a	1.92×10^{19}
			Powdered iron	t	1,212,000	8.55×10^{14} a	1.04×10^{21}
	Human economics system	Non-renewable resources (F_N)	Electricity	J	1.17×10^{15}	160,000 b	1.87×10^{20}
	Steel production system	Recycled materials	Blast furnace gas	J	2.72×10^{14}	66,000 a	1.79×10^{19}
	Output						
	Natural system	Pollutant	SO_2				2.43×10^{17}
			Dust				4.25×10^{17}
			NO_x				6.44×10^{17}
	Human economics system	Products	Pellet	t	1,200,000	1.09×10^{15} a	6.20×10^{18}
		Recycled materials	Desulphurizing Slag	t	6200	1.00×10^{15} a	
			Pellet dust removal ash	t	48	8.30×10^{14} a	3.98×10^{16}
Sintering	**Input**						
	Natural system	Renewable resources (R)	Fresh water	t	87,000	6.64×10^{11} a	5.78×10^{16}
		Non-renewable resources(N)	Domestic powdered iron	t	1,625,000	8.55×10^{14} a	1.39×10^{21}
			Powdered iron abroad	t	6,025,000	8.55×10^{14} a	5.15×10^{21}
			Limestone	t	1,464,000	1.00×10^{15} a	1.46×10^{21}
			High magnesium powder	t	85,900	1.00×10^{15} a	8.59×10^{19}
	Human economics system	Non-renewable resources (F_N)	Electricity	J	1.17×10^{15}	160,000 b	1.87×10^{20}
			Coke powder	J	1.11×10^{16}	10,600 a	1.18×10^{20}
	Steel production system	Recycled materials	Dust removal ash	t	48,200	8.30×10^{14} a	4.00×10^{19}
			Water treatment sludge	t	53,850	1.00×10^{15} a	5.39×10^{19}
			Sinter reentry	t	184,400	1.09×10^{15} a	2.01×10^{20}
			Pellet return	t	100,300	1.09×10^{15} a	1.09×10^{20}
			Blast furnace gas	J	2.07×10^{15}	66,000 a	1.37×10^{20}
			Use of steam	J	8.64×10^{14}	3090 a	2.67×10^{18}
	Output						
	Natural system	Pollutant	SO_2				1.90×10^{18}
			Dust				4.81×10^{18}
			NO_x				5.46×10^{18}
	Human economics system	Products	Sinter	t	9,150,000	1.09×10^{15} a	9.98×10^{21}
		Recycled materials	Desulphurized gypsum	t	23,600	1.00×10^{15} c	2.36×10^{19}
			Desulphurization waste ash	t	16,560	8.30×10^{14} a	1.38×10^{19}
			Sinter reentry	t	757,360	1.09×10^{15} a	8.26×10^{20}
			Sintering dust	t	75,900	8.30×10^{14} a	6.30×10^{19}
			Steam generation	J	6.66×10^{14}	3090 a	2.06×10^{18}

Table A1. Cont.

Production Process	System	Resource Type	Indicator	Unit	Raw Data	Emergy Conversion Rate sej/unit	Emergy sej
Puddling	**Input**						
	Natural system	Renewable resources (R)	Compressed air	t	1,217,695.5	5.16×10^{13} [a]	6.29×10^{19}
			Fresh water	t	1,260,000	6.64×10^{11} [a]	8.37×10^{17}
		Non-renewable resources (N)	Iron block	t	101,700	8.55×10^{14} [a]	8.70×10^{19}
			Coal	J	5.09×10^{15}	39,801 [a]	2.02×10^{20}
			Pulverized coal	J	6.59×10^{14}	39,801 [a]	2.62×10^{19}
	Human economics system	Non-renewable resources (F$_N$)	Electricity	J	7.70×10^{14}	160,000 [b]	1.23×10^{20}
			Coke	J	4.45×10^{16}	10,600 [a]	4.71×10^{20}
			Nut coke	J	7.32×10^{14}	10,600 [a]	7.76×10^{18}
	Steel production system	Intermediate products	Sinter	t	6,630,000	1.09×10^{15} [a]	7.23×10^{21}
			Pellet	t	1,200,000	1.09×10^{15} [a]	1.31×10^{21}
		Recycled materials	Use of steam	J	1.73×10^{14}	3090 [a]	5.34×10^{17}
			Blast furnace gas	J	1.43×10^{16}	66,000 [a]	9.43×10^{20}
	Output						
	Natural system	Pollutant	SO$_2$				1.47×10^{17}
			Dust				3.55×10^{18}
			NO$_x$				5.44×10^{17}
	Human economics system	Products	Pig iron	t	4,650,000	2.71×10^{15} [a]	1.26×10^{22}
		Recycled materials	Blast furnace slag	t	1,417,000	8.55×10^{14} [a]	1.21×10^{21}
			Gas ash	t	57,500	8.30×10^{14} [a]	4.77×10^{19}
			Blast furnace gas	J	1.79×10^{16}	66,000 [a]	2.12×10^{21}
			Hot blast stove flue gas	J	27,619,708.4	8.30×10^{14} [a]	2.29×10^{22}
			Dust and ash	t	39,100	8.30×10^{14} [a]	3.25×10^{19}
Steel-making	**Input**						
	Natural system	Renewable resources (R)	Fresh water	t	1,176,000	6.64×10^{11} [a]	7.81×10^{17}
			Nitrogen consumption	t	4,515,739.44	2.00×10^{14} [d]	9.03×10^{20}
			Oxygen consumption	t	2,667,180.15	5.16×10^{13} [b]	1.38×10^{20}
			Compressed air	t	1,178,415	5.16×10^{13}	6.08×10^{19}
		Non-renewable resources (N)	Iron block	t	110,000	8.55×10^{14}	9.41×10^{19}
			Ferroalloy	t	13,300	8.55×10^{14}	1.14×10^{19}
			Doomite	t	65,500	1.00×10^{15} [a]	6.55×10^{19}
			Flour	t	600	1.00×10^{15} [a]	6.00×10^{17}
	Human economics system	Non-renewable resources (F$_N$)	Electricity	J	7.85×10^{14}	160,000 [b]	1.26×10^{20}
			White ash	t	468,000	2.16×10^{15} [a]	1.01×10^{21}
	Steel production system	Intermediate products	Pig iron	t	4,650,000	2.17×10^{15} [a]	1.01×10^{22}
			Steel scrap	t	32,000	3.09×10^{15} [a]	9.89×10^{19}
		Recycled materials	Blast furnace gas	J	2.86×10^{14}	66,000 [a]	1.89×10^{19}
			Convertor gas	J	1.29×10^{15}	66,000 [a]	8.52×10^{19}

173

Table A1. Cont.

Production Process	System	Resource Type	Indicator	Unit	Raw Data	Emergy Conversion Rate sej/unit	Emergy sej
Steel-making	**Output** Natural system	Pollutant	Dust	t	4,500,000		2.63×10^{18}
	Human economics system	Products	Billet steel	t	427,600	3.09×10^{15} [a]	1.32×10^{21}
		Recycled materials	Steel slag	t	53,800	3.09×10^{15} [a]	5.38×10^{19}
			Dust mud	t	120	1.00×10^{15} [a]	9.96×10^{16}
			Dust	t	27,800	8.30×10^{14} [a]	2.38×10^{19}
			Iron oxide skin	t	19,500	8.55×10^{14} [a]	1.62×10^{19}
			Remainder residue			8.30×10^{14} [a]	
			Steam generation	J	1.01×10^{15}	3090 [a]	3.13×10^{18}
			Convertor gas	J	4.35×10^{15}	66,000 [a]	2.87×10^{20}
Rolling	**Input** Natural system	Renewable resources (R)	compressed air	t	104,472.48	5.16×10^{13} [a]	5.39×10^{18}
			Fresh water	t	670,000	6.64×10^{11} [a]	4.45×10^{17}
	Human economics system	Non-renewable resources (F_N)	Electricity	J	1.44×10^{15}	160,000 [b]	2.30×10^{20}
	Steel production system	Intermediate products	Billet steel	t	3,243,700	3.09×10^{15} [a]	1.00×10^{22}
		Recycled materials	Blast furnace gas	J	3.90×10^{15}	66,000 [a]	2.58×10^{20}
	Output Natural system	Pollutant	SO_2				6.95×10^{16}
			Dust				1.88×10^{17}
			NO_x				2.07×10^{17}
	Human economics system	Products	Rolled steel	t	3,200,000	3.09×10^{15} [a]	2.19×10^{19}
		Recycled materials	Iron oxide sludge	t	21,900	1.00×10^{15} [a]	8.74×10^{19}
			Steel scrap	t	28,300	3.09×10^{15} [a]	
			Steam generation	J	3.95×10^{14}	3090 [a]	1.22×10^{18}
Oxygen production	**Input** Natural system	Renewable resources (R)	air	t	1,964,117.14	5.16×10^{13} [a]	1.01×10^{20}
			Fresh water	t	286,651.81	6.64×10^{11} [a]	1.90×10^{17}
	Human economics system	Non-renewable resources (F_N)	Electricity	J	7.93×10^{14}	160,000 [b]	1.27×10^{20}
Power plant	**Input** Steel production system	Recycled materials	Blast furnace gas	J	1.14×10^{16}	65,999 [a]	7.50×10^{20}
			Convertor gas	J	3.06×10^{15}	66,000 [a]	2.02×10^{20}
			Steam generation	J	1.67×10^{16}	3090 [a]	5.15×10^{19}
			Use of steam	J	1.67×10^{16}	3090 [a]	5.15×10^{19}
			Electricity	J	1.89×10^{15}	160,000 [b]	3.03×10^{20}
	Output Natural system	Pollutant	SO_2				4.42×10^{17}
			Dust				4.34×10^{17}
			NO_x				5.15×10^{17}

Table A1. Cont.

Production Process	System	Resource Type	Indicator	Unit	Raw Data	Emergy Conversion Rate sej/unit	Emergy sej
Whole process	**Input**						
	Human economics system	Renewable resources (R)	Labor	λ	8500	3.10×10^{16} [a]	2.64×10^{20}
			Investment in fixed assets	$	46,937,658.6	4.94×10^{12} [e]	2.32×10^{20}
			Soil loss	g	4.43×10^{11}	1.70×10^{9} [c]	7.53×10^{20}

Note: ① The calculation of the emergy of pollutants is detailed in the Section 3.2.2; ② Conversion parameters of some raw data: Compressed air mass: volume × 1.239 g/L; steam heat: mass × 2817.2381 J/g; oxygen mass: volume × 10,470 g/m^3; nitrogen mass: volume × 9168.8 g/m^3; coke heat: mass × 28,470 J/g; coal heat: mass × 8374 J/g; electric heat: kWh × 3.6 × 10^6 J/kWh; blast furnace gas heat: volume × 3344 kJ/m^3; Convertor gas heat: volume × 7527 kJ/m^3; Lifetime of factory: 20 years. ③ Energy of soil loss = soil loss mass organic matter content × soil organic matter calorific value; soil organic matter calorific value is 106 kcal/t; surface soil thickness is 0.15 m; organic matter content is 5%; soil bulk density is 1.3 g/cm^3. ④ Labor, fixed assets investment and soil loss are only counted in the total production process of factory for the three indicators are not easy to collected in subsystems. ⑤ Reference: [a]: [35]; [b]: [23]; [c]: [36]; [d]: [37]; [e]: [38].

References

1. Renzulli, P.A.; Notarnicola, B.; Tassielli, G.; Arcese, G.; Di Capua, R. Life Cycle Assessment of Steel Produced in an Italian Integrated Steel Mill. *Sustainability* **2016**, *8*, 719. [CrossRef]
2. Kim, Y.; Worrell, E. International comparison of CO_2 emission trends in the iron and steel industry. *Energy Policy* **2002**, *30*, 827–838. [CrossRef]
3. Chen, B.; Yang, J.X.; Ouyang, Z.Y. Life Cycle Assessment of Internal Recycling Options of Steel Slag in Chinese Iron and Steel Industry. *J. Iron Steel Res. Int.* **2011**, *18*, 33–40. [CrossRef]
4. Michaelis, P.; Jackson, T. Material and energy flow through the UK iron and steel sector (Part 1: 1954–1994H). *Resour. Conserv. Recycl.* **2000**, *29*, 131–156. [CrossRef]
5. Michaelis, P.; Jackson, T. Material and energy flow through the UK iron and steel sector (Part 2: 1994–2019). *Resour. Conserv. Recycl.* **2000**, *29*, 209–230. [CrossRef]
6. Huang, Z.; Ding, X.; Sun, H.; Liu, S. Identification of main influencing factors of life cycle CO_2 emissions from the integrated steel works using sensitivity analysis. *J. Clean. Prod.* **2010**, *18*, 1052–1058. [CrossRef]
7. Lin, B.; Wang, X. Promoting energy conservation in China's iron & steel sector. *Energy* **2014**, *73*, 465–474.
8. Lin, B.Q.; Wang, X.L. Exploring energy efficiency in China's iron and steel industry: A stochastic frontier approach. *Energy Policy* **2014**, *72*, 87–96. [CrossRef]
9. Odum, H.T. *Environmental Accounting: Emergy and Environmental Decision Making*; John Wiley: New York, NY, USA, 1996; p. 370.
10. Cleveland, C.J.; Kaufmann, R.K.; Stern, D.I. Aggregation and the role of energy in the economy. *Ecol. Econ.* **2000**, *32*, 301–317. [CrossRef]
11. Giannetti, B.F.; Demétrio, J.F.C.; Bonilla, S.H.; Agostinho, F.; Almeida, C.M.V.B. Emergy diagnosis and reflections towards Brazilian sustainable development. *Energy Policy* **2013**, *63*, 1002–1012. [CrossRef]
12. Zhang, L.X.; Song, B.; Chen, B. Emergy-based analysis of four farming systems: Insight into agricultural diversification in rural China. *J. Clean. Prod.* **2012**, *28*, 33–44. [CrossRef]
13. Zhang, X.; Hu, H.; Zhang, R.; Deng, S. Interactions between China's economy, energy and the air emissions and their policy implications. *Renew. Sustain. Energy Rev.* **2014**, *38*, 624–638. [CrossRef]
14. Zhang, Y.R.; Liu, J.; Zhang, J.; Wang, R. Emergy-based evaluation of system sustainability and ecosystem value of a large-scale constructed wetland in North China. *Environ. Monit. Assess.* **2013**, *185*, 5595–5609. [CrossRef]
15. Li, Q.; Yan, M. Assessing the health of agricultural land with emergy analysis and fuzzy logic in the major grain-producing region. *Catena* **2012**, *99*, 9–17. [CrossRef]
16. Zhang, X.L.; Hu, Q.H.; Wang, C.B. Emergy evaluation of environmental sustainability of poultry farming that produces products with organic claims on the outskirts of mega-cities in China. *Ecol. Eng.* **2013**, *54*, 128–135. [CrossRef]
17. Brown, M.T.; Ulgiati, S. Emergy evaluations and environmental loading of electricity production systems. *J. Clean. Prod.* **2002**, *10*, 321–334. [CrossRef]
18. Geng, Y.; Zhang, P.; Ulgiati, S.; Sarkis, J. Emergy analysis of an industrial park: The case of Dalian, China. *Sci. Total Environ.* **2010**, *408*, 5273–5283. [CrossRef] [PubMed]
19. Yuan, F.; Shen, L.Y.; Li, M.Q. Emergy analysis of the recycling options for construction and demolition waste. *Waste Manag.* **2011**, *31*, 2503–2511. [CrossRef] [PubMed]
20. Ulgiati, S. A comprehensive energy and economic assessment of biofuels: When "green" is not enough. *Crit. Rev. Plant Sci.* **2001**, *20*, 71–106. [CrossRef]
21. Takahashi, F.; Ortega, E. Assessing the sustainability of Brazilian oleaginous crops—Possible raw material to produce biodiesel. *Energy Policy* **2010**, *38*, 2446–2454. [CrossRef]
22. Zhou, S.Y.; Zhang, B.; Cai, Z.F. Emergy analysis of a farm biogas project in China: A biophysical perspective of agricultural ecological engineering. *Commun. Nonlinear Sci.* **2010**, *15*, 1408–1418. [CrossRef]
23. Zhang, X.H.; Jiang, W.J.; Deng, S.H.; Peng, K. Emergy evaluation of the sustainability of Chinese steel production during 1998–2004. *J. Clean. Prod.* **2009**, *17*, 1030–1038. [CrossRef]
24. Pan, H.Y.; Zhang, X.H.; Wu, J.; Zhang, Y.; Lin, L.; Yang, G.; Deng, S.; Li, L.; Yu, X.; Qi, H.; et al. Sustainability evaluation of a steel production system in China based on emergy. *J. Clean. Prod.* **2016**, *112*, 1498–1509. [CrossRef]

25. Ulgiati, S.; Brown, M.T. Quantifying the environmental support for dilution and abatement of process emissions: The case of electricity production. *J. Clean. Prod.* **2002**, *10*, 335–348. [CrossRef]
26. World Health Organization (WHO). Metrics: Disability-adjusted Life Year (DALY). 2013. Available online: http://www.who.int/healthinfo/global_burden_disease/metrics_daly/en/ (accessed on 7 December 2014).
27. Ministry of Environmental Protection of the People's Republic China (MEPPRC). Ambient Air Quality Standard (GB3095-2012). 2012. Available online: http://kjs.mep.gov.cn/hjbhbz/bzwb/dqhjbh/dqhjzlbz/201203/t20120302_224165.htm (accessed on 7 December 2014). (In Chinese)
28. Ukidwe, N.U.; Bakshi, B.R. Industrial and ecological cumulative exergy consumption of the United States via the 1997 input-output benchmark model. *Energy* **2007**, *32*, 1560–1592. [CrossRef]
29. Liu, G.; Yang, Z.; Chen, B.; Zhang, Y.; Su, M.; Zhang, L. Emergy evaluation of the urban solid waste Handling in Liaoning Province, China. *Energies* **2013**, *6*, 5486–5506. [CrossRef]
30. Ulgiati, S.; Brown, M.T. Monitoring Patterns of Sustainability in Natural and Man-made Ecosystems. *Ecol. Model.* **1998**, *108*, 23–36. [CrossRef]
31. Burchart-Korol, D. Significance of Environmental Life Cycle Assessment (Lca) Method in The Iron And Steel Industry. *Metalurgija* **2011**, *50*, 205–208.
32. Iosif, A.M.; Hanrot, F.; Birat, J.P.; Ablitzer, D. Physicochemical modelling of the classical steelmaking route for life cycle inventory analysis. *Int. J. Life Cycle Assess.* **2010**, *15*, 304–310. [CrossRef]
33. Xiao, L.S.; Wang, R.; Chiang, P.C.; Pan, S.Y.; Guo, Q.H.; Chang, E.E. Comparative Life Cycle Assessment (LCA) of Accelerated Carbonation Processes Using Steelmaking Slag for CO_2 Fixation. *Aerosol Air Qual. Res.* **2014**, *14*, 892–904. [CrossRef]
34. Vadenbo, C.O.; Boesch, M.E.; Hellweg, S. Life Cycle Assessment Model for the Use of Alternative Resources in Ironmaking. *J. Ind. Ecol.* **2013**, *17*, 363–374. [CrossRef]
35. Chen, X. *Evaluation of the sustainability of Iron and Steel Eco-Industrial Parks Based on Emergy and Exergy Theory*; Dalian University of Technology: Shenyang, China, 2009.
36. Lan, S.F.; Qin, P.; Lu, H.F. *Emergy Analysis of Ecological Economic System*; Chemical Industry Press: Beijing, China, 2002; p. 328.
37. Brown, M.T.; Bardi, E. *Handbook of Emergy Evaluation: Folio #3 Emegy of Ecosystems*; University of Florida: Gainesville, FL, USA, 2001; Available online: www.emergysystems.org (accessed on 2 July 2010).
38. Ma, F.J.; Liu, J.T. Agricultural Ecosystem Services Assessment Based on Emergy Analysis—A Case Study of Luancheng County. *Resour. Sci.* **2014**, *36*, 1949–1957.

© 2018 by the authors. Licensee MDPI, Basel, Switzerland. This article is an open access article distributed under the terms and conditions of the Creative Commons Attribution (CC BY) license (http://creativecommons.org/licenses/by/4.0/).

Article

Risk Assessment Using Fuzzy TOPSIS and PRAT for Sustainable Engineering Projects

G.K. Koulinas *, O.E. Demesouka, P.K. Marhavilas, A.P. Vavatsikos and D.E. Koulouriotis

Department of Production and Management Engineering, Democritus University of Thrace, 12 Vas. Sofias st., 67100 Xanthi, Greece; odemesou@pme.duth.gr (O.E.D.); marhavil@ee.duth.gr (P.K.M.); avavatsi@pme.duth.gr (A.P.V.); jimk@pme.duth.gr (D.E.K.)
* Correspondence: gkoulina@pme.duth.gr

Received: 13 December 2018; Accepted: 22 January 2019; Published: 24 January 2019

Abstract: In this study, we propose a safety risk assessment process using the fuzzy extension of the technique for order of preference by similarity to ideal solution (TOPSIS) for assigning priorities to risks in worksites, in order to promote the health, safety and well-being of workers, issues that are embedded within the concept of sustainability, specifically belonging to the social sphere of sustainability. The multicriteria method works in cooperation with a simple quantitative risk analysis and assessment process, the proportional risk assessment technique (PRAT), the functionality of which is based on real data. The efficiency of this approach is validated through treating a construction project example in Greece, and the results are compared with real fatal and non-fatal accidents data for the years 2014–2016. This integrated multicriteria approach can be used by risk managers as a tool for assessing safety risks and making informed decisions about the manner that a constraint budget would be spent in order to maximize health and safety in workplace.

Keywords: risk assessment; Fuzzy TOPSIS; construction safety; PRAT method; sustainability

1. Introduction

Preserving the health, safety and well-being of labor on work sites is a key concern worldwide as part of the effort to increase the productivity and promote the sustainable growth of business. The World Health Organization (WHO) [1], refers to sustainable development as a strategy to "meet the needs of the present world population without causing adverse effect on health and on the environment, and without depleting or endangering the global resource base, hence without compromising the ability of future generations to meet their needs." Recently, the U.S. Occupational Safety and Health Administration (OSHA) [2] recognized that actions for ensuring occupational health and safety can be integrated into sustainability efforts and benefit from sustainability movement dynamics in order to make work sites safer and healthy.

In addition, reduced accidents costs can save private and state budget resources that can be used more efficiently in sectors that need them more. Reports of the International Labor Organization (ILO) state that in the region of Europe and Central Asia, for the period 2009–2015, there were 64,230,125 workdays lost in the manufacturing sector due to injuries caused by accidents in the workspace and 44,723,674 workdays lost in the construction sector, as well [3].

Moreover, the European Agency for Safety and Health at Work [4] and the ILO [3] reported that, according to estimates, the cost of work-related injuries and illnesses worldwide is up to 3.9% of gross domestic product (GDP), namely about €2680 billion. As for the European Union (EU), injuries and illnesses related to the workplace costs 3.3% of its GDP, namely €476 billion annually, a part of which could be saved using more efficient health and safety practices, and thus, redirected to help treating many serious economic and societal problems.

Today, OSH issues are considered to be very important for organizations for economic (e.g., decrease lost working days), environmental (e.g., environmental hazards for employees), and social issues (e.g., ethical working conditions). Actually, the idea of sustainability has been utilized as a frame for OSHA and also for the Occupational Safety and Health (OSH) standards and techniques [5]. So far, the majority of OSH topics have been associated with organizations' compliance with legislation requirements, while a great part of the literature has recently incorporated OSH issues into the concept of social responsibility of organizations, which should go beyond the law by adopting voluntary OSH standards. However, the voluntary trend of organizations has lately gained ground in the context of the social responsibility of organizations to contribute to sustainable development [5,6]. This is integrated into the context of organizations as a commitment to OSH issues beyond the law which should be achieved through voluntary OSH standards (e.g., OSHAS 18001, ISO 45001). By this logic, the concept of sustainability is utilized as a frame to classify current OSH standards and techniques.

The main contribution of this study is the integration of the fuzzy extension of the most popular compromise programming multicriteria method, the technique for order of preference by similarity to ideal solution (TOPSIS), with a proven efficient quantitative technique for risk assessing and analysis, the proportional risk assessment technique (PRAT). The proposed process aims to be a useful tool for making informed decisions and assist systematic knowledge transfer between managers and engineers and practitioners. This study is organized into six sections: (1) the introduction, (2) the review of recent scientific literature, (3) the description of the fuzzy TOPSIS method, (4) the description of the proposed approach, (5) the application and results discussion, and (6) the conclusions.

2. Literature Review

In recent years there has been rising interest in the research field of occupational health and safety hazards assessment. Marhavilas and Koulouriotis [7] developed the "proportional risk assessment technique" (PRAT) and a "decision matrix technique", for quantification of risks, and applied them to a Greek aluminum extrusion company. Marhavilas and Koulouriotis [8] used an assessment approach using both stochastic and deterministic processes, and applied it in the largest Greek industry, the Public Power Corporation (PPC) and Aneziris et al. [9] proposed a model for construction projects risk quantification. Aminbakhsh, Gunduz, and Sonmez [10] used an approach for robust construction risks assessment while allocating budget. Marhavilas et al. [11] proposed an approach based on time-series harmonic analysis of accidents in the workplace. Guo and Haimes [12] proposed a framework for precursor analysis to assist the development of a precursor monitoring and decision support system. A great recent review on quantitative risk analysis is presented in Goerlandt et al. [13]. In the study of Dehdasht et al. [14] used the Decision Making Trial and Evaluation Laboratory (DEMATEL) method in association with Analytical Network Process (ANP) for risk assessment, specifically in oil and gas construction projects. Also, [15] and [16] used ANP and fuzzy systems for e-procurement risk factors estimation in a manufacturing company. The study of Jo et al. [17] analyzed data from accidents in Korean construction sector between years 2011 and 2015, in order to provide crucial information for defining policies to reduce construction accidents. Ghodrati, Yiu, Wilkinson, and Shahbazpour [18] proposed models to predict the safety outcome in the construction industry. Wu et al. [19] used neutrosophic sets for modifying operators that used with multiple attribute decision-making methods to assess risks in engineering construction projects.

In addition, great reviews of research directions and contributions in the occupational safety and health field with applications in the construction industry were undertaken by [20], and [21]. Recently, [5] conducted a comprehensive review to map the field of the interaction of sustainability and health-safety management systems, and provides an analytic description of the usage of occupational health and safety management standards in work sites. Finally, [22] proposed a framework for incorporating the fuzzy extension of a popular multicriteria decision making method, such as AHP, with a quantitative method for prioritizing risks in the workplace.

TOPSIS, was introduced by Hwang, C.L., and Yoon [23] as a multicriteria method for ranking and selection of alternatives using distance measures. The fuzzy-TOPSIS which is the extension of classic TOPSIS to fuzzy logic, has been introduced by [24], including ratings and weightings description by triangular fuzzy numbers. There is a wide range of applications for TOPSIS. Jozi and Majd [25], used TOPSIS for identifying risks from pollutants in the steel production process. Mahdevari, Shahriar, and Esfahanipour [26] proposed a process based on fuzzy TOPSIS for assessing health and safety risks and produce balanced budget plans, and Jozi, Shoshtary, and Zadeh [27] used AHP and TOPSIS for prioritization of hazardous factors identified by a Delphi method application in a construction project. Cococcioni, Lazzerini, and Pistolesi [28] proposed a multi-objective learning evolutionary algorithm for the classification of workers according to their risk perception for assigned tasks. TOPSIS is used for selecting the best Pareto-optimal solution. The TOPSIS method is used to select the best Pareto-optimal solution, among those generated by a non-dominated sorting artificial bee colony (NSBC) algorithm.

3. The Technique for Order of Preference by Similarity to Ideal Solution (TOPSIS) and Fuzzy TOPSIS Methods

3.1. TOPSIS Method

The functionality of the TOPSIS method is based on the consideration of two ideal solutions, the positive and the negative one. Then, the method works to find the shortest distance from the positive ideal solution and the longest distance from the negative ideal solution. TOPSIS belongs to the compromise programming methods, as its main principle is that the feasible solution sets' ranking depends on both their proximity to the positive ideal solution (PIS) (Equation (1)) and the negative ideal solution (NIS) (Equation (2)) [29].

$$L^{PIS} = \left(\sum_{j=1}^{m} \left(\left| f_j^+ - f_j(x) \right| \right)^2 \right)^{\frac{1}{2}} \quad (1)$$

$$L^{NIS} = \left(\sum_{j=1}^{m} \left(\left| f_j^- - f_j(x) \right| \right)^2 \right)^{\frac{1}{2}} \quad (2)$$

where $x \in A$, and m = number of alternatives,

Taking into consideration the fact that in TOPSIS the Euclidean distance measure is applied ($p = 2$) for the calculation of the distance from the positive ideal solution and the negative ideal solution, this comprises a special case of the compromise programming methods.

The method's model can further be simplified by expressing the distances from the optimal and the worst points of the analysis for the closeness coefficient of each alternative in a function form, as shown for the benefit criteria (f_t^+) and the cost criteria (f_k^-) respectively (Equations (3) and (4)).

$$f_j^+ = \{X = \max(or\ min) f_t(x) (or\ f_k(x)),\ \forall t(or\ k)\ \epsilon B(and\ K)\ \} \quad (3)$$

$$f_j^- = \{X = \min(or\ \max) f_t(x) (or\ f_k(x)),\ \forall t(or\ k)\ \epsilon B(and\ K)\ \} \quad (4)$$

The basic concept of this method is based on the monotonicity (increasing or decreasing) of each criterion, implying the easiness of defining the positive ideal solution and the negative ideal solution. Therefore, the alternatives' ranking depends on their distance from the best and the worst points of the analysis. The shorter the distance from the optimal solution and the farther the distance from the worst solution, the better the alternative.

The closeness coefficient (c_i), for the alternatives' ranking, is measured according to Equation (5) based on their distance from the positive ideal solution (S_i^+) and the negative ideal solution (S_i^-) [23,29].

$$C_i = \frac{S_i^-}{S_i^+ + S_i^-}, \ 0 \leq C_i \leq 1 \tag{5}$$

However, the uncertainty that real-world problems involve, increase the complexity of the decision making process, and as a result exact numeric values cannot represent decision maker's preferential system. For this reason, the use of linguistic variables introduced, to facilitate decision making ranking process.

3.2. Fuzzy TOPSIS Method

Initial efforts for extending the TOPSIS method to fuzzy sets were made by the study of Rebai [30] in which proposed the first fuzzy extension of TOPSIS method to rank alternatives using types of non-cardinal measures for measuring attributes' performance in a framework named "BBTOPSIS". However, the fuzzy extension of the TOPSIS method, used in the present study was, firstly, introduced by [24] to express the uncertainty existing in multicriteria decision support methods and specifically in the experts' judgements. Given the fact that the TOPSIS method aims at the alternatives ranking based on their distance from the PIS and NIS, these two points have to be identified first. Hereafter, each alternative's distance is measured from PIS and NIS and the alternative with the smallest distance from the PIS and the largest distance from the NIS is considered the best [29].

At first, as the analysis alternatives have been evaluated according to their severity and probability through the use of linguistic variables, they converted into fuzzy numbers according to the selected fuzzy scale. The normalization process is applied so as fuzzy numbers ranging between 0 to 1 (Equations (6) and (7)).

$$\tilde{r}_{ij} = \left(\frac{l_{ij}}{u_j^+}, \frac{m_{ij}}{u_j^+}, \frac{u_{ij}}{u_j^+} \right), 0 \leq C_i \leq 1 \tag{6}$$

$$\tilde{r}_{ij} = \left(\frac{l_j^-}{u_j^+}, \frac{l_j^-}{m_j^+}, \frac{l_j^-}{l_j^+} \right), \ l_j^- = \min_i l_{ij} \ \forall j^- \tag{7}$$

The normalized fuzzy decision matrix, in case of m alternatives and n criteria, is constructed as follows:

$$\tilde{R} = [\tilde{r}_{ij}]_{m \times n}, \ u_j^+ = \max_i u_{ij} \ \forall j^+ \tag{8}$$

where $\tilde{r}_{ij}$ refers to the normalized values of (l_{ij}, m_{ij}, u_{ij})

The weighted normalized value $\tilde{v}_{ij}$ is the product of the multiplication of weights ($\tilde{w}_j$) with the normalized fuzzy decision matrix $\tilde{r}_{ij}$. The weighted normalized decision matrix is obtained as:

$$\tilde{V} = [\tilde{w}_j \tilde{r}_{ij}] = [\tilde{v}_{ij}]_{m \times n} \ i = 1, 2, \ldots, m \ j = 1, 2, \ldots, n \tag{9}$$

Thereafter the fuzzy PIS and NIS are estimated as:

$$A^+ = (\tilde{v}_1^+, \tilde{v}_2^+, \ldots, \tilde{v}_n^+) = \left\{ \max_i v_{ij} \middle| (i = 1, 2, \ldots, m; j = 1, 2, \ldots, n) \right\} \tag{10}$$

$$((A^- = (\tilde{v}_1^-, \tilde{v}_2^-, \ldots, \tilde{v}_n^-) = \left\{ \min_i v_{ij} \middle| (i = 1, 2, \ldots, m; j = 1, 2, \ldots, n) \right\} \tag{11}$$

Finally each alternative's distance from the PIS (d_i^+) and NIS (d_i^-) points is calculated to be ranked using the closeness coefficient C_i as follows:

$$d_i^+ = \sum_{j=1}^{n} d\left(\tilde{v}_{ij}, \tilde{v}_j^+\right) \qquad (12)$$

$$d_i^- = \sum_{j=1}^{n} d\left(\tilde{v}_{ij}, \tilde{v}_j^-\right) \qquad (13)$$

Given that, the distance between two triangular fuzzy numbers (TFN) is obtained as

$$d_v(\widetilde{m},\widetilde{n}) = \sqrt{\tfrac{1}{3}\left[(l_1 - l_2)^2 + (m_1 - m_2)^2 + (u_1 - u_2)^2\right]}$$
$$C_i = \frac{d_i^-}{(d_i^+ + d_i^-)} \qquad (14)$$
$$\text{With } d_i^+,\ d_i^- \geq 0 \text{ and } C_i \in [0,1]$$

4. The Proposed Approach

The proposed risk assessment framework, works as follows: the well-known multicriteria compromise programming method, the fuzzy TOPSIS, is used for corresponding the decision maker's values and opinions regarding the importance of safety risk factors. In cooperation with this, PRAT [7] has been selected, due to its simplicity and effectiveness, to handle the real accidents' data and evaluate safety risks. This TOPSIS-PRAT mixed approach aims to provide to the risk manager a tool that can produce results using real-world accident data and the decision maker's value system and specific experience, as well. The flowchart of the proposed approach is shown in Figure 1.

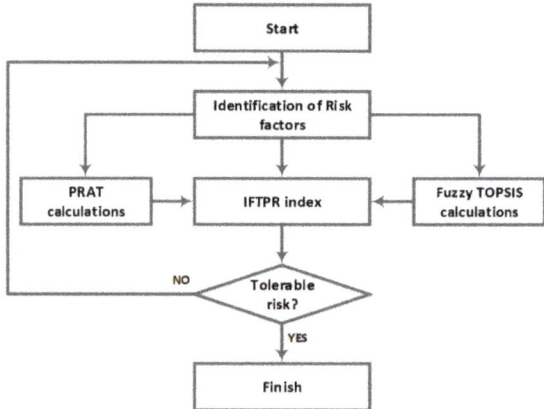

Figure 1. Flowchart of the proposed approach.

As illustrated, after identifying the safety risk factors for a given project, calculations for TOPSIS and PRAT realized separately. Next, the integrated fuzzy TOPSIS and PRAT (IFTPR) index is constructed as a part of the risk management process. Finally, the risk factors are ranked according to their merged index score, and the risk manager makes the decision about the tolerance of the total risk of the project. Figure 2 focuses on showing in more detail the cooperation framework of the two approaches.

Note that indexes P, S, Fr, R, TFN, PIS, NIS, Ci and the scores TPS and IFTPR are explained in details in the next Section 5.

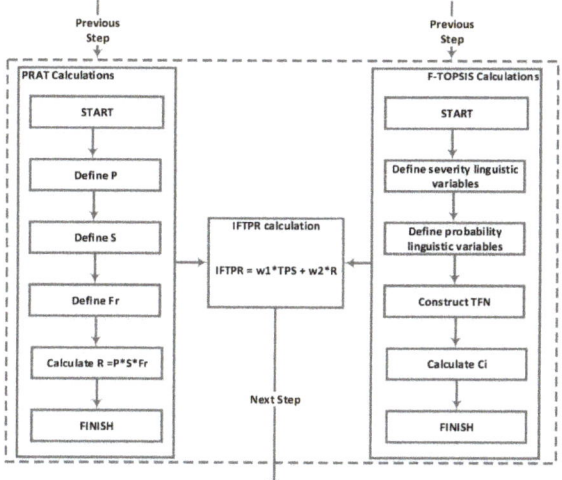

Figure 2. The fuzzy technique for order of preference by similarity to ideal solution (TOPSI)S and proportional risk assessment technique (PRAT) methods' integration steps.

5. Application in a Construction Project

The proposed method was applied in a construction project of a total building renovation in the region of Halkidiki, Greece. The project was the reconstruction of a luxurious 5-star hotel with about 200 rooms, suites and bungalows. In addition, the facilities include a large swimming area with a complex of pools, and many rooms with private gardens and exclusive pools, as well.

The building area is up to 10,000 m^2 including foyers, halls, a conference center and a fitness room. More specifically, the building was initially constructed in the 1980s and with only some minor improvements of the furniture had been used since then. It is worth mentioning that reconstruction activities included strengthening the stability of the building, a complete replacement of all electromechanical installations and the plumbing networks, as well. Furthermore, all the window frames of the building were replaced while its energy efficiency was upgraded. In addition, decoration and furnishings were also completely renewed. According to the TOPSIS methodology, an expert site manager is needed for making the prerequisite judgments that the method uses to build its output rankings of the safety risk factors. In our case, the supervising project manager of the construction company plays the role of the decision maker for making the judgements required by the multicriteria method employed.

For validating the functionality of the method, we used real accident data for the years 2014, 2015 and 2016 from the Hellenic Statistical Authority (ELSTAT), which is an independent authority being the national statistics representative of Greece in the service of the EU and in any other international organization. These data are collected constantly every year from ELSTAT and conform to the ESAW methodology [31,32]. The main goal of our approach is to create a hybrid ranking of risk factors consisting of the ranking of the most frequent and severe safety risk factors using the real data and PRAT method, and the preferential system corresponding to the judgements of the expert using the fuzzy TOPSIS method.

The present study aims to provide project managers a tool for ranking risks and consequently, lead to efficient accident preventing investments, in order to minimize total risk.

5.1. TOPSIS Method Calculations

The nine risk factors included in this study (Table 1) are these used by the ELSTAT's methodology, which follows the European Statistics on Accidents at Work (ESAW) methodology [31,32].

After that, the linguistic variables' scale defined according to the decision maker's preferential system. Tables 2 and 3 present the linguistic variables for evaluating the analysis criteria regarding severity and probability respectively.

Table 1. The risk factors of the analysis and their codes.

Description of Risk Factors	Code
Contact with electrical voltage, temperature, hazardous substances	F1
Drowned, buried, enveloped	F2
Horizontal or vertical impact with or against a stationary object (the victim is in motion)	F3
Struck by object in motion collision with	F4
Contact with sharp, pointed, rough, coarse material agent	F5
Trapped, crushed etc.	F6
Physical or mental stress	F7
Bite, Kick etc. (animal or human)	F8
Other contacts-modes of injury not listed in this classification	F9

Table 2. Probability-frequency scale, including ten gradation levels, in order to estimate the probability factor.

Probability Factor (P)	Description of Undesirable Event	Frequency of Events Occurring
10	Unavoidable	1 event during a time period of $\Delta t \ll 10^3$ h
9	Almost assured	1 event during a time period of $\Delta t < 10^3$ h
8	Frequent	1 event during a time period of Δt almost equal to 10^3 h
7	Probable	1 event during a time period of $10^3 < \Delta t < 10^{-4}$ h
6	Probability slightly greater than 50%	1 event during a time period of $10^4 < \Delta t < 10^5$ h
5	Probability 50%	1 event during a time period of Δt almost equal to 10^5 h
4	Probability slightly less than 50%	1 event during a time period of $10^5 < \Delta t < 10^6$ h
3	Almost improbable (or remote)	1 event during a time period of $10^6 < \Delta t < 10^7$ h
2	Improbable	1 event during a time period of Δt almost equal to 10^7 h
1	Impossible	1 event during a time period of $\Delta t > 10^7$ h

Table 3. Gradation severity of harm factor in association with the undesirable event.

Severity of Harm Factor (S)	Description of Undesirable Event
10	Death
9	Permanent total inefficiency
8	Permanent serious inefficiency
7	Permanent slight inefficiency
6	Absence from work > 3 weeks, and return with health problems
5	Absence from work > 3 weeks, and return after full recovery
4	Absence from work > 3 days and < 3 weeks, and return after full recovery
3	Absence from work < 3 days, and return after full recovery
2	Slight injuring without absence from the work, and with full recovery
1	No human injury

Then the TFN scale is constructed expressing the smallest possible value (l), the most promising value (m), and the largest possible value (u) that describes each fuzzy event. In the present paper, the following fuzzy scale is applied (Table 4).

According to the TOPSIS process, the construction expert selected as the decision maker has to evaluate each factor of this analysis, and make judgements regarding their severity and probability. These judgements are presented in Table 5. Next, the total risk for each factor is estimated by multiplying the TFN probability with the TFN severity.

Table 4. Fuzzy scale for linguistic variables evaluation.

Probability Factor (P)	Severity Factor (S)	Triangular Fuzzy Scale
P10	S10	(0.80, 0.90, 1.00)
P9	S9	(0.75, 0.85, 0.95)
P8	S8	(0.70, 0.80, 0.85)
P7	S7	(0.55, 0.65, 0.75)
P6	S6	(0.50, 0.60, 0.65)
P5	S5	(0.35, 0.55, 0.60)
P4	S4	(0.25, 0.40, 0.45)
P3	S3	(0.20, 0.30, 0.35)
P2	S2	(0.15, 0.20, 0.25)
P1	S1	(0.00, 0.10, 0.15)

Table 5. Analysis factors' evaluation.

Factor	P	TFN (P)	S	TFN (S)	TFN (P x S)
F1	P5	(0.35, 0.55, 0.60)	S8	(0.70, 0.80, 0.85)	(0.245, 0.440, 0.510)
F2	P5	(0.35, 0.55, 0.60)	S2	(0.15, 0.20, 0.25)	(0.053, 0.110, 0.150)
F3	P7	(0.55, 0.65, 0.75)	S8	(0.70, 0.80, 0.85)	(0.385, 0.520, 0,638)
F4	P7	(0.55, 0.65, 0.75)	S7	(0.55, 0.65, 0.75)	(0.303, 0.423, 0,563)
F5	P3	(0.20, 0.30, 0.35)	S3	(0.20, 0.30, 0.35)	(0.040, 0.090, 0.123)
F6	P6	(0.50, 0.60, 0.65)	S7	(0.55, 0.65, 0.75)	(0.275, 0.390, 0.488)
F7	P5	(0.35, 0.55, 0.60)	S3	(0.20, 0.30, 0.35)	(0.070, 0.165, 0.210)
F8	P6	(0.50, 0.60, 0.65)	S2	(0.15, 0.20, 0.25)	(0.075, 0.120, 0.163)
F9	P2	(0.15, 0.20, 0.25)	S2	(0.15, 0.20, 0.25)	(0.023, 0.040, 0.063)

From Table 5 the PIS (0.385, 0.520, 0.638) and NIS (0.023, 0.040, 0.063) points are estimated according to Equations (10) and (11). Then, the closeness coefficient Ci is calculated with the use of Equations (12)–(14) (Table 6). Given the fact that the factors' ranking is based on their risk, when the coefficient Ci is more close to 1 the factor is considered to be of high-risk and when the coefficient Ci is more close to 0 the factor is considered to be of low risk.

Table 6. The C_i estimation.

Factor	d+	d−	Ci	Normalized Ci
F1	0.119	0.369	0.757	19.20%
F2	0.415	0.067	0.138	3.50%
F3	0.000	0.480	1.000	25.36%
F4	0.086	0.397	0.823	20.87%
F5	0.436	0.046	0.095	2.41%
F6	0.131	0.349	0.727	18.44%
F7	0.369	0.115	0.237	6.01%
F8	0.401	0.080	0.166	4.21%
F9	0.481	0.000	0.000	0.00%
			3.943	

5.2. Proportional Risk Assessment Technique (PRAT) Method Calculations

In parallel, the risk estimation using the PRAT method proceeds. This simple process uses real accidents data included in Tables 7 and 8. For each safety risk factor, a risk value (R) is calculated by:

$$R = P \times S \times Fr \quad (15)$$

where P is the probability of an accident type occurring, S is the severity of harm for a given risk factor, and Fr is the frequency of an accident type occurrence. According to Tables 2, 3 and 9, these three factors, can take values in the scale of 1–10, besides the quantity R that can be expressed in the scale of 1–1000.

Table 7. Non-fatal accidents at work by contact/mode of injury. (Source: Hellenic Statistics Authority, Press Release Survey on Accidents at Work (2014), (2015), (2016)).

	Mode of Injury	2014	2015	2016	Total %	Events/Hour	Average Time for an Event (Hours)
F1	Contact with electrical voltage, temperature, hazardous substances	112	142	132	2.88%	0.0335	29.84
F2	Drowned, buried, enveloped	0	1	1	0.01%	0.0002	5760.00
F3	Horizontal or vertical impact with or against a stationary object (the victim is in motion)	1534	1702	1778	37.44%	0.4352	2.30
F4	Struck by object in motion collision with	980	975	1027	22.27%	0.2589	3.86
F5	Contact with sharp, pointed, rough, coarse material agent	514	555	561	12.17%	0.1415	7.07
F6	Trapped, crushed etc.	410	422	484	9.83%	0.1142	8.75
F7	Physical or mental stress	458	408	446	9.80%	0.1139	8.78
F8	Bite, Kick etc. (animal or human)	44	59	55	1.18%	0.0137	72.91
F9	Other contacts/modes of injury not listed in this classification	4	3	2	0.07%	0.0335	1280.00
	No information	139	192	253	4.36%	0.0507	19.73
	Sum	4195	4459	4739	100.0		

Table 8. Fatal accidents at work by contact/mode of injury. (Source: Hellenic Statistics Authority, Press Release Survey On Accidents At Work (2014), (2015), (2016)).

	Mode of Injury	2014	2015	2016	Total %	Events/Hour	Average Time for an Event (Hours)
F1	Contact with electrical voltage, temperature, hazardous substances	4	7	3	10.14%	0.0009	822.86
F2	Drowned, buried, enveloped	4	1	1	4.35%	0.0012	1920.00
F3	Horizontal or vertical impact with or against a stationary object (the victim is in motion)	17	17	11	32.61%	0.0005	256.00
F4	Struck by object in motion collision with	11	17	17	32.61%	0.0039	256.00
F5	Contact with sharp, pointed, rough, coarse material agent	0	0	1	0.72%	0.0039	11520.00
F6	Trapped, crushed etc.	5	2	7	10.14%	0.0001	822.86
F7	Physical or mental stress	1	0	0	0.72%	0.0012	11520.00
F8	Bite, Kick etc. (animal or human)	2	0	0	1.45%	0.0001	5760.00
F9	Other contacts/modes of injury not listed in this classification	0	0	0	0.00%	0.0002	0.00
	No information	2	1	7	7.25%	0.0009	1152.00
	Sum	46	45	47	100.0		

Table 9. Gradation of the frequency (or the exposure) factor in association with the undesirable event.

Frequency Factor (Fr)	Description of Undesirable Event
10	Permanent presence of damage
9	Presence of damage every 30 s
8	Presence of damage every 1 min
7	Presence of damage every 30 min
6	Presence of damage every 1 h
5	Presence of damage every 8 h
4	Presence of damage every 1 week (182.5 h)
3	Presence of damage every 1 month (730 h)
2	Presence of damage every 1 year (8760 h)
1	Presence of damage every 5 years (43800 h)

The real data included in Tables 7 and 8 were used to define the values for P, S, and F factors, which are illustrated in Table 10.

Table 10. Classification of important hazard sources after PRAT.

Description of Risk Factors		Probability Factor (P)	Severity of Harm Factor (S)	Frequency Factor (Fr)	Risk Value (R)	Normalized Risk Value
		Tables 7 and 8	Tables 7 and 8	Tables 7 and 8		
Contact with electrical voltage, temperature, hazardous substances	F1	10	10	3	300	14.08%
Drowned, buried, enveloped	F2	8	10	2	160	7.51%
Horizontal or vertical impact with or against a stationary object (the victim is in motion)	F3	10	10	4	400	18.78%
Struck by object in motion collision with	F4	10	10	4	400	18.78%
Contact with sharp, pointed, rough, coarse material agent	F5	10	7	2	140	6.57%
Trapped, crushed etc.	F6	10	10	3	300	14.08%
Physical or mental stress	F7	10	7	1	70	3.29%
Bite, Kick etc. (animal or human)	F8	10	8	2	160	7.51%
Other contacts/modes of injury not listed in this classification	F9	7	0	0	0	0.00%
No information	-	10	10	2	200	9.39%
Sum					2130	100.00%

In this analysis, the level of each factor is considered as follows:

For estimating the level of factor P, we use the average time for an accident to occur (Table 7, column 8), and for severity factor (S) we considered that if a risk factor resulted to at least one fatal incident for each year, it is assigned the highest level (10). The rest of the factors were assigned levels according to the expert's experience. Regarding factor Fr, we have assigned different levels according to the "average time for a (fatal) accident to occur" (Table 8, column 8) data. For computing the values of the columns "Events/hour" and "Average time for an event (hours)" in Table 7, we have considered the working weeks being 48 per year, having 5 working days each, and 16 working hours per day, resulting in a total of D = 1 × 48 × 5 × 16 = 3840 hours/year. It is worth mentioning that the construction company uses two shifts per day, namely 16 working hours. After computing the R values, we sum these values for every factor. The normalized risk value for each factor defined by dividing each factor's R value with the total (Table 10).

5.3. Integrated TOPSIS with PRAT

The IFTPR index consist of the TOPSIS score (TPS) with a weighting of w1 = 50%, and the risk value (R) calculated with PRA technique, with a weighting of w2 = 50%. Initially, we considered these factors (TPS and R) as equally important for computing the IFTPR.

$$\text{IFTPR} = 0.5 \times \text{TPS} + 0.5 \times \text{R} \tag{16}$$

The calculations leading to this index are shown in Table 11, while ranking of risk factors according to the IFTPR index are illustrated in the last column of Table 12. In addition, Table 12 summarizes rankings with solely fuzzy TOPSIS, PRAT, and real fatal and non-fatal accidents.

Table 11. Calculations for the integrated fuzzy TOPSIS and PRAT (IFTPR) method ranking.

Description of Risk Factors	PRAT (R)	F-TOPSIS (TPS)	IFTPR (0.5 × R + 0.5 × TPS)
Contact with electrical voltage, temperature, hazardous substances	14.08%	19.20%	16.64%
Drowned, buried, enveloped	7.51%	3.50%	5.51%
Horizontal or vertical impact with or against a stationary object (the victim is in motion)	18.78%	25.36%	22.07%
Struck by object in motion collision with	18.78%	20.87%	19.83%
Contact with sharp, pointed, rough, coarse material agent	6.57%	2.41%	4.49%
Trapped, crushed etc.	14.08%	18.44%	16.26%
Physical or mental stress	3.29%	6.01%	4.65%
Bite, Kick etc. (animal or human)	7.51%	4.21%	5.86%
Other contacts-modes of injury not listed in this classification	0.00%	0.00%	0.00%
No Information	9.39%	0.00%	4.69%

Table 12. Rankings of risk factors for the different approaches used.

Real Fatal Accidents Data	Real Non Fatal Accidents Data	Fuzzy TOPSIS	PRAT	F-TOPSIS & PRAT
a	b	d	e	f
F3 (32.61%)	F3 (37.44%)	F3 (25.36%)	F3 (18.78%)	F3 (22.07%)
F4 (32.61%)	F4 (22.27%)	F4 (20.87%)	F4 (18.78%)	F4 (19.83%)
F1 (10.14%)	F5 (12.17%)	F1 (19.20%)	F1 (14.08%)	F1 (16.64%)
F6 (1.14%)	F6 (9.83%)	F6 (18.22%)	F6 (14.08%)	F6 (16.26%)
F2 (4.35%)	F7 (9.80%)	F7 (6.01%)	F2 (7.51%)	F8 (5.86%)
F8 (1.45%)	F1 (2.88%)	F8 (4.21%)	F8 (7.51%)	F2 (5.51%)
F5 (0.72%)	F8 (1.18%)	F2 (3.50%)	F5 (6.57%)	F7 (4.65%)
F7 (0.72%)	F9 (0.07%)	F5 (2.41%)	F7 (3.29%)	F5 (4.49%)
F9 (0.00%)	F2 (0.01%)	F9 (0.00%)	F9 (0.00%)	F9 (0.00%)

5.4. Results

Table 12 summarizes rankings with applying solely the Fuzzy TOPSIS, and PRAT methods and rankings extracted by the data of real fatal and non-fatal accidents.

According to the TOPSIS results (Tables 6 and 12), the decision maker evaluates the F3 (horizontal or vertical impact with or against a stationary object) as the most important risk factor in the project. Next, he ranks F4 (Struck by object in motion-collision with), and third F1 (Contact with electrical voltage, temperature, hazardous substances), while F6 (Trapped, crushed, etc.) has ranked fourth. This ranking corresponding the decision maker's previous experience is in accordance with the ranking from the real data of fatal accidents, which means that the manager is informed about the four most important risks that workers treat in their workplace in a construction project. As for the remaining five factors, the decision maker ranks F7 (Physical or mental stress), and then F8 (Bite, Kick, etc. (animal or human)), F2 (Drowned, buried, enveloped) and F5 (Contact with sharp, pointed, rough, coarse material agent), respectively. Note that the F9 (No information), although existing in the statistical data, cannot be evaluated by the decision maker. The rankings of the rest four factors illustrate that the manager has no clear view about how important these risks are, and so, a supplementary process is needed to support the decision maker's opinion in order to make better informed decisions.

According to the PRAT results, most important factors are F3 and F4, followed by F1 and F6, which have a relative small difference from the first two factors. In the next three places are the F2, F8 and F5 factors, having clearly smaller risk value than the previous four factors. The next two factors are F7 and F9 that due to their relatively small frequency of appearance have the last two places in the list. Note that the PRAT ranking does not include the "No information" factor since it only exists for statistical reasons.

The cooperation of the fuzzy TOPSIS and PRAT methods results in clearly more efficient output than applying each method separately. Also, it can be stated that this ranking is an indicator about which safety measures should the manager chose with a priority when he allocates the constrained budget.

More specifically, the factors "Horizontal or vertical impact with or against a stationary object" (F3), "Struck by object in motion collision with" (F4), "Contact with electrical voltage, temperature, hazardous substances"(F1) and "Trapped, crushed etc."(F6) are proved to be much more important risk sources than the other factors since they cause about 74.8% of the total accidents.

The next group of factors including "Bite, Kick etc." (F8), "Drowned, buried, enveloped" (F2), "Physical or mental stress"(F7) and "Contact with sharp, pointed, rough, coarse material agent"(F5) are ranked to have smaller importance due to the small numbers of real accidents caused during the last three years. In addition, as illustrated by the given judgements, the decision maker has no such previous experience in the work site of a construction project since the rankings are extracted by fuzzy TOPSIS.

Grouping the risk factors in such a manner, illustrates that the decision maker is relatively experienced since he recognizes the four most important risk factors. As for the rest of them, probably because he rarely treats such a kind of risks, the ranking of the expert is quite different than reality. The informed ranking list after applying PRAT and merging with fuzzy TOPSIS results is, as expected, slightly different than the original constructed with the multicriteria method alone, and surely could be used for allocating constrained budget more efficiently to measures that can prevent accidents in the workplace. The contribution of merging these two powerful methods is that the ranking of the decision maker is modified and transformed to obtain facts about accidents that are closer to reality.

6. Conclusions

As OSHA [2] states in a recently published report "A building, no matter how energy efficient or healthy for occupants, is not sustainable if a construction worker is killed while building it". Furthermore, an employer can really be considered as sustainable only if it ensures the safety and health of their employees [2]. Conforming to these statements, an assessment of risk factors in the workplace is a key aspect in every project. Accidents can surely lead to exceeding budget and time constraints and often can cause lengthy and costly trials and compensation payments for a company. On the other hand, investments in safety measures can reduce risks to a tolerable amount, but in order to do this, near optimal allocation of available budget is crucial. The main contribution of the proposed approach is the cooperation of the fuzzy extension of TOPSIS multicriteria method, the fuzzy TOPSIS, with a simple quantitative process (PRAT) the function of which is based on real accident data. The TOPSIS method is based in linear programming and works with distances from the positive and negative ideal solution. More specifically, it is considered an ideal and a non-ideal solution, and the process aims to find the shortest distance from the positive ideal solution and the longest distance from the negative ideal solution. This synergic framework is employed for assessment and prioritization of risks that construction workers often treat in their work sites. Generally, fuzzy TOPSIS is used to express the decision maker's experience and PRAT to update knowledge and experience of the expert, using real accident data.

Regarding the implications of the present approach, in practice it can be a very useful framework for risk managers and management practitioners for supporting semi-automatic and fast safety risk prioritization that benefits both from the expertise of the manager, and from real data that the decision maker may not know when treating emergency instances, and then allocating constrained budget in such a manner as to maximize health and safety and minimize the total risk in the workplace. Also, the proposed approach can be used as a knowledge and experience transfer tool from more experienced risk managers to less experienced practitioners in order to assist them in learning how to make informed decisions. In addition, the framework of the present study, and more specific data, can be used to develop hybrid multicriteria methods with other existing quantitative techniques for specific sectors of an economy such as manufacturing and heavy industry where frequently serious accidents occur. Nevertheless, the present process could be improved in the future, by employing sensitivity analysis on the weights of TPS and R that fuzzy TOPSIS and PRAT contribute to the integrated index IFTPR in order to understand if there are some circumstances under which it is needed as a method to override others.

Author Contributions: Conceptualization, G.K.K., O.E.D., P.K.M., A.P.V., D.E.K.; methodology, G.K.K., O.E.D., P.K.M., A.P.V., D.E.K.; software, G.K.K., O.E.D., A.P.V.; validation, G.K.K., O.E.D., P.M., A.P.V., D.E.K.; formal analysis, G.K.K., O.E.D.; investigation, G.K.K., O.E.D.; resources, G.K.K.; data curation, O.E.D., writing—original draft preparation, G.K.K.; writing—review and editing, G.K.K., O.E.D., P.K.M., A.P.V., D.E.K.; visualization, G.K.K., O.E.D., P.K.M., A.P.V., D.E.K.; supervision, D.E.K.; project administration, G.K.K., D.E.K.; funding acquisition, D.E.K.

Funding: This research received no external funding.

Conflicts of Interest: The authors declare no conflict of interest.

References

1. WHO Global Strategy on Occupational Health for All: The Way to Health at Work. Available online: https://www.who.int/occupational_health/publications/globstrategy/en/index3.html (accessed on 1 July 2018).
2. OSHA Sustainability in the Workplace: A New Approach for Advancing Worker Safety and Health. Available online: https://www.osha.gov/sustainability/docs/OSHA_sustainability_paper.pdf (accessed on 1 July 2018).
3. ILO Days Lost Due to Cases of Occupational Injury with Temporary Incapacity for Work by Economic Activity. Available online: http://www.ilo.org/ilostat/faces/oracle/webcenter/portalapp/pagehierarchy/Page27.jspx?indicator=INJ_DAYS_ECO_NB&subject=OSH&datasetCode=A&collectionCode=YI&_adf.ctrl-state=f16q5nvmt_4&_afrLoop=516942705047488&_afrWindowMode=0&_afrWindowId=f16q5nvmt_1#!%40 (accessed on 20 Jun 2018).
4. EU-OSHA Work-Related Accidents and Injuries Cost EU €476 Billion a Year According to New Global Estimates. Available online: https://osha.europa.eu/en/about-eu-osha/press-room/eu-osha-presents-new-figures-costs-poor-workplace-safety-and-health-world (accessed on 20 June 2018).
5. Marhavilas, P.; Koulouriotis, D.; Nikolaou, I.; Tsotoulidou, S. International Occupational Health and Safety Management-Systems Standards as a Frame for the Sustainability: Mapping the Territory. *Sustainability* **2018**, *10*, 3663. [CrossRef]
6. European Commission Communication from the Commission Concerning Corporate Social Responsibility: A Business Contribution to Sustainable Development. Available online: https://eur-lex.europa.eu/LexUriServ/LexUriServ.do?uri=COM:2002:0347:FIN:EN:PDF (accessed on 26 September 2018).
7. Marhavilas, P.K.; Koulouriotis, D.E. A risk-estimation methodological framework using quantitative assessment techniques and real accidents' data: Application in an aluminum extrusion industry. *J. Loss Prev. Process Ind.* **2008**, *21*, 596–603. [CrossRef]

8. Marhavilas, P.K.; Koulouriotis, D.E. Developing a new alternative risk assessment framework in the work sites by including a stochastic and a deterministic process: A case study for the Greek Public Electric Power Provider. *Saf. Sci.* **2012**, *50*, 448–462. [CrossRef]
9. Aneziris, O.N.; Topali, E.; Papazoglou, I.A. Occupational risk of building construction. *Reliab. Eng. Syst. Saf.* **2012**, *105*, 36–46. [CrossRef]
10. Aminbakhsh, S.; Gunduz, M.; Sonmez, R. Safety risk assessment using analytic hierarchy process (AHP) during planning and budgeting of construction projects. *J. Saf. Res.* **2013**, *46*, 99–105. [CrossRef] [PubMed]
11. Marhavilas, P.K.; Koulouriotis, D.E.; Spartalis, S.H. Harmonic analysis of occupational-accident time-series as a part of the quantified risk evaluation in worksites: Application on electric power industry and construction sector. *Reliab. Eng. Syst. Saf.* **2013**, *112*, 8–25. [CrossRef]
12. Guo, Z.; Haimes, Y.Y. Risk Assessment of Infrastructure System of Systems with Precursor Analysis. *Risk Anal.* **2016**, *36*, 1630–1643. [CrossRef]
13. Goerlandt, F.; Khakzad, N.; Reniers, G. Validity and validation of safety-related quantitative risk analysis: A review. *Saf. Sci.* **2017**, *99*, 127–139. [CrossRef]
14. Dehdasht, G.; Mohamad Zin, R.; Ferwati, S.M.; Mohammed Abdullahi, M.; Keyvanfar, A.; McCaffer, R. DEMATEL-ANP Risk Assessment in Oil and Gas Construction Projects. *Sustainability* **2017**, *9*, 1420. [CrossRef]
15. Ramkumar, M.; Schoenherr, T.; Jenamani, M. Risk assessment of outsourcing e-procurement services: Integrating SWOT analysis with a modified ANP-based fuzzy inference system. *Prod. Plan. Control* **2016**, *27*, 1171–1190. [CrossRef]
16. Ramkumar, M. A modified ANP and fuzzy inference system based approach for risk assessment of in-house and third party e-procurement systems. *Strateg. Outsourc. Int. J.* **2016**, *9*, 159–188. [CrossRef]
17. Jo, W.B.; Lee, S.Y.; Kim, H.J.; Khan, M.R. Trend Analysis of Construction Industrial Accidents in Korea from 2011 to 2015. *Sustainability* **2017**, *9*, 1297. [CrossRef]
18. Ghodrati, N.; Yiu, T.W.; Wilkinson, S.; Shahbazpour, M. A new approach to predict safety outcomes in the construction industry. *Saf. Sci.* **2018**, *109*, 86–94. [CrossRef]
19. Wu, S.; Wang, J.; Wei, G.; Wei, Y. Research on Construction Engineering Project Risk Assessment with Some 2-Tuple Linguistic Neutrosophic Hamy Mean Operators. *Sustainability* **2018**, *10*, 1536. [CrossRef]
20. Sousa, V.; Almeida, N.M.; Dias, L.A. Risk-based management of occupational safety and health in the construction industry—Part 1: Background knowledge. *Saf. Sci.* **2014**, *66*, 75–86. [CrossRef]
21. Sousa, V.; Almeida, N.M.; Dias, L.A. Risk-based management of occupational safety and health in the construction industry—Part 2: Quantitative model. *Saf. Sci.* **2015**, *74*, 184–194. [CrossRef]
22. Koulinas, G.K.; Marhavilas, P.K.; Demesouka, O.E.; Vavatsikos, A.P.; Koulouriotis, D.E. Risk analysis and assessment in the worksites using the fuzzy-analytical hierarchy process and a quantitative technique—A case study for the Greek construction sector. *Saf. Sci.* **2019**, *112*, 96–104. [CrossRef]
23. Hwang, C.L.; Yoon, K. *Multiple Attribute Decision Making Methods and Applications*; Springer: Berlin, Germany, 1981; ISBN 978-3-642-48318-9.
24. Chen, C.-T. Extensions of the TOPSIS for group decision-making under fuzzy environment. *Fuzzy Sets Syst.* **2000**, *114*, 1–9. [CrossRef]
25. Jozi, S.A.; Majd, N.M. Health, safety, and environmental risk assessment of steel production complex in central Iran using TOPSIS. *Environ. Monit. Assess.* **2014**, *186*, 6969–6983. [CrossRef]
26. Mahdevari, S.; Shahriar, K.; Esfahanipour, A. Human health and safety risks management in underground coal mines using fuzzy TOPSIS. *Sci. Total Environ.* **2014**, *488–489*, 85–99. [CrossRef]
27. Jozi, S.A.; Shoshtary, M.T.; Zadeh, A.R.K. Environmental Risk Assessment of Dams in Construction Phase Using a Multi-Criteria Decision-Making (MCDM) Method. *Hum. Ecol. Risk Assess.* **2015**, *21*, 1–16. [CrossRef]
28. Cococcioni, M.; Lazzerini, B.; Pistolesi, F. A semi-supervised learning-aided evolutionary approach to occupational safety improvement. In Proceedings of the 2016 IEEE Congress on Evolutionary Computation, CEC 2016, Vancouver, BC, Canada, 24–29 July 2016; pp. 3695–3701.
29. Hwang, C.-L.; Lai, Y.-J.; Liu, T.-Y. A new approach for multiple objective decision making. *Comput. Oper. Res.* **1993**, *20*, 889–899. [CrossRef]
30. Rebai, A. BBTOPSIS: A bag based technique for order preference by similarity to ideal solution. *Fuzzy Sets Syst.* **1993**, *60*, 143–162. [CrossRef]

31. Eurostat European Statistics on Accident at Work (ESAW) Methodology. Available online: http://ec.europa.eu/eurostat/ramon/statmanuals/files/ESAW_2001_EN.pdf (accessed on 20 June 2018).
32. Eurostat European Statistics on Accident at Work (ESAW) Summary Methodology. Available online: http://ec.europa.eu/eurostat/documents/3859598/5926181/KS-RA-12-102-EN.PDF/56cd35ba-1e8a-4af3-9f9a-b3c47611ff1c (accessed on 20 June 2018).

© 2019 by the authors. Licensee MDPI, Basel, Switzerland. This article is an open access article distributed under the terms and conditions of the Creative Commons Attribution (CC BY) license (http://creativecommons.org/licenses/by/4.0/).

Article

Critical Success Factors for Project Planning and Control in Prefabrication Housing Production: A China Study

Long Li [1], Zhongfu Li [1], Guangdong Wu [2,*] and Xiaodan Li [1]

[1] Department of Construction Management, Dalian University of Technology, Dalian 116024, China; qdlgll2012@163.com (L.L.); lizhongfu@dlut.edu.cn (Z.L.); gstslxd@126.com (X.L.)
[2] Department of Construction Management, Jiangxi University of Finance & Economics, Nanchang 330013, China
* Correspondence: gd198410@163.com; Tel.: +86-791-8384-2078

Received: 14 February 2018; Accepted: 9 March 2018; Published: 16 March 2018

Abstract: The process of prefabrication housing production (PHP) has been inevitably faced with diverse challenges. A number of factors affect the successful implementation of PHP. However, the critical success factors (CSFs) remain unrevealed. This paper aims to examine the CSFs for the planning and control of PHP projects. A total of 23 factors were identified as a result of literature review, in-depth interviews and pilot studies with experts in the construction industry. A questionnaire survey was conducted with designers, manufacturers, and contractors in China. The result showed that the top five CSFs were: (1) designers' experience of PHP, (2) manufacturer's experience of PHP, (3) project manager's ability to solve problems, (4) maturity of techniques used in the detailed design phase, and (5) persistent policies and incentives. The 23 CSFs were further categorized into five groups via exploratory factor analysis, namely: (1) technology and method, (2) information, communication and collaboration, (3) external environment, (4) experience and knowledge, and (5) competence of the project manager. In particular, "technology and method" played the dominant role. This study contributes to the existing body of knowledge via a holistic approach covering the key actors of PHP such as designers, manufacturers as well as contractors to examine CSFs of PHP. These findings provided designers and project managers with a useful set of criteria for the effective project planning and control of PHP and facilitated the successful implementation PHP.

Keywords: prefabrication housing production; planning and control; critical success factors; China

1. Introduction

Prefabrication housing production (PHP) has been widely promoted all over the world to improve quality, reduce waste and energy consumption, and provide a safer work environment [1–3]. Similar terms used in different countries or regions include "off-site production or off-site construction", "industrialized building/housing/construction", "prefabrication, preassembly, modular and off-site fabrication", and "prefabrication construction" [4–6]. PHP is an emerging construction method in which building components can be produced in a controlled environment and assembled quickly on site. According to previous PHP practices in different regions in the world such as Sweden, Singapore, and Malaysia, additional complexity is introduced into PHP because of more requirements for cooperation and coordination [7–9]. In addition, PHP has a higher demand for information delivery across all stakeholders in the supply chain and is more complex in terms of project planning, organization, coordination, and communication [10]. Compared to the construction process of non-prefabricated houses, the organization of PHP is more complicated because of the inclusion of more new stakeholders, such as offshore manufacturers, transporters, and local authorities [11]. Consequently, these kinds of complexity will lead to inefficiencies and uncertainties in PHP and will present new challenges for the planning and control of projects [12]. Therefore, the planning and control of PHP become more difficult due to the increased challenges and complexity.

In addition to traditional project objectives (i.e., time, cost, quality, and safety), productivity and sustainability are project objectives of PHP [13]. This is due to the rapid urbanization process and the growing demand to achieve a resource-conscious and environmentally friendly society [14,15]. Due to available resources and various project features, PHP's performance is not always better than on-site construction methods. For example, previous studies have identified a large number of barriers to PHP practices (e.g., lack of policies and regulations, technical difficulties, high costs, and fragmented industry structure and supply chain) [5,16–18]. The majority of issues that lead to delay, poor quality, and cost overrun in PHP are in the construction stage where the process of installation and erection is carried out [19]. Stricter planning and control measures should be adopted in the design, components production, and assembly process in order to achieve these objectives These measures include fewer design changes, more effective materials logistics management, a higher degree of standardization, and a more reliable components production schedule [20–23]. If not employed appropriately, PHP may suffer from a series of issues such as production delays, substantial cost overruns, and order change [24,25].

A considerable number of studies have been conducted on project critical success factors (CSFs). However, there is no consensus regarding the factors that influence the success of PHP projects [26,27]. Previous studies have shown that project success factors are not common to all types of projects. Many researchers had identified CSFs that are specific to certain kinds of projects such as public–private partnerships (PPP) infrastructure projects, six sigma projects, international construction projects, and green building projects [28–31]. However, few studies have explored CSFs to improve their project planning and control (PP&C) outcomes in PHP. The study conducted by Ismail et al. (2012) on management factors for PHP is an exception. Only management-related factors were highlighted in Ismail's study [32]. Other factors such as supply chain-related factors, technical factors, and industry-related factors were largely overlooked. PHP has undergone rapid growth in China in recent years, thus providing the opportunity for a comprehensive investigation. As such, the overall aim of this study is to fill this research gap in the field of PHP by identifying CSFs. The specific objectives are to (1) identify the CSFs of PHP projects, (2) explore the underlying relationships among factors related to the successful implementation of PHP, and (3) provide a useful reference for key stakeholders of effective PP&C of PHP. The next section provides a literature review.

2. Literature Review

2.1. PHP in China

PHP has been gradually adopted in China since the middle 2000s [14,15,33]. This is due to common factors such as the sustainable development of the national economy, the growth of labor costs, and the increasing demand for sustainability. Over the past half-century, PHP in China went through various stages: an initial development stage (1950–1970s), an exploratory development stage (1980–early 2000s), and an expansion development stage (middle 2000s–now) [15]. With the growing demand for environmental protection and labor shortages in China, PHP has been developed and expanded gradually since the middle 2000s. Especially in the past three years, the government has released a lot of policies and initiatives in order to promote the adoption of PHP. An increasing number of developers such as Vanke Corporation, Beijing Uni-construction Real Estate Development Corporation, and Country Garden have entered the market. More than 100 manufacturing plants specifically designed for PHP have been invested in by companies such as the China Construction Science & Technology Group (Beijing, China), the China Mingsheng Drawing Technology Group (Changsha, China), and Yuhui Construction Corporation (Harbin, China). A variety of building systems such as "prefabricated shear wall structure systems" and "prefabricated reinforced concrete shear wall structural system" have been successfully implemented in China. Under the current situation, all the stakeholders in the industrial chain, such as the government, developers, designers, manufacturers, and general contractors, have started to devote enormous resources to prefabricated buildings. A huge number of PHP projects have started or have been implemented. Research into critical success factors for project planning and control in prefabrication housing production has become crucial. Although many scholars have fully explored CSFs in relation to construction projects, CSFs for PHP have not yet been systematically discussed. Therefore, this research plugs a gap.

2.2. PP&C and Project Success in PHP

Effective planning and control of construction activities are essential to achieving exceptional performance [34,35]. PP&C involve a systematic and iterative process of defining directives, executing and adjusting them according to project feedback [36]. Various project objectives such as delivery on time, keeping the project costs within budget, and meeting the quality requirements must be satisfied to deliver a successful project. Additional objectives such as safety, sustainability, and reliability are equally or even more important in PHP.

Various factors can affect the performance of construction projects. Previous studies have identified many critical success factors (CSFs) for construction projects. Chan et al. (2004) identified five groups of factors influencing the success of construction project implementation, namely project-related factors, project procedures, project management actions, human-related factors, and the external environment. Zwikael and Globerson (2006) explored the impact of 16 planning processes and identified the most sensitive processes for a successful project [37]. Ling et al. (2009) established 24 project practices that were significantly correlated with Singaporean firms' project performances in China, especially those relating to risk management [30]. Li et al. (2011) grouped the factors that affected the successful delivery of green building projects into five components, namely project manager's competence, technical and innovation-oriented factors, human resource–oriented factors, coordination of designers, and contractors' support from designers and senior management [31]. Liu et al. (2014) identified CSFs such as sound feasibility analysis, effective interface management, and effective conflict management that contribute to the success of public–private partnership (PPP) infrastructure projects in different phases. O'Connor et al. (2014) presented 21 CSFs including owner's planning resources, timely design freeze, capability of the fabricator, and heavy lifting equipment for the successful implementation of modularization in projects [38]. Heravi et al. (2015) explored the influence of project stakeholders and identified four critical stakeholder groups including the project owner, developer, designers, and contractors [27].

3. Research Methodology

3.1. Overall Research Framework

The research goal of this paper is to explore the CSFs of PHP. To achieve this purpose, this study follows the research framework proposed by Deng et al. [39] and Arif et al. [40]. Arif et al. identified 17 key variables related to political risk management of international construction companies through literature reviews and pilot studies. In order to research the rank and relationships of affecting factors, the factor analysis and average score methods were introduced in their study [40]. Deng et al. explored the factors that inhibit the promotion of SI system building by conducting an investigation through a questionnaire [39]. Then the mean score method was used to explore critical factors, and factor analysis was applied to explore the potential relationship between initial variables in the questionnaire. The research methodology of this study is based on a literature review, in-depth interviews, a questionnaire survey, ranking, and exploratory factor analysis (Figure 1). Statistical Product and Service Solutions software 19.0 provides professional factor analysis and in this study was also applied to conduct mean score ranking.

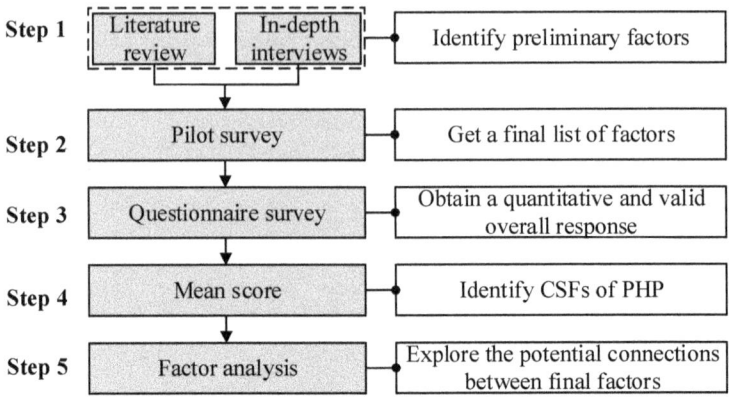

Figure 1. Research methodology flow chart.

3.2. Questionnaire

The preliminary list of variables was formulated based on (1) a literature review, and (2) in-depth interviews with PHP experts. In order to ensure the reliability of the survey, the research group selected five experts with long-term prefabricated construction experience in authoritative enterprises as respondents. The five experts cover all types of related companies. Before conducting a full investigation, a pilot study was conducted with selected experts in the field of construction management to verify the initial list of variables. Five experts who had over 10 years of working experience and participated in more than eight PHP projects were invited to revise the initial list of variables. The profiles of these experts are listed in Table 1. The questionnaire was refined based on the feedback received from the pilot survey. Finally, 23 variables that influence the implantation of PHP were obtained (Table 2).

Table 1. The profiles of the five experts.

Experts	Type	Company	Working Experience	Major	The Number of PHP Projects Participated In
1	Manufacturer	Company A: One of the earliest companies in China to manufacture the prefabricated components, it is very representative as a manufacturer	12	Civil engineering	12
2	Design	Company B: The Company B is one of the largest six design institutes in China and is authoritative in the design of prefabricated buildings.	15	Architecture	20
3	Contractor	Company C: A company with the highest qualification and level in this field of prefabricated construction	12	Civil engineering	10
4	consulting company	Company D: Company D specializes in the consulting of prefabricated housing production	15	Civil engineering	25
5	Contractor	Company C: A company with the highest qualification and level in this field of prefabricated construction	12	Civil engineering	8

Table 2. The preliminary list of 23 factors.

Code	Factors	Sources
F1	Well-developed specifications and regulations	[15,39]
F2	Persistent policies and incentives	[5,41,42]
F3	Sustainability request by the local government	[43,44]
F4	Difficulty to obtain planning permission by the local government	[11]
F5	Designers' experience of PHP	[26,45]
F6	Involvement of the designer during the production and construction stage	[46]
F7	Involvement of contractors and manufacturers during the design stage	[9]
F8	Project manager's proportion of time spent on planning and control	[27,47,48]
F9	Design processes management method	[11,49]
F10	Manufacturer's experience of PHP	[5,23]
F11	Sufficiency of manufacturers and suppliers of prefabricated components	[11,39]
F12	The quality management method of prefabricated components	[15,50]
F13	Rationality of the transportation method of prefabricated components	[11,51]
F14	The maturity of manufacture technology	[52]
F15	Skills and knowledge of labors	[15,17,39,53]
F16	Project manager's ability to solve problems	[54]
F17	Effective communication among participants	[9,55,56]
F18	Project manager's attitude towards planning and control	[27,47,56]
F19	Information sharing among participants	[10,57]
F20	The maturity of techniques used in the detailed design phase	[11,58]
F21	Efficient coordination between off-site and on-site	[59]
F22	Adoption of Information and communication technology (ICT) such as building information modeling (BIM), enterprise resource planning (ERP), and radio frequency identification (RFID)	[60–63]
F23	Rationality of the assembly planning method	[52,64]

The survey was conducted in March–May 2017. A snowball sampling technique was adopted because of the lack of sampling framework. A similar technique was used in studies by other scholars [5,39]. By means of the snowball sampling technique, the questionnaire can be shared through social networks by the initial respondents (e.g., 25 initial respondents from popular design, manufacturing, and construction firms in this research) to approach a wider range of respondents. This questionnaire was distributed to 400 professionals (designers, manufacturers, and contractors). They are the main actors in the supply chain of PHP and have a significant influence on PP&C of PHP [11,13]. A total of 136 valid responses were obtained, including 42 from designers, 43 from manufacturers, and 51 from contractors. Therefore, the response rate of this study is 34%, which was higher than in similar studies (28.2%) conducted by Yuan [65] and Liu et al. [66] (25.8%) in the building and construction industry in China. Additionally, this rate was higher than the average response rate, ranging from 20% to 30% in the construction industry [67,68]. Therefore, this sample was adequate for data analysis. The profile of the sample group indicated that 12.5% of the respondents were senior managers, 34.56% were middle managers, and 52.94% were engineers and technicians. The majority of the respondents (83.8%) have more than three years of experience in PHP. In addition, 36% of the respondents have more than five years of experience. Considering that China's prefabricated houses were built on a large scale after 2015, the survey participants are reliable.

In addition, the average years of working experience of the respondents was 5.4 years, which is higher than in a similar study conducted in Cao et al. [39] (2.4 years); the maximum years of experience in PHP of the respondents was 15. This result was acceptable as most of the PHPs in China were completed in the last 10 years, as stated by Zhang et al. [15]. This research result was consistent with previous studies (2014) as the precast concrete frame was the main form of PHP. This survey covers most of the regions in China that have developed PHP. The geographical distribution of the respondents is shown in Figure 2.

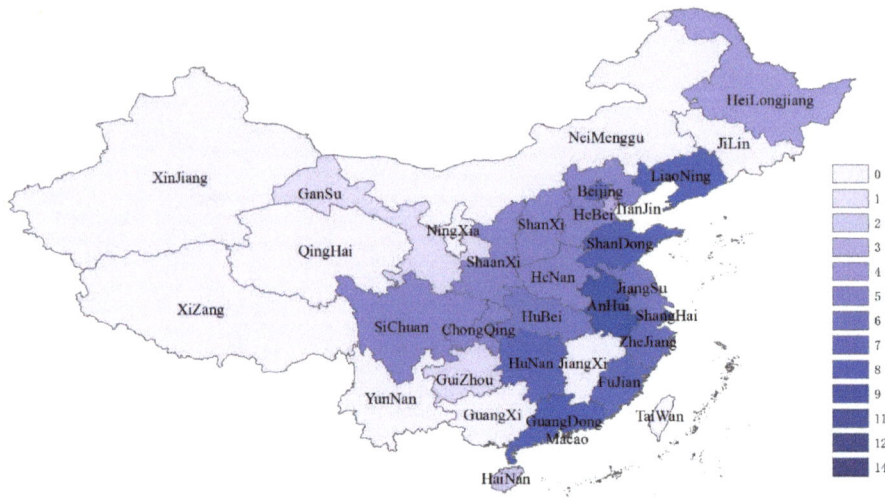

Figure 2. Geographical distribution of the respondents.

4. Data Analysis Method

In order to rank the importance of variables, the mean score (MS) method is introduced into this research. The five-point Likert scale (1 = Least important, 2 = Slightly important, 3 = Important, 4 = Very important, and 5 = Most important) was used to calculate the MS of each factor. To check whether respondents in three key stakeholders (designers, manufacturers, and contractors) gave the same ranking, the MS of the individual variable under the same category was analyzed. MS showed the relative importance of each factor. The MS method is usually considered as a research method to rank the relative importance of factors [40,66]. If the MS of several factors is exactly the same, the factor with the lower standard deviation (SD) is assigned a higher level. Furthermore, factor analysis, which is a statistical method to detect clusters of related variables [39], was used to group variables into a small number of underlying factors. In this study, factor analysis was used to explore the interrelation of 23 variables so that critical factors contributing to the implementation of PHP can be obtained. The frequencies of the responses and their percentages are shown in Table 3.

To test whether each variable was significantly important to the implementation of PHP, a one-sample t-test was conducted. Considering the five-point Likert scale, 3.00 is the mid value, that is, the test value [5]. As shown in Table 4, all 23 variables had significant importance, meaning that the p-values were below 0.05 and the mean scores were above 3.00. Several issues and tests have been widely considered and used in previous studies to determine whether each group of data is suitable for performing factor analysis [39,40]. In this study, the procedure recommended by Deng et al. was followed [40]. The ratio of the sample size to the number of variables is 6, which was higher than the ideal ratio of 5:1. So, for the factor analysis, the sample size is already sufficient. The Cronbach's alpha is 0.875, which is higher than the desired 0.80. This indicates that all data are reliable. The Kaiser–Meyer–Olkin (KMO) index should be greater than or equal to 0.5, while the Bartlett's test of sphericity ($p < 0.05$) should be used to verify whether factor analysis is suitable for data analysis.

Table 3. The frequencies of the responses and their percentages.

F	Least Important	Slightly Important	Important	Very Important	Most Important	Mean Value
F1	2 (1.47%)	7 (5.15%)	33 (24.26%)	66 (48.53%)	28 (20.59%)	3.82
F2	0 (0%)	3 (2.21%)	25 (18.38%)	67 (49.26%)	41 (30.15%)	4.07
F3	2 (1.47%)	7 (5.15%)	34 (25%)	61 (44.85%)	32 (23.53%)	3.84
F4	3 (2.21%)	6 (4.41%)	27 (19.85%)	73 (53.68%)	27 (19.85%)	3.85
F5	0 (0%)	1 (0.74%)	6 (4.41%)	36 (26.47%)	93 (68.38%)	4.63
F6	1 (0.74%)	1 (0.74%)	41 (30.15%)	63 (46.32%)	30 (22.06%)	3.88
F7	1 (0.74%)	2 (1.47%)	39 (28.68%)	58 (42.65%)	36 (26.47%)	3.93
F8	0 (0%)	1 (0.74%)	53 (38.97%)	59 (43.38%)	23 (16.91%)	3.76
F9	3 (2.21%)	8 (5.88%)	31 (22.79%)	65 (47.79%)	29 (21.32%)	3.8
F10	0 (0%)	4 (2.94%)	5 (3.68%)	44 (32.35%)	83 (61.03%)	4.51
F11	1 (0.74%)	6 (4.41%)	26 (19.12%)	77 (56.62%)	26 (19.12%)	3.89
F12	3 (2.21%)	17 (12.5%)	60 (44.12%)	32 (23.53%)	24 (17.65%)	3.42
F13	4 (2.94%)	7 (5.15%)	59 (43.38%)	40 (29.41%)	26 (19.12%)	3.57
F14	1 (0.74%)	9 (6.62%)	57 (41.91%)	55 (40.44%)	14 (10.29%)	3.53
F15	3 (2.21%)	13 (9.56%)	24 (17.65%)	29 (21.32%)	67 (49.26%)	4.06
F16	0 (0%)	2 (1.47%)	18 (13.24%)	77 (56.62%)	39 (28.68%)	4.13
F17	0 (0%)	1 (0.74%)	48 (35.29%)	59 (43.38%)	28 (20.59%)	3.84
F18	0 (0%)	0 (0%)	40 (29.41%)	67 (49.26%)	29 (21.32%)	3.92
F19	0 (0%)	3 (2.21%)	47 (34.56%)	58 (42.65%)	28 (20.59%)	3.82
F20	3 (2.21%)	5 (3.68%)	17 (12.5%)	61 (44.85%)	50 (36.76%)	4.1
F21	0 (0%)	2 (1.47%)	47 (34.56%)	54 (39.71%)	33 (24.26%)	3.87
F22	5 (3.68%)	9 (6.62%)	39 (28.68%)	65 (47.79%)	18 (13.24%)	3.6
F23	2 (1.47%)	7 (5.15%)	33 (24.26%)	66 (48.53%)	28 (20.59%)	3.82

Table 4. Results of MS method and factor analysis.

Code	Mean Value	SD	p-Value	Rank	Components				
					1	2	3	4	5
F23	3.897	0.913	<0.001 [a]	9	0.852	-	-	-	-
F14	3.566	0.956	<0.001 [a]	22	0.842	-	-	-	-
F22	3.603	0.929	<0.001 [a]	20	0.738	-	-	-	-
F20	4.103	0.913	<0.001 [a]	4	0.689	-	-	-	-
F12	3.419	0.993	<0.001 [a]	23	0.670	-	-	-	-
F9	3.801	0.917	<0.001 [a]	18	0.646	-	-	-	-
F13	3.581	0.970	<0.001 [a]	21	0.537	-	-	-	-
F6	3.882	0.780	<0.001 [a]	11	-	0.789	-	-	-
F7	3.926	0.822	<0.001 [a]	7	-	0.747	-	-	-
F19	3.816	0.781	<0.001 [a]	16	-	0.740	-	-	-
F21	3.868	0.796	<0.001 [a]	12	-	0.733	-	-	-
F17	3.838	0.752	<0.001 [a]	14	-	0.698	-	-	-
F3	3.838	0.896	<0.001 [a]	15	-	-	0.805	-	-
F11	3.890	0.786	<0.001 [a]	10	-	-	0.760	-	-
F4	3.846	0.868	<0.001 [a]	13	-	-	0.654	-	-
F1	3.816	0.871	<0.001 [a]	17	-	-	0.651	-	-
F2	4.074	0.757	<0.001 [a]	5	-	-	0.641	-	-
F5	4.625	0.608	<0.001 [a]	1	-	-	-	0.847	-
F10	4.515	0.710	<0.001 [a]	2	-	-	-	0.845	-
F15	4.059	1.121	<0.001 [a]	6	-	-	-	0.768	-
F16	4.132	0.686	<0.001 [a]	3	-	-	-	-	0.764
F18	3.919	0.633	<0.001 [a]	8	-	-	-	-	0.737
F8	3.765	0.733	<0.001 [a]	19	-	-	-	-	0.720
Cronbach alpha					0.883	0.803	0.824	0.798	0.708
Initial eigenvalues					6.567	3.861	2.160	1.464	1.405
Variance (%)					28.553	16.786	9.392	6.366	6.111
Cumulative variance (%)					28.553	45.339	54.732	61.097	67.208

[a] The one-sample t-test result is significant at the 0.05 level (two-tailed).

5. Research Findings and Discussion

5.1. Ranking of the Factors

This section aims to research those factors that have the greatest impact on the PP&C of PHP on the basis of the MS method. The mean values in Table 4 range from 3.419 for the factor quality of prefabricated components to 4.625 for the factor designers' experience of PHP. This result proved that all the respondents considered these 23 factors to be critical to PHP. As for the ranks of the 23 factors in three categories, as respondents in different stakeholders had different knowledge and roles, their ranks were not completely consistent. The top five factors according to the overall ranking are discussed further.

5.2. Designers' Experience of PHP

The most critical factor to the successful implementation of PHP is designers' experience of PHP (mean value = 4.625). Respondents in the three categories all agreed that this variable was the most significant. Designers' performance, from inception to completion, is critical to a successful project [26]. The design of PHP is very different from the conventional design. On the one hand, the design must be largely finished prior to the production. The fast production and assembly of building components need accurate design. On the other hand, the design changes can reduce the benefits of PHP because of the high rework cost, as indicated in a PHP project in Sweden [45]. Incompetent design may ultimately lead to production quality problems such as joint failure, poor thermal insulation, and water vapor penetration. In addition, PHP has a greater advantage in terms of component standardization and modularity [16]. If designers have adequate experience on the standardization of design, there will be an improvement in both the project constructability and the speed of construction.

The architectural design and the detailed design of prefabricated components in PHP projects are clearly separate in China. This is evidenced by the fact that the manufacturer is only a subcontractor. Designers' experience of PHP would have a significant influence on the subsequent detailed design of prefabricated components. As the detailed design takes inputs from customer requirements and architectural design, it is important for both manufacturing and on-site construction [61]. An experienced designer would consider requests from manufacturing and construction before going ahead. The design result completed by experienced designers would eventually bring value to the construction, e.g., fewer design errors, more efficient manufacturing productivity, and lower cost.

5.3. Manufacturers' Experience of PHP

Manufacturers' experience of PHP was ranked the second most important factor (mean value = 4.515). The ranking of this variable in the three categories was consistent. The importance of this factor to the implementation of PHP has been recognized in various PHP markets such as Singapore, Taiwan, the USA, and Turkey. The manufacturing phase is an extra stage compared with the traditional construction method [18,51]. However, in terms of prefabricated structural components such as precast walls, the advantage of quick installation would be undermined by work delays caused by poor management of manufacturing. It has been recognized in PHP projects in Singapore that late delivery by the precast components' manufacturers is the most common issue for the main contractors [59]. The knowledge and experience of production managers are crucial to consider when making production plans to achieve on-time delivery [69]. For projects located in the downtown, the potential delays caused by traffic congestion and strict size and load restrictions on transportation [70] must be considered when delivering prefabricated components from manufacturers' plants to the construction site.

The quality of the prefabricated components significantly influenced the installation productivity. Physical damage to the components frequently occurred, especially during the storage and transportation process. Without rich experience in practice, physical damage to the components, such as corners and broken ribs, would happen without using appropriate battens during stacking.

Moreover, the probability of damage to finished components increases when conducting loading and unloading tasks [71].

5.4. Project Manager's Ability to Solve Problems

The project manager's ability to solve problems was ranked the third most important factor (mean value = 4.132). This result is consistent with the findings of Jabar et al. (2013a), whose study was based on a survey of the Malaysia PHP market [54]. As the problem-solving skill was the most important competency of a project manager apart from technical knowledge. Only the ranking of this variable in the designers' group was not consistent. "F6: Involvement of the designer during the production and construction stage" was ranked third by the respondents in the designers group. This result indicates that the respondents in the designers group are more concerned with the association between designers and subsequent activities. The project manager is the person in the PHP project who coordinates on-site and off-site activities and is responsible for the project objectives. The time taken for vertical transportation of prefabricated components depends on the weight and size of the prefabricated components and the loading capacity of available hoists and cranes [17]. In this circumstance, the project manager needs to reduce the duration of the subsequent activities to control the overall construction schedule.

5.5. The Maturity of Techniques Used in the Detailed Design Phase

The maturity of techniques used in the detailed design phase is ranked the fourth most important factor (mean value = 4.103). The ranking by manufacturers and contractors is consistent with the overall ranking, as their activities are all affected by the upstream stakeholder. The detailed design is a multi-disciplinary design that includes assembly design and analysis, mold design, and piece and connection design. The detailed design phase is essential in PHP as the role of detailed design is to transform construction drawings into assembly drawings, in which the dimension of each component and the connection method are labeled. In China, the detailed design process is very time-consuming as the detailed design process is based on two-dimensional computer-aided design (CAD) drawings. Taking the external wall as an example, this process is dependent on the designer's knowledge to determine which part of the wall can be prefabricated. Then the detailed assembly drawings of the wall are produced based on the location of the wall, the design specification, and the production constraints. If there are design changes, the detailed design needs to be reprocessed to keep consistent with the overall design of the building.

Sacks et al. (2004) had proposed that the design and detailing tasks of precast concrete should be automated [72]. With the development of three-dimensional (3D) modeling software and building information modeling (BIM) technology, the way building information is represented and managed has the potential to be revolutionized [52]. Although the value of BIM has been recognized by designers in the construction industry, the applications of BIM are limited to visualization, collision detection, and construction simulation [60,73–75]. In terms of the detailed design of a PHP project, numerous limitations related to the information exchange in both geometric shape information and other semantically meaningful information between architects (the Industry Foundation Classes (IFC) format) and fabricators (the Standard ACIS Text (SAT) format) have impeded the BIM technology in the detailed design phase [58]. The data interoperability between the software widely used by designers (e.g., Autodesk Revit, Tekla Structures, ArchiCAD, and Graphisoft) in the construction industry and the software that was popular in manufacturing (e.g., Catia, Solidworks, and Unigraphics (UG)) has not yet been achieved.

5.6. Persistent Policies and Incentives

The persistent incentive policies factor took fifth place (mean value = 4.074). The ranking of this variable in the three categories is consistent as the policies have a common influence on all stakeholders. The adoption of prefabrication involves capital investment and technology innovation when no mature technical systems exist, especially in developing countries such as China and Malaysia. When the PHP is still in its infancy, contractors are more willing to choose a mature method rather than new ones without incentives from local government [39]. In European countries, China, and Australia, the government plays a crucial role in the construction industry. Government policies should be favorable to the prefabrication initiative in order to increase the adoption of prefabrication [41,76]. In the last five years, a growing number of contractors in China have invested in precast concrete plants due to the government's preferential policies. It is apparent that persistent incentive policies have promoted the development of PHP in China.

5.7. Exploratory Factor Analysis

There are three steps in factor analysis: a test for suitability of data, factor extraction, and factor rotation. The value of KMO of this study is 0.806, which is higher than the minimum desirable value of 0.5, and Bartlett's test of sphericity was statistically significant ($x^2 = 1733.497$, $df = 253$, $p < 0.001 < 0.05$). This result proved that factor analysis can be used in this study.

This study uses a combination of varimax rotation method and principal component analysis to analyze all 23 factors. After that, five clusters with eigenvalues greater than 1 were extracted, accounting for 67.208% of the variance (Table 4). Each of the 23 CSFs belongs to only one cluster, with a factor loading value greater than 0.5. The number of initial variables in all five clusters is greater than or equal than three. Therefore, the factors in each cluster can accurately reflect the features [39].

All the affecting factors are divided into five clusters through principal component analysis. The five clusters can be labeled as (1) technology and method; (2) information, communication, and collaboration; (3) the external environment; (4) experience and knowledge; and (5) the competence of the project manager.

5.7.1. Cluster 1: Technology and Method

The cluster "technology and method" consists of seven CSFs, namely (1) the rationality of the assembly planning method; (2) the maturity of the manufacturing technology; (3) the adoption of information communication technology (ICT) such as BIM, enterprise resource planning (ERP), and radio frequency identification (RFID); (4) the maturity of the techniques used in the detailed design phase; (5) the quality management method of prefabricated components; (6) the design processes management method; and (7) the rationality of the transportation method of prefabricated components. This cluster reveals 28.553% of the total variables.

This cluster involves factors related to technologies and methods used in different stages of a PHP project such as design, manufacture, transportation, and assembly. As the design stage accounts for more than 70% of the cost and has a significant influence on a project, it is the earliest stage at which new technologies have been applied. Accurate design is required in the PHP to avoid design changes in the production and construction stage. There is a need for technological enhancement in the design, manufacturing, and construction process, especially for improvement in the design and construction technology. As for the design technology, some Chinese design firms are still based on two-dimensional (2D) application tools such as AutoCAD for drafting, while some design firms are trying to transition to a BIM platform such as Autodesk Revit. In addition, during the production process, BIM can be combined with an automatic production line by providing component dimensional information for the detailed component diagram. The automatic production line uses a computer-aided design to computer-aided manufacturing (CAD-CAM) controlled concrete distributor to spread the right amount of concrete according to CAD and robots to place the mold and reinforcement [77].

It is recognized that design automation dramatically improves the productivity of the PHP project as a variety of design rules and constraints were imposed by the manufacturer and the client. The configuration system proposed by Jensen et al. [78] can support and facilitate the design automation by adding rules in SolidWorks (a manufacturing CAD tool) and sharing information through Extensible Markup Language (XML) with Autodesk Revit. Design processes management methods are also important to PHP. One well-known design processes management method is the design structure matrix (DSM) method, which represents the complex design processes in a matrix form. DSM can expose the driver of rework clearly and early to avoid potential risks during project planning [79]. The sequence deviation quotient (SDQ) is another design processes management method proposed by Haller et al. to detect and control superfluous iteration as well as control the risk of time overruns and quality issues within the design phase of PHP [45].

The assembly planning has influences on the manufacturing process as the manufacturer must deliver the prefabricated components to the construction site prior to assembly. Push and pull are two familiar planning methods. In contrast with the traditional planning method, which is "push"-based, lean construction advocates for a planning method that is "pull"-based [64]. Most contractors in China adopted the push method. Consequently, all the prefabricated components have been produced beforehand and delivered to the construction site. As the prefabricated components such as precast concrete components are usually bulky, they will block the construction site [59]. Meanwhile, it is difficult to search for the appropriate component from the storage stacks, which increases the assembly cost. When using the pull method, the components are delivered to the construction site only if the contractor needs it.

The manufacturing technology involves mechanization and automation. The level of mechanization is higher than the level of automation in China. Investment in various machines (e.g., casting machines that pour the concrete onto tables with high accuracy, tilting tables designed for the manufacture of large reinforced concrete elements, and floating machines to smooth the surface finish of precast products) in the precast concrete plant improves productivity.

Although the level of mechanization is relatively high in China, sometimes the quality of prefabricated components cannot satisfy the requirement of customers. Low precision of junction and broken nibs are the two main quality issues. Such quality issues often caused repair or rework. As these issues were detected at the construction site, the resulting costs are higher than be detected in the plant. Six Sigma theory and total quality management have been used in the manufacturing industry to improve the quality of products [80]. The precast concrete firms should use these quality management methods to avoid these quality issues.

The transportation cost of precast components is jointly determined by the number of truckloads used in the delivery process and the unit cost of delivery. The weight and size of the precast concrete components to be transported mainly affect the number of truckloads. The unit cost of delivery is directly affected by the distance between the manufacturer's plant and the construction site [16].

5.7.2. Cluster 2: Information, Communication, and Collaboration

The "information, communication, and collaboration" cluster consists of five CSFs: (1) involvement of the designer during the production and construction stage, (2) involvement of contractors and manufacturers during the design stage, (3) information sharing among participants, (4) efficient coordination between off-site and on-site, and (5) effective communication among participants. This cluster reveals 16.786% of the total variables.

The information required in the management of PHP includes materials, prefabricated components, quality inspection, inventory, and transportation. Traditionally, the information was recorded in notebooks. Data will be entered into the computer for control processing until engineers return to the office. There are spatial and time gaps between the plant (on-site) and the office, which increases the difficulty of communication [81].

The precast concrete systems have two major advantages, namely low cost and speedy erection. Only when good coordination is achieved among all the key stakeholders involved in a project can problems such as delays in production, erection schedules, and constructability be avoided [16]. A huge amount of rework may also be needed in the design stage due to the inefficiency of communication between the designer and manufacturer. Moreover, the contractor's good communication with the designer and the manufacturer is also essential to the success of the project in the erection stage [16].

Prefabrication also required advanced coordination between professionals [51]. Close coordination between prefabrication and construction processes is needed as these two groups of activities run in parallel [61]. Stakeholders in the supply chain should communicate with each other as early as possible in order to improve business efficiency [4].

5.7.3. Cluster 3: External Environment

The cluster "external environment" consists of five CSFs: (1) sustainability request by the local government, (2) sufficiency of manufacturers and suppliers of prefabricated components, (3) difficulty of obtaining planning permission from the local government, (4) well-developed specifications and regulations, and (5) persistent policies and incentives. This cluster reveals 9.392% of the total variables. Various studies suggest that the environment includes many aspects such as economy, politics, society, and industry [26,30] as external factors affecting the project's success. The factors belonging to this "external environment" can be classified into political environment and industrial relation environment.

Well-developed specifications and regulations are also key to the successful implementation of PHP [5]. The specifications and regulations provide the basis for the design of prefabricated buildings, prefabricated components production, quality checking, and evaluation. In the last three years, these specifications and regulations have steadily improved in China. One of the important reasons to promote PHP is the requirement for environmental protection and energy savings [14,82,83], as sustainability is a long-term goal of rapid urbanization. Although some regions in China have created incentives such as tax exemptions and rewards, the process of obtaining planning permission is slow. The design and construction planning must be approved by several agencies.

Designers, manufacturers, and contractors involved in the PHP are the three main stakeholders in the supply chain [11,13]. For the contractor, the manufacturer is the supplier that provides prefabricated components. Therefore, the supplier selection process is important for the planning and control of appropriate components [84]. The manufacturer also needs to order materials such as cement, reinforcement, molds, insulating panels, and concrete admixtures from other upstream suppliers. The precast fabricators must preorder materials before confirming an order to reduce the production time span and decrease the risk of late delivery [22]. If the qualified precast manufacturers are few, a lack of serious competition may lead to higher prices. In the context of China's new urbanization, government incentives of have been the driving force for the advantages of PHP [85]. Government policies have affected market demand and supply as well as the technological update.

5.7.4. Cluster 4: Experience and Knowledge

The "experience and knowledge" cluster consists of three CSFs: (1) designers' experience of PHP, (2) manufacturers' experience of PHP, and (3) skills and knowledge of laborers. This cluster reveals 6.366% of the total variables.

The manufacturers' performance is crucial to the successful promotion of PHP. One of the main factors that hinder performance improvement is a lack of expertise and experience, which may lead to poor design and practices [16]. The above factor may eliminate the advantages of PHP such as predictable schedules. For example, a lack of expertise in components design may cause severe conflicts between manufacturers and designers; a lack of manufacturing experience can cause delays in the flow of deliveries to the construction site; a lack of competence on the part of the contractor can lead to delays in the installation schedule [86]. Furthermore, the on-site assembly and joining of prefabricated components require skilled workers, especially those with machine-oriented skills, both on-site and in the factory [17]. The transition from on-site construction to prefabrication construction requires workers to master new knowledge related to machine operation and maintenance.

5.7.5. Cluster 5: Competence of the Project Manager

The cluster "competence of the project manager" consists of three CSFs, namely (1) the project manager's ability to solve problems, (2) the project manager's attitude towards planning and control, and (3) the project manager's proportion of time spent on planning and control. This cluster accounts for 6.111% of the total variance among all the critical factors (Table 4).

A critical factor affecting communication and planning is the competence of the project manager [26,87]. The conception of competence includes not only knowledge and skills, but also attitudes, behaviors, and work habits [87]. In a PHP project, the project manager needs to coordinate and communicate with relevant stakeholders to achieve the objectives. The project manager should have problem-solving competence, which means being able to identify, analyze, and solve problems [29]. The project manager should invest adequate time to ensure that activities are executed according to the plan or make the plan reliable.

6. Conclusions

The promotion of PHP is complex as it involves a variety of stakeholders as well as activities executed in different locations. These kinds of complexity present significant challenges for PP&C of PHP projects. This study provides a comprehensive list of factors that affect the successful implementation of PHP, based on in-depth interviews, a literature review, and the questionnaire method. This study identified 23 CSFs using the mean score method. The relationships between these critical factors were examined and analyzed by means of factor analysis. Although empirical evidence for this study is from the Chinese PHP market, the methodology derived from this study may provide a reference for similar studies in other PHP markets, according to the characteristics and situations faced in these different environments. Hence, this study contributes to the existing body of knowledge via a holistic approach covering the key actors of PHP such as designers, manufacturers, and contractors and the factors that inhibit the promotion of PHP in the broader global community. The findings of this study can provide stakeholders involved in PHP with a useful set of criteria. Also, the findings would enable PHP practitioners to possess a deeper understanding of the factors that are critical to successful implementation of PHP.

This study identified 23 factors and ranked them in terms of relative importance. All 23 factors were critical as they all have mean scores above 3. The top five CSFs are designers' experience of PHP, manufacturers' experience of PHP, project manager's ability to solve problems, the maturity of the techniques used in the detailed design phase, and consistent policies and incentives. Although these five CSFs were identified based on the Chinese PHP market, significant influences on the implementation of PHP have also been recognized in other countries such as the USA, Australia, Sweden, Turkey, and Malaysia. Factor analysis was used to determine the main factors that affect the PP&C of PHP. The results revealed five clusters that account for 67.208% of the overall factors. The five clusters are (1) technology and method; (2) information, communication, and collaboration; (3) the external environment; (4) experience and knowledge; and (5) the competence of the project manager.

The primary limitation of this study is that not every form of PHP practices was covered. The survey sample mainly covered precast concrete frames as these are the main form of PHP practices in China in recent years. The form of PHP practices may be different according to the characteristics of different regions. Thus, future investigations of various PHP practices in different regions or different stages of development of PHP should be considered in future research.

Acknowledgments: This work was supported by the National Natural Science Foundation of China under grant nos. 71371041, 71561009).

Author Contributions: Long Li conceived the research and completed the research paper in English; Zhongfu Li and Xiaodan Li participated in designing the research and drafting the article; Guangdong Wu revised it critically for important intellectual content.

Conflicts of Interest: The authors declare no conflicts of interest.

References

1. Pons, O.; Wadel, G. Environmental impacts of prefabricated school buildings in Catalonia. *Habitat Int.* **2011**, *35*, 553–563. [CrossRef]
2. Nadim, W.; Goulding, J.S. Offsite production in the UK: The way forward? A UK construction industry perspective. *Constr. Innov.* **2010**, *10*, 181–202. [CrossRef]
3. Li, Z.; Shen, G.Q.; Xue, X. Critical review of the research on the management of prefabricated construction. *Habitat Int.* **2014**, *43*, 240–249. [CrossRef]
4. Pan, W.; Gibb, A.G.F.; Dainty, A.R.J.; Asce, M. Strategies for Integrating the Use of Off-Site Production Technologies in House Building. *J. Constr. Eng. Manag.* **2012**, *138*, 1331–1340. [CrossRef]
5. Mao, C.; Shen, Q.; Pan, W.; Ye, K. Major Barriers to Off-Site Construction: The Developer's Perspective in China. *J. Manag. Eng.* **2015**, *31*, 4014043. [CrossRef]
6. Malmgren, L. *Industrialized Construction—Explorations of Current Practice and Opportunities*; Lund University Publications: Lund, Sweden, 2014; ISBN 9789197954389.
7. Höök, M.; Stehn, L. Connecting lean construction to prefabrication complexity in Swedish volume element housing. In Proceedings of the 13th Annual Conference International Group Lean Construction, Sydney, Australia, 19–21 July 2005; pp. 317–325.
8. Pheng, L.S.; Chuan, C.J. Just-in-time management in precast concrete construction: A survey of the readiness of main contractors in Singapore. *Integr. Manuf. Syst.* **2001**, *12*, 416–429. [CrossRef]
9. Ismail, F.; Yusuwan, N.M.; Baharuddin, H.E.A. Management Factors for Successful IBS Projects Implementation. *Procedia Soc. Behav. Sci.* **2012**, *68*, 99–107. [CrossRef]
10. Ergen, E.; Akinci, B. Formalization of the Flow of Component-Related Information in Precast Concrete Supply Chains. *J. Constr. Eng. Manag.* **2008**, *134*, 112–121. [CrossRef]
11. Li, C.Z.; Hong, J.; Xue, F.; Shen, G.Q.; Xu, X.; Mok, M.K. Schedule risks in prefabrication housing production in Hong Kong: A social network analysis. *J. Clean. Prod.* **2016**, *134*, 482–494. [CrossRef]
12. Park, M.; Pena-Mora, F. Dynamic Planning and Control of Construction Projects. *Comput. Civ. Build. Eng.* **2000**, 930–939. [CrossRef]
13. Teng, Y.; Mao, C.; Liu, G.; Wang, X. Analysis of stakeholder relationships in the industry chain of industrialized building in China. *J. Clean. Prod.* **2017**, *152*, 387–398. [CrossRef]
14. Cao, X.; Li, X.; Zhu, Y.; Zhang, Z. A comparative study of environmental performance between prefabricated and traditional residential buildings in China. *J. Clean. Prod.* **2014**, *109*, 131–143. [CrossRef]
15. Zhang, X.; Skitmore, M.; Peng, Y. Exploring the challenges to industrialized residential building in China. *Habitat Int.* **2014**, *41*, 176–184. [CrossRef]
16. Polat, G. Factors affecting the use of precast concrete systems in the United States. *J. Constr. Eng. Manag.* **2008**, *134*, 169–178. [CrossRef]
17. Chiang, Y.H.; Chan, E.H.W.; Lok, L.K.L. Prefabrication and barriers to entry-a case study of public housing and institutional buildings in Hong Kong. *Habitat Int.* **2006**, *30*, 482–499. [CrossRef]
18. Pan, W.; Gibb, A.G.F.; Dainty, A.R.J. *Leading UK Housebuilders' Utilisation of Offsite Modern Methods of Construction*; Taylor & Francis: Abingdon, UK, 2008; Volume 36, ISBN 0961321070120.
19. Jabar, I.L.; Ismail, F.; Mustafa, A.A. Issues in Managing Construction Phase of IBS Projects. *Procedia Soc. Behav. Sci.* **2013**, *101*, 81–89. [CrossRef]
20. Ajayi, S.O.; Oyedele, L.O.; Bilal, M.; Akinade, O.O.; Alaka, H.A.; Owolabi, H.A. Critical management practices influencing on-site waste minimization in construction projects. *Waste Manag.* **2017**, *59*, 330–339. [CrossRef] [PubMed]
21. Khalili, A.; Chua, D.K. Integrated prefabrication configuration and component grouping for resource optimization of precast production. *J. Constr. Eng. Manag.* **2014**, *140*. [CrossRef]
22. Ko, C.-H. Material Transshipment for Precast Fabrication. *J. Civ. Eng. Manag.* **2013**, *19*, 335–347. [CrossRef]
23. Wang, C.; Liu, M.; Asce, A.M.; Hsiang, S.M.; Leming, M.L.; Asce, M. Causes and Penalties of Variation: Case Study of a Precast Concrete Slab Production Facility. *J. Constr. Eng. Manag.* **2013**, *138*, 1–12. [CrossRef]
24. Chen, Y.; Okudan, G.E.; Riley, D.R. Decision support for construction method selection in concrete buildings: Prefabrication adoption and optimization. *Autom. Constr.* **2010**, *19*, 665–675. [CrossRef]
25. Pan, W.; Dainty, A.R.J.; Gibb, A.G.F. Establishing and Weighting Decision Criteria for Building System Selection in Housing Construction. *J. Constr. Eng. Manag.* **2012**, *138*, 1239–1250. [CrossRef]

26. Chan, A.; Scott, D.; Chan, A. Factors affecting the success of a construction project. *J. Constr. Eng. Manag.* **2004**, *130*, 153–155. [CrossRef]
27. Heravi, A.; Coffey, V.; Trigunarsyah, B. Evaluating the level of stakeholder involvement during the project planning processes of building projects. *Int. J. Proj. Manag.* **2015**, *33*, 985–997. [CrossRef]
28. Liu, J.; Love, P.E.D.; Smith, J.; Regan, M. Life Cycle Critical Success Factors for Public-Private Partnership Infrastructure Projects. *J. Manag. Eng.* **2015**, *31*, 1–7. [CrossRef]
29. Marzagão, D.S.L.; Carvalho, M.M. Critical success factors for Six Sigma projects. *Int. J. Proj. Manag.* **2016**, *34*, 1505–1518. [CrossRef]
30. Ling, F.Y.Y.; Low, S.P.; Wang, S.Q.; Lim, H.H. Key project management practices affecting Singaporean firms' project performance in China. *Int. J. Proj. Manag.* **2009**, *27*, 59–71. [CrossRef]
31. Li, Y.Y.; Chen, P.-H.; Chew, D.A.S.; Teo, C.C.; Ding, R.G. Exploration of critical external partners of architecture/engineering/construction (AEC) firms for delivering green building projects in Singapore. *J. Constr. Eng. Manag.* **2011**, *7*, 193–209. [CrossRef]
32. Rahim, A.A.; Hamid, Z.A.; Zen, I.H.; Ismail, Z.; Kamar, K.A.M. Adaptable Housing of Precast Panel System in Malaysia. *Procedia Soc. Behav. Sci.* **2012**, *50*, 369–382. [CrossRef]
33. Xue, X.; Zhang, X.; Wang, L.; Skitmore, M.; Wang, Q. Analyzing collaborative relationships among industrialized construction technology innovation organizations: A combined SNA and SEM approach. *J. Clean. Prod.* **2016**. [CrossRef]
34. Rodrigues, J.S.; Costa, A.R.; Gestoso, C.G. Project Planning and Control: Does National Culture Influence Project Success? *Procedia Technol.* **2014**, *16*, 1047–1056. [CrossRef]
35. Thomas, M.; Jacques, P.H.; Adams, J.R.; Kihneman-Wooten, J. Developing an effective project: Planning and team building combined. *Proj. Manag. J.* **2008**, *39*, 105–113. [CrossRef]
36. Son, J.; Rojas, E.M. Impact of Optimism Bias Regarding Organizational Dynamics on Project Planning and Control. *J. Constr. Eng. Manag.* **2011**, *137*, 147–157. [CrossRef]
37. Zwikael, O.; Globerson, S. From Critical Success Factors to Critical Success Processes. *Int. J. Prod. Res.* **2006**, *44*, 3433–3449. [CrossRef]
38. O'Connor, J.T.; O'Brien, W.J.; Choi, J.O. Critical success factors and Eenablers for optimum and maximum industrial modularization. *J. Constr. Eng. Manag.* **2014**, *140*, 1–11. [CrossRef]
39. Cao, X.; Li, Z.; Liu, S. Study on factors that inhibit the promotion of SI housing system in China. *Energy Build.* **2015**, *88*, 384–394. [CrossRef]
40. Deng, X.; Lw, S.P.; Li, Q.; Zhao, X. Developing Competitive Advantages in Political Risk Management for International Construction Enterprises. *J. Constr. Eng. Manag.* **2014**, *140*, 1–10. [CrossRef]
41. Arif, M.; Egbu, C.; Mohammed, A.; Egbu, C. Making a case for offsite construction in China. *Eng. Constr. Archit. Manag.* **2010**, *17*, 536–548. [CrossRef]
42. Azam Haron, N.; Abdul-Rahman, H.; Wang, C.; Wood, L.C. Quality function deployment modelling to enhance industrialised building system adoption in housing projects. *Total Qual. Manag. Bus. Excell.* **2015**, *26*, 703–718. [CrossRef]
43. Lu, W.; Yuan, H. Investigating waste reduction potential in the upstream processes of offshore prefabrication construction. *Renew. Sustain. Energy Rev.* **2013**, *28*, 804–811. [CrossRef]
44. Wu, P.; Low, S.P.; Jin, X. Identification of non-value adding (NVA) activities in precast concrete installation sites to achieve low-carbon installation. *Resour. Conserv. Recycl.* **2013**, *81*, 60–70. [CrossRef]
45. Haller, M.; Lu, W.; Stehn, L.; Jansson, G. An indicator for superfluous iteration in offsite building design processes. *Archit. Eng. Des. Manag.* **2015**, *11*, 360–375. [CrossRef]
46. Nadim, W.; Goulding, J.S. Offsite production: A model for building down barriers: A European construction industry perspective. *Eng. Constr. Archit. Manag.* **2011**, *18*, 82–101. [CrossRef]
47. Faniran, O.O.; Oluwoye, J.O.; Lenard, D.J. Interaction between construction planning and influence factors. *J. Constr. Eng. Manag.* **1998**, *124*, 245–256. [CrossRef]
48. Cheng, Y.M. An exploration into cost-influencing factors on construction projects. *Int. J. Proj. Manag.* **2014**, *32*, 850–860. [CrossRef]
49. Olawale, Y.; Sun, M. PCIM: Project Control and Inhibiting-Factors Management Model. *J. Manag. Eng.* **2013**, *29*, 60–70. [CrossRef]

50. Kim, M.-K.; Cheng, J.C.P.; Sohn, H.; Chang, C.-C. A framework for dimensional and surface quality assessment of precast concrete elements using BIM and 3D laser scanning. *Autom. Constr.* **2015**, *49*, 225–238. [CrossRef]
51. Jaillon, L.; Poon, C.S. The evolution of prefabricated residential building systems in Hong Kong: A review of the public and the private sector. *Autom. Constr.* **2009**, *18*, 239–248. [CrossRef]
52. Goulding, J.S.; Pour Rahimian, F.; Arif, M.; Sharp, M.D. New offsite production and business models in construction: Priorities for the future research agenda. *Archit. Eng. Des. Manag.* **2015**, *11*, 163–184. [CrossRef]
53. Tam, V.W.Y.; Tam, C.M.; Zeng, S.X.; Ng, W.C.Y. Towards adoption of prefabrication in construction. *Build. Environ.* **2007**, *42*, 3642–3654. [CrossRef]
54. Jabar, I.L.; Ismail, F.; Aziz, N.M.; Janipha, N.A.I. Construction Manager's Competency in Managing the Construction Process of IBS Projects. *Procedia Soc. Behav. Sci.* **2013**, *105*, 85–93. [CrossRef]
55. Xue, X.; Shen, Q.; Ren, Z. Critical review of collaborative working in construction projects: Business environment and human behaviors. *J. Manag. Eng.* **2010**, *26*, 196–208. [CrossRef]
56. Puddicombe, M.S. The Limitations of Planning: The Importance of Learning. *J. Constr. Eng. Manag.* **2006**, *132*, 949–955. [CrossRef]
57. Zhao, X.; Hwang, B.-G.; Low, S.P. Critical success factors for enterprise risk management in Chinese construction companies. *Constr. Manag. Econ.* **2013**, *31*, 1199–1214. [CrossRef]
58. Jeong, Y.-S.; Eastman, C.M.; Sacks, R.; Kaner, I. Benchmark tests for BIM data exchanges of precast concrete. *Autom. Constr.* **2009**, *18*, 469–484. [CrossRef]
59. Pheng, L.S.; Chuan, C.J. Just-in-Time Management of Precast Concrete Components. *J. Constr. Eng. Manag.* **2001**, *127*, 494–501. [CrossRef]
60. Liu, H.; Al-Hussein, M.; Lu, M. BIM-based integrated approach for detailed construction scheduling under resource constraints. *Autom. Constr.* **2015**, *53*, 29–43. [CrossRef]
61. Babič, N.C.; Podbreznik, P.; Rebolj, D. Integrating resource production and construction using BIM. *Autom. Constr.* **2010**, *19*, 539–543. [CrossRef]
62. Lu, Y.; Li, Y.; Skibniewski, M.J.; Wu, Z.; Wang, R.; Le, Y. Information and Communication Technology Applications in Architecture, Engineering, and Construction Organizations: A 15-Year Review. *J. Manag. Eng.* **2014**, *31*, 1–19. [CrossRef]
63. Ergen, E.; Akinci, B.; Sacks, R. Tracking and locating components in a precast storage yard utilizing radio frequency identification technology and GPS. *Autom. Constr.* **2007**, *16*, 354–367. [CrossRef]
64. Tommelein, I.D. Journey toward Lean Construction: Pursuing a Paradigm Shift in the AEC Industry. *J. Constr. Eng. Manag.* **2015**, *141*, 1–12. [CrossRef]
65. Yuan, H. Critical Management Measures Contributing to Construction Waste Management: Evidence From Construction Projects in China. *Proj. Manag. J.* **2013**, *80*, 39. [CrossRef]
66. Liu, J.; Zhao, X.; Li, Y. Exploring the Factors Inducing Contractors' Unethical Behavior: Case of China. *J. Prof. Issues Eng. Educ. Pract.* **2017**, *143*, 4016023. [CrossRef]
67. Akintoye, A. Analysis of factors influencing project cost estimating practice. *Constr. Manag. Econ.* **2000**, *18*, 77–89. [CrossRef]
68. Hwang, B.; Zhao, X.; Ong, S. Value Management in Singaporean Building Projects: Implementation Status, Critical Success Factors, and Risk Factors. *J. Manag. Eng.* **2014**, *31*, 4014094. [CrossRef]
69. Ko, C. An integrated framework for reducing precast fabrication inventory. *J. Civ. Eng. Manag.* **2010**, *16*, 418–427. [CrossRef]
70. Polat, G. Precast concrete systems in developing vs. industrialized countries. *J. Civ. Eng. Manag.* **2010**, *16*, 85–94. [CrossRef]
71. Wu, P.; Low, S.P. Barriers to achieving green precast concrete stock management—A survey of current stock management practices in Singapore. *Int. J. Constr. Manag.* **2014**, *14*, 78–89. [CrossRef]
72. Sacks, R.; Eastman, C.M.; Lee, G. Parametric 3D modeling in building construction with examples from precast concrete. *Autom. Constr.* **2004**, *13*, 291–312. [CrossRef]
73. Hu, Z.; Zhang, J.; Yu, F.; Tian, P.; Xiang, X. Construction and facility management of large MEP projects using a multi-Scale building information model. *Adv. Eng. Softw.* **2016**, *100*, 215–230. [CrossRef]
74. Manrique, J.D.; Al-Hussein, M.; Telyas, A.; Funston, G. Constructing a Complex Precast Tilt-Up-Panel Structure Utilizing an Optimization Model, 3D CAD, and Animation. *J. Constr. Eng. Manag.* **2007**, *133*, 199–207. [CrossRef]

75. Zhao, X. A scientometric review of global BIM research: Analysis and visualization. *Autom. Constr.* **2017**, *80*, 37–47. [CrossRef]
76. Mohsin, W. *Offsite Manufacturing as a Means of Improving Productivity in New Zealand Construction Industry: Key barriers to Adoption and Improvement Measures*; Massey University: Palmerston North, New Zealand, 2011.
77. Vähä, P.; Heikkilä, T.; Kilpeläinen, P.; Järviluoma, M.; Gambao, E. Extending automation of building construction—Survey on potential sensor technologies and robotic applications. *Autom. Constr.* **2013**, *36*, 168–178. [CrossRef]
78. Jensen, P.; Olofsson, T.; Johnsson, H. Configuration through the parameterization of building components. *Autom. Constr.* **2012**, *23*, 1–8. [CrossRef]
79. Danilovic, M.; Browning, T.R. Managing complex product development projects with design structure matrices and domain mapping matrices. *Int. J. Proj. Manag.* **2007**, *25*, 300–314. [CrossRef]
80. Tchidi, M.F.; He, Z.; Li, Y.B. Process and Quality Improvement Using Six Sigma in Construction Industry. *J. Civ. Eng. Manag.* **2012**, *18*, 158–172. [CrossRef]
81. Yin, S.Y.L.; Tserng, H.P.; Wang, J.C.; Tsai, S.C. Developing a precast production management system using RFID technology. *Autom. Constr.* **2009**, *18*, 677–691. [CrossRef]
82. Hong, J.; Shen, G.Q.; Mao, C.; Li, Z.; Li, K. Life-cycle energy analysis of prefabricated building components: An input-output-based hybrid model. *J. Clean. Prod.* **2016**, *112*, 2198–2207. [CrossRef]
83. Ji, Y.; Li, K.; Liu, G.; Shrestha, A.; Jing, J. Comparing greenhouse gas emissions of precast in-situ and conventional construction methods. *J. Clean. Prod.* **2018**, *173*, 124–134. [CrossRef]
84. Safa, M.; Shahi, A.; Haas, C.T.; Hipel, K.W. Supplier selection process in an integrated construction materials management model. *Autom. Constr.* **2014**, *48*, 64–73. [CrossRef]
85. Jiang, R.; Mao, C.; Hou, L.; Wu, C.; Tan, J. A SWOT analysis for promoting off-site construction under the backdrop of China's new urbanisation. *J. Clean. Prod.* **2017**, 1–10. [CrossRef]
86. Zhao, X.; Hwang, B.G.; Gao, Y. A fuzzy synthetic evaluation approach for risk assessment: A case of Singapore's green projects. *J. Clean. Prod.* **2016**, *115*, 203–213. [CrossRef]
87. Yang, L.-R.; Chen, J.-H.; Chang, S.-P. Testing a Framework for Evaluating Critical Success Factors of Projects. *J. Test. Eval.* **2016**, *44*, 20140074. [CrossRef]

 © 2018 by the authors. Licensee MDPI, Basel, Switzerland. This article is an open access article distributed under the terms and conditions of the Creative Commons Attribution (CC BY) license (http://creativecommons.org/licenses/by/4.0/).

Article

Cost Forecasting of Substation Projects Based on Cuckoo Search Algorithm and Support Vector Machines

Dongxiao Niu, Weibo Zhao *, Si Li and Rongjun Chen

School of Economics and Management, North China Electric Power University, Beijing 102206, China; ndx@ncepu.edu.cn (D.N.); 1172206202@ncepu.edu.cn (S.L.); crj713@ncepu.edu.cn (R.C.)
* Correspondence: zhaoweibo@ncepu.edu.cn; Tel.: +86-10-6177-3472

Received: 4 December 2017; Accepted: 31 December 2017; Published: 5 January 2018

Abstract: Accurate prediction of substation project cost is helpful to improve the investment management and sustainability. It is also directly related to the economy of substation project. Ensemble Empirical Mode Decomposition (EEMD) can decompose variables with non-stationary sequence signals into significant regularity and periodicity, which is helpful in improving the accuracy of prediction model. Adding the Gauss perturbation to the traditional Cuckoo Search (CS) algorithm can improve the searching vigor and precision of CS algorithm. Thus, the parameters and kernel functions of Support Vector Machines (SVM) model are optimized. By comparing the prediction results with other models, this model has higher prediction accuracy.

Keywords: cost prediction of substation project; Ensemble Empirical Mode Decomposition; Cuckoo Search; Support Vector Machines

---

1. Introduction

The prediction of the cost level of the power grid project is an important part of the economic evaluation of the power system. Grid projects are often capital-intensive and have high technical requirements [1]. It is of great significance to predict the cost level of grid projects effectively for improving the investment efficiency and sustainability. Grid projects mainly include transmission projects and substation projects, and these two types of projects are directly related to the safety production, normal operation, and economic benefits of the power grid. At present, there are many researches on the cost prediction of transmission project [2,3]. Whereas, few scholars have studied the cost prediction of the substation project [4,5]. In the construction of substation projects, a reasonable cost prediction can provide decision support and reference for the power grid companies, which is also helpful to promote the sustainable development of the investment in substation projects [6]. However, due to the impact of regional economic development, the surrounding natural environment and project management level, the cost of substation projects often tends to be non-linear, irregular, and difficult to predict [7].

Scholars around the world have conducted in-depth studies on the cost of transmission and substation projects. The researches mainly include the construction and prediction of transmission project cost index [8,9], the analysis of transmission and substation project cost affecting factors [10–12], and the prediction of substation project cost [13], etc. In terms of constructing the cost index, Liu et al. [8] built the cost index of power grid projects by considering different technologies and voltage classes, and obtained the total project cost index by weighting the typical program, in which the weight could be determined by the Paasche index analysis method or Laspeyres index analysis method. Tao et al. [3] selected more than 300 cost indicators of transmission projects from 2002 to 2010, and pointed out that changes of transmission project cost were affected by the previous period.

The Markov chain prediction model was applied to describe the changes of each period. In addition, the total investment of transmission project was composed of the construction cost, the installation cost, the equipment cost, and other expenses. From the four parts, nine key indicators were selected as the comprehensive cost index to be predicted. Hua et al. [9] constructed cost index analysis and prediction model based on Autoregressive Integrated Moving Average Model (ARIMA) and exponential smoothing models, and then, realized the process of modeling and obtained forecast results through SPSS software, which have some references meaning for the application of prediction method.

In terms of the analysis of affecting factors of project cost, Wang et al. [12] set up the evaluation system of cost index of 500 kV transmission project according to the samples of transmission project cost. Activity Based Classification (ABC) analysis, key parameter simplification, and principal component analysis were applied to deal with the samples. Three level indexes and their calculation formulas were obtained through calculation. Furthermore, the Least Squares Support Vector Machine (LSSVM) based on particle swarm optimization was used to calculate and verify the evaluation index system; the results proved that the proposed model has high prediction accuracy. The influence of various factors on the cost of substation engineering was studied from the aspects of technology, organization, external environment, and cost parameters in [13].

In the prediction of project cost, the prediction methods mainly include time series method, multiple regression model, and intelligent prediction method, and the selection of prediction methods could affect the accuracy of prediction results directly. Xu et al. [14] pointed out that the building cost index had been widely used to measure the cost level of the construction industry. However, to improve the accuracy of measurement, the interaction between the cost indices and other variables (such as consumer price index) should be considered. Therefore, the cointegration theory and Vector Auto Regression (VAR) model was proposed for predicting the changes of construction cost, which could assess the risk and uncertainty of rising costs. Zhu et al. [15] took the situation of the region, transaction date, transaction conditions, and individual factors into account firstly to construct a hierarchical structure system of real estate price factors. Then, the 1–9 scale was used to build the comparison judgment matrix and the ranking weight of case level relative to price could be calculated layer by layer. Moreover, the stochastic fuzzy regression analysis method was introduced to predict the cost of residential buildings accurately. Shahandashti et al. [16] pointed that the cost of highway construction could vary greatly over time, which was directly related to the income of highway contractors. Therefore, the research selected sixteen indices from the National Highway Construction Cost Index (NHCCI) as candidate indices through the literature research. Based on the results of the co-integration test, a Vector Error Correction (VEC) model was established to predict the construction cost index of national highway and the results showed that the multivariate time series model was more accurate than the single variable model.

In the field of intelligent prediction model, scholars had explored a series of high-precision intelligent prediction methods. Qin et al. [17] considered qualitative and quantitative cost index as an input set and a single cost as an output set of indicators. The correlation of housing project cost input indicators could be eliminated by means of principal component analysis. Support Vector Machines (SVM) and LSSVM were applied, respectively, to predict the cost of 25 residential projects in Hangzhou, in which the prediction error was within 7%. Zhou et al. [18] predicted the cigarette sales based on LSSVM and optimized the LSSVM parameters on the basis of the improved Cuckoo Search (CS) algorithm. The introduction of inertia weights in the path and location updates of the cuckoo nest helped to avoid falling into local optimum. When considering the after-effectiveness of cigarette sales, the best time delay was determined by comparing the prediction accuracy under different delay numbers, and the cigarette sales at the current time and five time periods ahead was proved to be corresponding by calculating the actual data. Besides, this article made multi-step prediction through the iterative method and predicted the sales volume of cigarettes in different periods in the future, which improved the dynamics of the prediction. Shao et al. [19] proposed that it was of great practical significance to explain the dependence between medium-term demand and external variables scientifically. He solved the problem of nonlinear power demand prediction by combining Ensemble Empirical Mode Decomposition (EEMD) method with semi-parametric model. The model could capture the potential important features, including the climate and economic development from the original electricity demand data. Reliable confidence interval of longer term fluctuation trend was obtained by means of predicting the actual monthly electricity demand data from Suzhou and Guangzhou.

However, the accuracy of the existing prediction methods cannot meet the requirements of the modern management of power grid projects. In view of the above problems based on the traditional prediction methods, this paper proposes an optimized SVM model based on the improved CS algorithm that decomposed by EEMD to predict the substation projects' cost.

2. Basic Theory

2.1. EEMD

Empirical Mode Decomposition (EMD) method can decompose the irregular signal into a number of well-characterized Intrinsic Mode Function (IMF) components, and EMD algorithm is essentially a stationary signal process [20–24]. However, due to the existence of modal aliasing in the EMD algorithm, the precision and breadth of the EMD algorithm are restricted. In this paper, it is chosen to add the uniformly distributed Gaussian white noise signal to the EMD to decompose the signal, in order to avoid the EMD algorithm modal aliasing influences. This decomposition method is the EEMD algorithm, and the steps of the algorithm are as follows [25–28]:

1. Determine the number of decomposed IMFs and the number of decompositions
2. Add Gaussian white noise sequence to the input signals
3. Normalize the signals after adding the white noise sequence
4. Decompose the normalized signals to obtain multiple IMF components and one surplus variable:

$$X(t) = \sum_{i=1}^{N} \sum_{j=1}^{M} c_{j,i}(t), \ i = 1 \ldots N, \ j = 1 \ldots M \tag{1}$$

$$x(t) = \frac{1}{N} \sum_{i=1}^{N} \sum_{j=1}^{M} c_{j,i}(t) = \sum_{i=1}^{N-1} \sigma_i(t) + r(t), \ i = 1 \ldots N, \ j = 1 \ldots M \tag{2}$$

Among them $r(t)$ is the remainder. The method is applied to predict the cost level of substation project. The original irregular cost data can be decomposed into a number of stationary IMF components through the EEMD processing, which are then input into the SVR model for prediction. Finally, the predictive value of substation project cost level can be obtained by adding the total amount.

2.2. SVM

SVM was proposed by Cortes and Vapnik in 1995 [29–31], which can solve the small sample and complex nonlinear regression problems effectively. It maps the data X_i into high-dimensional space F by nonlinear mapping ϕ, and performs linear regression in high-dimensional space. The mapped linear function is $f(x) = \omega\phi + b$, which is used to solve the optimal function Equation (1) by finding the weight ω and threshold b in the linear function, according to the SVM criterion [32–35].

$$\min \frac{1}{2} \|\omega\|^2 + C \sum_{i=1}^{l} \xi_i \tag{3}$$

$$s.t. \begin{cases} y_i(\omega\phi + b) \geq 1 - \xi_i \\ \xi_i \geq 0 \\ C > 0 \end{cases} \tag{4}$$

The above problem can be transformed into a dual problem by introducing Lagrange multiplier $\alpha_i \geq 0$, $\beta_i \geq 0$, thus the classification decision function of SVM becomes:

$$f(x) = \text{sgn}(\sum_{i=1}^{l} y_i \alpha_i K(x_i, x_j) + b) \tag{5}$$

$K(x_i, x_j)$ in the formula is a kernel function. This paper selects radial basis function (RBF) as a kernel function.

$$K(x_i, x_j) = \exp(-\frac{\|x_i - x_j\|^2}{2\sigma^2}) \tag{6}$$

where σ is the width of the kernel function.

The key to the accuracy of SVM regression model is the penalty factor C and the width of kernel function σ [36]. Therefore, this paper chooses the improved CS algorithm to optimize C and σ in order to improve the generalization ability of SVM.

2.3. Optimized CS Algorithm

The CS algorithm can search the optimization much faster and more accurately by simulating the random walking process of cuckoo in the search for the suitable egg laying hosts [37–39]. According to existing research, the CS algorithm has the following three rules [40,41]:

1. The number of eggs produced by a cuckoo per time is 1.
2. The host bird's nest where high-quality eggs are located is the optimal solution and will be retained for the next generation.
3. The number of host nests is certain, and the probability that cuckoo eggs are found by nest owners is $P_a \in [0,1]$.

During the search, cuckoo's flight search path follows the Lévy distribution, namely:

$$x_i^{(t+1)} = x_i^{(t)} + \alpha \oplus L(\lambda) \quad i = 1, 2, \cdots, n \tag{7}$$

where $x_i^{(t+1)}$ and $x_i^{(t)}$ are the bird's nest positions of the $(t+1)$th and the tth generation, n is the number of cuckoo, $\oplus$ is the point to point multiplication, and $L(\lambda)$ is the Lévy flight path. The relationship between searching path and time is as follows:

$$L(\lambda) \sim u = t^{-\lambda}(1 < \lambda \leq 3) \tag{8}$$

In the traditional CS algorithm, the probability of finding cuckoo eggs and the step size α of position updating are fixed values, which leads to the problems of the weak global searching ability, slow convergence speed, and low precision. Therefore, an improved cuckoo algorithm is proposed in this paper to update the values of P_a and α dynamically, as follows [42,43]:

$$P_a(t) = P_{a\max} - \frac{t(P_{a\max} - P_{a\min})}{N} \tag{9}$$

$$\alpha(t) = \alpha_{\max} e^{\frac{\ln(\frac{\alpha_{\max}}{\alpha_{\min}}) \times t}{N}} \tag{10}$$

where t and N are the number of current iterations and the total number of iterations, $P_{a\max}$ and $P_{a\min}$ are the maximum and minimum values of the detection probability, $\alpha_{\max}$ and $\alpha_{\min}$ are the maximum and minimum step coefficients.

However, CS algorithm has defects of lacking of search vitality and slow speed of. The optimization ability of CS algorithm can be improved effectively by adding Gauss perturbation [44]. Assuming that the optimal location of the nest $x_i^{(i)}$, $(i = 1, 2, \cdots, n)$, is obtained after the calculation of t

times CS iterations. In order to prevent the next iteration of $x_i^{(i)}$ and maintain the Gaussian disturbance, the next phase of $x_i^{(i)}$ is searched. Suppose that the matrix $p_t = [x_1^{(t)}, x_2^{(t)}, \cdots, x_n^{(t)}]^T$ is made up of the better position $x_i^{(i)}$ of the bird's nest, $x_i^{(i)}$ is a d-dimensional vector, and the dimension of p_t is $d \times n$. Matrix p_t combined with Gaussian perturbation is the basic step of GCS algorithm, namely:

$$p_t = p_t + a \oplus \varepsilon \quad (11)$$

where ε is a random matrix with the same order of p_t, which follows $N(0,1)$ distribution, and a is constant. In the search for a better nest position vitality at the same time, the position of the bird's nest can be overextended easily because of the large random range of ε. Therefore, the selection of suitable a is particularly important. After obtaining a reasonable set of p_t and comparing it with each nest in p_t, only a better nest position is reserved to obtain a better set of nest positions p_t [45].

3. Substation Project Cost Prediction Model Based on EEMD-GCS-SVM

The cost prediction of substation project is affected by many factors, and the cost level is non-stationary and irregular. The specific process of substation project cost prediction model based on EEMD-GCS-SVM is shown in the Figure 1. Specific steps are:

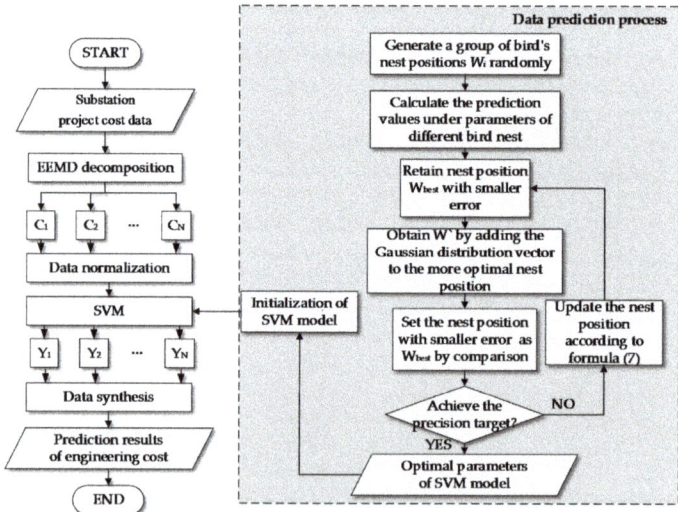

Figure 1. Flow chart of the Ensemble Empirical Mode Decomposition (EEMD)-GCS- Support Vector Machines (SVM) prediction model.

1. Decompose the substation cost data to obtain the IMF components and surplus variables through the EEMD method, and normalize the data.
2. Initialize the parameters and kernel functions of SVM model, input the normalized decomposed variables into SVM model, and find and determine the optimal parameters and kernel function of SVM model by using GCS algorithm. In order to search for the best parameters of the prediction model faster, the range of c, ε are set as [0.01, 100], [0.01, 100], respectively. Then, train the prediction model by plugging the historical data into the model and search the best parameter by using the GCS Algorithm. Firstly, set the N_{nest} (number of birds' nest) as 20, while P_a (probability of bird's eggs by bird's nest owner) is 0.45, and N (number of iterations) is 200. After that, randomly generate N_{nest} bird nest location $W = (W_1, W_2, ..., N_{nest})^T$. Each bird nest W_i has s parameters (s = the number of weights between input layer and hidden layer + the number of

weights between hidden layer and output layer + the number of translation factors + the number of expansion factors). The predicted values of each bird's nest were calculated, and the nest which has the smallest error in the 20 nests is found, marked as W_{best}. Then W_{best} retain to the next generation.
3. Train the SVM model by using the training set, and then input the test set data to obtain the predictive value of the cost data.

4. Example Analysis

4.1. Basic Data

There are a lot of voltage levels in the substation projects. 220 kV is a widely used voltage level of substation projects, and its cost data is more easily obtained at the same time. Thus, take the cost level data of 220 kV new outdoor substation projects at certain place in 2014–2016 (as shown in Table 1) as an example to validate the model. The cost level of substation projects is represented by the cost per kVA as the research data sample. Starting from the data of the first sample project, taking approximately equal time as the interval between the selected samples, we get the cost data of 72 samples, and sort them according to the completion time of the projects.

Table 1. The cost level of 220 kV new outdoor substation project at a certain place.

Serial Number	Cost (Yuan/kV·A)	Serial Number	Cost (Yuan/kV·A)	Serial Number	Cost (Yuan/kV·A)	Serial Number	Cost (Yuan/kV·A)
1	284.22	19	396.62	37	438.26	55	285.86
2	337.44	20	279.78	38	314.15	56	288.38
3	369.32	21	257.53	39	299.14	57	321.98
4	347.03	22	259.8	40	454.53	58	466.19
5	419.99	23	290.07	41	342.54	59	398.09
6	358.64	24	275	42	344.12	60	498.41
7	449.02	25	320.91	43	370.65	61	425.52
8	383.35	26	343.77	44	304.73	62	342.21
9	308.3	27	306.23	45	329.5	63	423.47
10	346.53	28	335	46	355.37	64	367.21
11	262.64	29	381.5	47	296.73	65	330.49
12	297.97	30	394.85	48	319.74	66	275.96
13	257.35	31	400.6	49	371.6	67	297.36
14	235.51	32	430.9	50	298.83	68	364.17
15	339.83	33	354.63	51	361.29	69	292.85
16	407.24	34	390.91	52	427.77	70	354.06
17	348.25	35	371.08	53	283.38	71	347.54
18	284.35	36	372.2	54	294.15	72	278.54

First of all, decompose the cost level data by EEMD into five IMF components, and the decomposition results are shown in Figure 2.

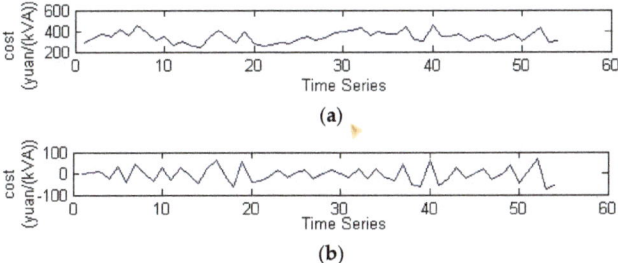

Figure 2. Cont.

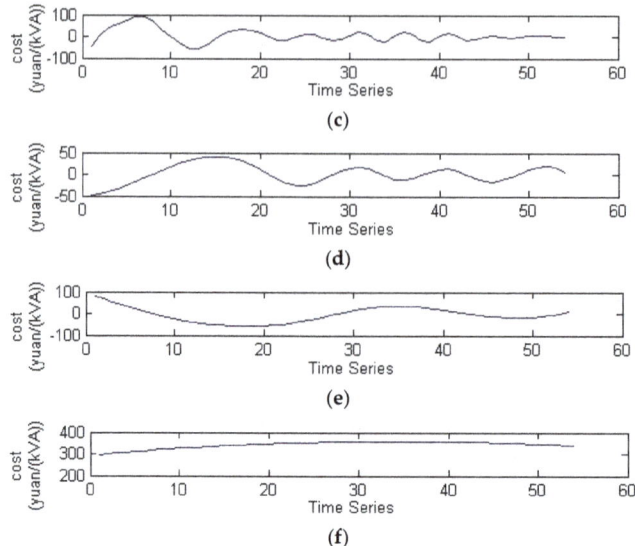

Figure 2. The result of the EEMD. (**a**) Original data; (**b**) IMF1; (**c**) IMF2; (**d**) IMF3; (**e**) IMF4; and (**f**) IMF5.

From the decomposition results, each IMF component obtained after the EEMD algorithm shows obvious regularity and periodicity, the algorithm helps to improve the prediction accuracy of subsequent SVM model. Then, the first 54 sets of cost data are taken as training group, the decomposed IMF components and the surplus components are input into the GCS-SVM prediction model, respectively, for training. The total cost of each component is reduced to the predictive value after the prediction results of each component are obtained. The latter 18 sets of cost data are used as test group to verify the prediction effect of the model.

4.2. Results Analysis and Comparison

After training the forecasting model, the best value of c and ε are 29.8425 and 0.4871, respectively. In order to verify the accuracy of the model, the GCS-SVM model without EEMD algorithm, the EEMD-CS-SVM model with the non-optimized CS algorithm, and the EEMD-GCS-SVM model are used to predict the samples in this paper. The results are shown in Figure 3a–c.

It can be seen from Figure 3a–c that the prediction result by EEMD-GCS-SVM model, which has been decomposed by EEMD and optimized by GCS algorithm, has the highest fitting and prediction accuracy. The results indicate that the EEMD and GCS algorithm is helpful to improve the accuracy of the model.

In addition, BP neural network model, SVM model, GCS-SVM model, EEMD-CS-SVM model, and EEMD-GCS-SVM model are, respectively, used to predict the data of the test set in this paper, and the error comparison results are shown in Table 2 and Figure 4.

According to the above table, the RMSE of the EEMD-GCS-SVM model is 0.51, MAE is 0.43, and MAPE is 0.13%, which indicates that the prediction accuracy is higher than EEMD-CS-SMV and GCS-SVM, and is significantly better than the BP neural network model and the SVM model.

The boxplot can directly reflect the accuracy of each model. It can be seen from the boxplot that the prediction error of EEMD-GCS-SVM is smaller. It means that this model is more accurate than other models, and is more suitable for the prediction of the cost level of 220 kV substation project.

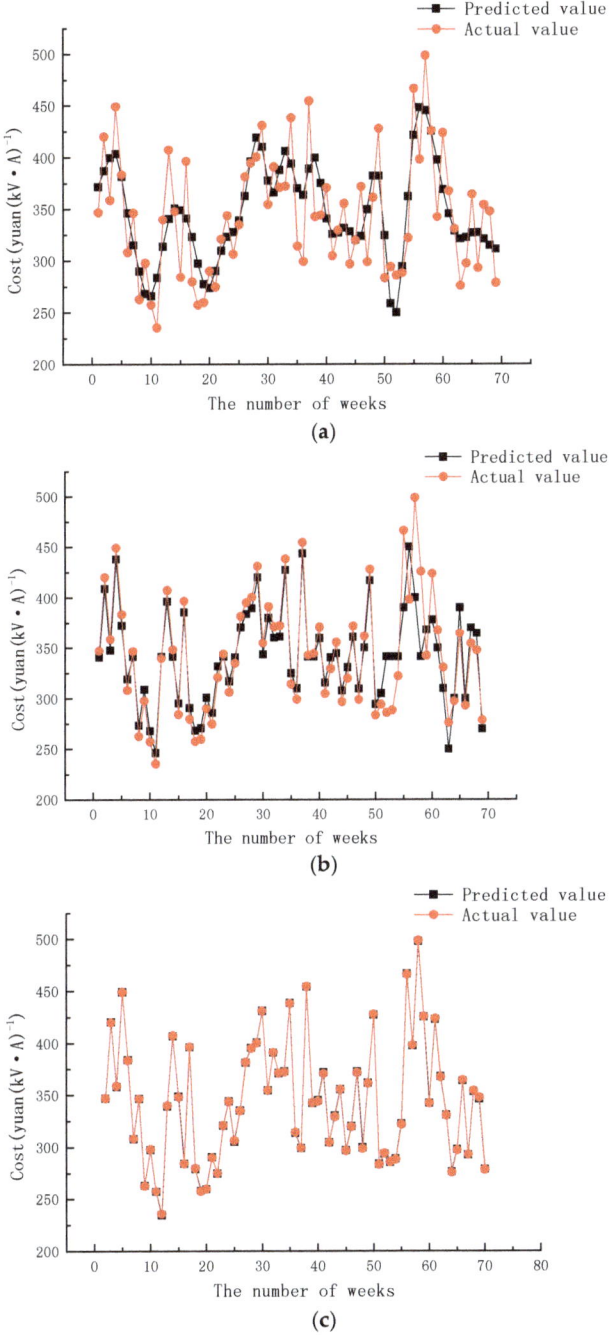

Figure 3. Prediction results. (**a**) The prediction result by EEMD-CS-SVM; (**b**) The prediction result by GCS-SVM; and (**c**) The prediction result by EEMD-GCS-SVM.

Table 2. Comparison of the model errors.

Model	BP	SVM	GCS-SVM	EEMD-CS-SVM	EEMD-GCS-SVM
RMSE (yuan/kV·A)	78.47	53.45	24.76	36.02	0.51
MAE (yuan/kV·A)	57.85	42.10	16.71	31.38	0.43
MAPE	16.42%	12.36%	4.72%	9.28%	0.13%

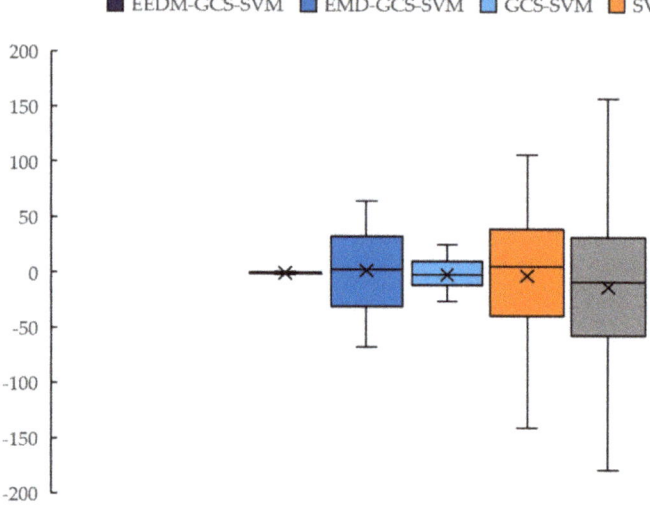

Figure 4. Comparison of the model errors.

5. Conclusions

With the development of power system reform, it is urgent to improve the cost management level of power grid project. Thus, it is of great significance to control the cost level of grid projects effectively for improving the investment efficiency and sustainability. However, the prediction of substation project cost level is difficult, and the prediction accuracy of the traditional method is insufficient.

(1) The EEMD-GCS-SVM model established in this paper can effectively improve the prediction accuracy of substation project cost with a MAPE value of only 0.13%, which is much better than that of the un-optimized and EEMD models.
(2) EEMD method can decompose irregular and non-stationary sequence signals into multiple IMF components and surplus components. The decomposed signals show regularity and periodicity obviously, which improves the prediction accuracy of the model.
(3) On the basis of CS optimized SVM parameters and kernel function, adding Gauss perturbation can effectively improve the search vitality and range of CS algorithm. The optimal SVM parameters are obtained, the calculation of kernel function is faster, and the computational efficiency and prediction accuracy is improved in the model.

However, the research only analyzed the cost prediction of the 220 kV substation due to the limited availability of data. Besides, there is no verification and analysis of more regions. Thus, more types of data should be widely used to verify the ability of the modified model in the further study. In future research, we will continue to apply the model to more prediction fields and explore more scientific and accurate prediction methods.

Acknowledgments: This research was supported by the National Natural Science Fund (71471059).

Author Contributions: All authors designed the research, established the forecasting model, completed the data arrangement and calculation, created the tables and figures, and finished the writing of the paper. All authors read and approved the final manuscript.

Conflicts of Interest: The authors declare no conflict of interest.

References

1. Rao, R.; Zhang, X.; Shi, Z.; Luo, K.; Tan, Z.; Feng, Y. A Systematical Framework of Schedule Risk Management for Power Grid Engineering Projects' Sustainable Development. *Sustainability* **2014**, *6*, 6872–6901. [CrossRef]
2. Zhao, H.; Li, N. Risk Evaluation of a UHV Power Transmission Construction Project Based on a Cloud Model and FCE Method for Sustainability. *Sustainability* **2015**, *7*, 2885–2914. [CrossRef]
3. Tao, Y.; Cui, S.; Ke, X. Forecasting Tendency of the Transmission Line Project Cost Index Using Markov Chain. *Open Electr. Electron. Eng. J.* **2014**, *8*, 56–63. [CrossRef]
4. Zhu, J.; Yao, X.; Zhao, Z. Empirical Research on Cost Prediction Model of Technical Renovation for Electric Power Substation Project. *Eng. Econ.* **2017**, *3*, 5–9.
5. Pei, Q.F.; Zhang, H.M.; Yang, Y.K. Study on the Management of Construction Cost in Substation Project. *Adv. Mater. Res.* **2014**, *1030–1032*, 2639–2642. [CrossRef]
6. Jaleel, J.A.; Sadina, M.F. Reliability Evaluation of 220 kV Substation Using Fault Tree Method and Its Prediction Using Neural Networks. *Environ. Manag.* **2013**, *54*, 951–970.
7. Ling, Y.; Yan, P.; Han, C.; Yang, C. Prediction model of transmission line engineering cost based on BP neural network. *China Power* **2012**, *45*, 95–99.
8. Liu, W.; Shi, H.; Lu, Y.; Hua, F.; Qiu, J. Study on Grid Engineering Cost Index Forecasting Based on ARIMA-ES Hybrid Model. *Manag. Rev.* **2016**, *28*, 45–53.
9. Niu, D.X.; Hua, F.Y. Research on Prediction of Transmission and Transformation Project Cost Index Based on ARIMA and Exponential Smoothing Models. In *Proceedings of the 22nd International Conference on Industrial Engineering and Engineering Management 2015*; Atlantis Press: Paris, France, 2016; pp. 771–779.
10. Zhou, M.; Zhao, S.; Zhu, J. Index Evaluation System of Power Transmission Project Cost Based on Support Vector Machine Method. *Electr. Power Constr.* **2014**, *35*, 102–106.
11. Huang, J. Substation project cost management and control measures. *Sci. Res.* **2016**, *15*, 00102–00103.
12. Wang, N.N.; Wang, F.; Yin, Y.T.; Li, H.; Hou, Y. Research on Cost Predicting of Power Transformation Projects Based on SVM. *Const. Econ.* **2016**, *37*, 48–52.
13. Song, Z.; Niu, D.; Xiao, X.; Zhu, L. Substation Engineering Cost Forecasting Method Based on Modified Firefly Algorithm and Support Vector Machine. *Electr. Power* **2017**, *50*, 168–173.
14. Xu, J.W.; Moon, S. Stochastic Forecast of Construction Cost Index Using a Cointegrated Vector Autoregression Model. *J. Manag. Eng.* **2013**, *29*, 10–18. [CrossRef]
15. Zhu, X.W.; Li, W.J. Prediction Model of Urban Real Estate Project Cost. *Stats. Decis.* **2010**, *14*, 47–49.
16. Shahandashti, S.M.; Ashuri, B. Highway Construction Cost Forecasting Using Vector Error Correction Models. *J. Manag. Eng.* **2016**, *32*. [CrossRef]
17. Qin, Z.F.; Lei, X.L.; Zhai, D.; Jin, L.Z. Forecasting the Costs of Residential Construction Based on Support Vector Machine and Least Squares-Support Vector Machine. *J. Zhejiang Univ. Sci. A* **2016**, *43*, 357–363.
18. Zhou, J.Y.; Zhang, K.W. Cigarette Sales Forecasting Based on Hybrid Kernels Least Square Support Vector Machine Optimized by Modified Cuckoo Search Algorithm. *Comput. Eng. Appl.* **2015**, *51*, 250–254.
19. Shao, Z.; Gao, F.; Yang, S.L.; Yu, B. A new semiparametric and EEMD based framework for mid-term electricity demand forecasting in China: Hidden characteristic extraction and probability density prediction. *Renew. Sustain. Energy Rev.* **2015**, *52*, 876–889. [CrossRef]
20. Huang, Y.; Wu, D.; Zhang, Z.; Chen, H.; Chen, S. EMD-based pulsed TIG welding process porosity defect detection and defect diagnosis using GA-SVM. *J. Mater. Process. Technol.* **2017**, *239*, 92–102. [CrossRef]
21. He, K.; Wang, H.; Du, J.; Zou, Y. Forecasting Electricity Market Risk Using Empirical Mode Decomposition (EMD)—Based Multiscale Methodology. *Energies* **2016**, *9*, 931. [CrossRef]
22. Amjady, N.; Abedinia, O. Short Term Wind Power Prediction Based on Improved Kriging Interpolation, Empirical Mode Decomposition, and Closed-Loop Forecasting Engine. *Sustainability* **2017**, *9*, 2104. [CrossRef]
23. Malik, H.; Sharma, R. EMD and ANN based intelligent fault diagnosis model for transmission line. *J. Intell. Fuzzy Syst.* **2017**, *32*, 3043–3050. [CrossRef]

24. Acharya, U.R.; Fujita, H.; Sudarshan, V.K.; Oh, S.L.; Muhammad, A.; Koh, J.E.W.; Tan, J.H.; Chua, C.K.; Chua, K.P.; Tan, R.S. Application of empirical mode decomposition (EMD) for automated identification of congestive heart failure using heart rate signals. *Neural Comput. Appl.* **2016**, *28*, 3073–3094. [CrossRef]
25. Helske, J.; Luukko, P. Ensemble Empirical Mode Decomposition (EEMD) and Its Complete Variant (CEEMDAN). *Int. J. Public Health* **2016**, *60*, 1–9.
26. Lee, D.H.; Ahn, J.H.; Koh, B.H. Fault Detection of Bearing Systems through EEMD and Optimization Algorithm. *Sensors* **2017**, *17*, 2477. [CrossRef] [PubMed]
27. Montalvo, C.; Gavilán-Moreno, C.J.; García-Berrocal, A. Cofrentes nuclear power plant instability analysis using ensemble empirical mode decomposition (EEMD). *Ann. Nucl. Energy* **2017**, *101*, 390–396. [CrossRef]
28. Li, W.; Yang, X.; Li, H.; Su, L. Hybrid Forecasting Approach Based on GRNN Neural Network and SVR Machine for Electricity Demand Forecasting. *Energies* **2017**, *10*, 44. [CrossRef]
29. Cortes, C.; Vapnik, V. Support-vector networks. *Mach. Learn.* **1995**, *20*, 273–297. [CrossRef]
30. Arabloo, M.; Ziaee, H.; Lee, M.; Bahadori, A. Prediction of the properties of brines using least squares support vector machine (LS-SVM) computational strategy. *J. Taiwan Inst. Chem. Eng.* **2015**, *50*, 123–130. [CrossRef]
31. Fan, J.; Wu, J.; Kong, W.; Zhang, M.H. Predicting Bio-indicators of Aquatic Ecosystems Using the Support Vector Machine Model in the Taizi River, China. *Sustainability* **2017**, *9*, 892. [CrossRef]
32. Liang, Y.; Niu, D.; Ye, M.; Hong, W.-C. Short-Term Load Forecasting Based on Wavelet Transform and Least Squares Support Vector Machine Optimized by Improved Cuckoo Search. *Energies* **2016**, *9*, 827. [CrossRef]
33. Liu, J.-P.; Li, C.-L. The Short-Term Power Load Forecasting Based on Sperm Whale Algorithm and Wavelet Least Square Support Vector Machine with DWT-IR for Feature Selection. *Sustainability* **2017**, *9*, 1188. [CrossRef]
34. Kachoosangi, F.T. How Reliable Are ANN, ANFIS, and SVM Techniques for Predicting Longitudinal Dispersion Coefficient in Natural Rivers? *J. Hydraul. Eng.* **2016**, *142*. [CrossRef]
35. Giorgi, M.G.D.; Ficarella, A.; Lay-Ekuakille, A. Cavitation Regime Detection by LS-SVM and ANN with Wavelet Decomposition Based on Pressure Sensor Signals. *IEEE Sens.* **2015**, *15*, 5701–5708. [CrossRef]
36. Tian, J.; Li, C.; Liu, J.; Yu, F.L.; Cheng, S.H.; Zhao, N.; Jaafar, W.Z.W. Groundwater Depth Prediction Using Data-Driven Models with the Assistance of Gamma Test. *Sustainability* **2016**, *8*, 1076. [CrossRef]
37. Sun, W.; Sun, J. Daily PM$_{2.5}$ concentration prediction based on principal component analysis and LSSVM optimized by cuckoo search algorithm. *J. Environ. Manag.* **2017**, *188*, 144–152. [CrossRef] [PubMed]
38. Xiao, L.; Shao, W.; Yu, M.; Ma, J.; Jin, C. Research and application of a hybrid wavelet neural network model with the improved cuckoo search algorithm for electrical power system forecasting. *Appl. Energy* **2017**, *198*, 203–222. [CrossRef]
39. Mellal, M.A.; Williams, E.J. Cuckoo optimization algorithm with penalty function for combined heat and power economic dispatch problem. *Energy* **2015**, *93*, 1711–1718. [CrossRef]
40. Kaboli, S.H.A.; Selvaraj, J.; Rahim, N.A. Long-term electric energy consumption forecasting via artificial cooperative search algorithm. *Energy* **2016**, *115*, 857–871. [CrossRef]
41. Makhloufi, S.; Mekhaldi, A.; Teguar, M. Three powerful nature-inspired algorithms to optimize power flow in Algeria's Adrar power system. *Energy* **2016**, *116*, 1117–1130. [CrossRef]
42. Wang, Z.; Wang, C.; Wu, J. Wind Energy Potential Assessment and Forecasting Research Based on the Data Pre-Processing Technique and Swarm Intelligent Optimization Algorithms. *Sustainability* **2016**, *8*, 1191. [CrossRef]
43. Cai, X.; Nan, X.; Gao, B. Application of SVM Optimized ICS in Fault Diagnosis of Analog Circuit. *Bull. Sci. Technol.* **2017**, *33*, 79–82.
44. Zheng, H.; Zhou, Y. A novel Cuckoo Search optimization algorithm base on gauss distribution. *J. Comput. Inf. Syst.* **2012**, *8*, 4193–4200.
45. Lai, J.; Liang, S.; Center, C. Application of GCS-SVM model in network traffic prediction. *Comput. Eng. Appl.* **2013**, *49*, 75–78.

© 2018 by the authors. Licensee MDPI, Basel, Switzerland. This article is an open access article distributed under the terms and conditions of the Creative Commons Attribution (CC BY) license (http://creativecommons.org/licenses/by/4.0/).

Article

Integrating a Procurement Management Process into Critical Chain Project Management (CCPM): A Case-Study on Oil and Gas Projects, the Piping Process

Sung-Hwan Jo [1], Eul-Bum Lee [2,*] and Kyoung-Youl Pyo [2]

[1] Dae-Woo Engineering and Construction, Engineering Management Team, Division of Plant Engineering, 75 Saemunan-Ro, Jongro-Ku, Seoul 03182, Korea; shjo81@postech.ac.kr
[2] Graduate Institute of Ferrous Technology & Graduate School of Engineering Mastership, Pohang University of Science and Technology (POSTECH), 77 Cheongam-Ro, Nam-Ku, Pohang 37673, Korea; kypyo@postech.ac.kr
* Correspondence: dreblee@postech.ac.kr; Tel.: +82-54-279-0136

Received: 8 April 2018; Accepted: 29 May 2018; Published: 31 May 2018

Abstract: Engineering, Procurement, and Construction (EPC) of oil and gas megaprojects often experience cost overruns due to substantial schedule delays. One of the greatest causes of these overruns is the mismanagement of the project schedule, with the piping works (prefabrication and installation) occupying a majority of that schedule. As such, an effective methodology for scheduling, planning, and controlling of piping activities is essential for project success. To meet this need, this study used the Critical Chain Project Management (CCPM) to develop a piping construction delay prevention methodology, incorporating material procurement processes for EPC megaprojects. Recent studies indicate that the traditional scheduling method used on oil and gas mega projects has critical limitations regarding resource scarcity, calculation of activity duration, and dealing with uncertainties. To overcome these limitations, the Theory of Constraints-based CCPM was proposed and implemented to provide schedule buffers management. Nonexistent in literature, and of critical importance, is this paper's focus on the resource buffer, representing material uncertainty and management. Furthermore, this paper presents a step-by-step process and flow chart for project, construction, and material managers to effectively manage a resource buffer through the CCPM process. This study extends the knowledge of traditional resource buffers in CCPM to improve material and procurement management, thus avoiding the shortage of piping materials and minimizing delays. The resultant process was validated by both deterministic and probabilistic schedule analysis through two case studies of a crude pump unit and propylene compressor installation at a Middle Eastern Refinery Plant Installation. The results show that the CCPM method effectively handles uncertainty, reducing the duration of piping works construction by about a 35% when compared to the traditional method. Furthermore, the results show that, in not considering material uncertainty (resource buffers), projects schedules have the potential for approximately a 5% schedule growth with the accompanying delay charges. The findings have far-reaching applications for both oil and gas and other sectors. This CCPM case-study exemplifies that the material management method represents an opportunity for industry to administrate pipeline installation projects more effectively, eliminate project duration extension, develop schedule-based risk mitigation measures pre-construction, and enable project teams to efficiently manage limited human and material resources.

Keywords: CCPM; piping construction; material procurement management; resource competition; buffer management; PERT/CPM; stochastic simulation; construction delay

1. Introduction

1.1. EPC Megaprojects Schedule Under-Performance

Over the last decade, EPC (Engineering, Procurement, and Construction) megaprojects (>1 billion USD), especially within the petrochemical and oil and gas industries, have experienced substantial cost overrun and prolonged schedule delay. Frequently, the burden of these overruns and delays fall on the contractor. This is most commonly the case in lump-sum turn-key contracts where the EPC contractors manage the contractual liabilities of the project. In these projects, the contractors are at risk of incurring considerable financial damages for poor project performance [1]. The frequency of project overruns experienced in the petrochemical and oil and gas industry, and the resultant contractor losses, suggests a major need for managerial process improvement.

Poor schedule management is one of the greatest causes of these overruns on mega-EPC projects [2]. The traditional scheduling management process, Program Evaluation and Review Technique/Critical Path Method (PERT/CPM), is found to be lacking, namely in its ability to handle resource constraints and project uncertainty. Therefore, this study proposes a Critical Chain Project Management (CCPM) which is better equipped to handle resource constraints and provides a more transparent uncertainty management process. While certain authors are dedicated to this topic, no existing literature discusses the management of the resource buffer (material availability uncertainty). Therefore, this paper presents a supplementary procurement management process, including schedule management of the uncertainties, to both fill the research gap and aid practitioners in effectively estimating project schedules. This process was developed for piping installation (one of the major petrochemical and oil and gas activities) and validated through an executed Korean EPC project.

1.2. Problem Background: Fluctuation of Korean EPC Contractors' Revenue in Overseas Market

As already stated, the focus of this paper is the oil and gas industry. In practice, the global oil and gas projects market consists of: (1) upstream, development of oil and gas fields for production; (2) midstream, transfer oil and gas through pipelines; and (3) downstream, converting oil and gas to petrochemical products. Globally, spending on oil and gas (EPC) projects has been in decline, with approximately USD 437 billion worth of work performed in 2016, a 42% decrease from 2014. Within the last few years, this market has experienced severe decline in the investments and cancellation of capital projects due to this steep downturn. In response to this downturn, oil and gas EPC companies have taken great lengths to improve their overall business portfolio including effective partnering, complementing skillsets [3].

Similar to the global trends, the Korean oil and gas EPC contractors working internationally have experienced profit fluctuation and project volatility. This has been caused by variables, including oil price drop, global market recession, etc. The total annual overseas onshore and offshore contracts executed by Korean EPC contractors has substantially reduced over the last few years, as shown in Figure 1. In 2014, the total value for mainly oil and gas projects awarded to Korean EPC contractors was about USD 39 billion (approximately 80% of total overseas plant projects of USD 52 billion). This was substantially reduced to only USD 9 billion in 2016 with only a slight bounce-up in 2017 (USD 13 billion) potentially due to the crude oil price recovery [4].

This downturn has resulted in major Korean EPC contractors' hardships such as earning shock due to insolvency at overseas construction sites. In a study of six Korean oil and gas EPC contractor's overseas experience, the losses experienced ranged from USD 80 million to 1.2 billion from 2013 to 2016 [2]. This is not an isolated occurrence. To date, most international overseas oil and gas (mainly plant) EPC contractors are experiencing cost-overruns, schedule delays, and outstanding disputes and claims on their projects [1]. This problem is further intensified by the loss of work in floating or fixed-platform oil and gas production facilities. While these projects once encompassed a major portion of overseas profit, most Korean EPC contractors now experience large deficits and losses in performing mega offshore EPC projects (fabrication of oil and gas production facilities). According to some media reports, major Korean shipbuilders' cumulative financial damages were about USD 12 billion over the last few years due to their under-performance in megaprojects [2].

To exacerbate low profits, these contractors often experience poor project performances [1]. The main causes of the losses and damages include: (a) poor estimation resulting in bid/contract amount insufficient to cover project expenses; (b) poor project management, control, and engineering due to inexperience/lack of knowledge; resulting in (c) schedule delays and associated penalties; and (d) poorly managed claims, disputes, legal remediations, etc. [2]. Among these, industry experts believe that the most significant is schedule delays. These delays come with significant penalties, upward of USD 1–2 million per day of delay (so-called delay liquidated damages) [5].

With low profits caused by low project availability and poor project performance, it is ever more important for EPC contractors to prevent the delay of the schedule and reduce the length of the project duration. Although many activities exist, none is more critical than piping installation, namely the procurement of piping materials. As such, this paper presents the Critical Chain Project Management (CCPM) used to develop a piping construction delay prevention methodology, incorporating material procurement processes for EPC megaprojects to prevent project delay on mega-EPC projects, leading to higher contractor profits.

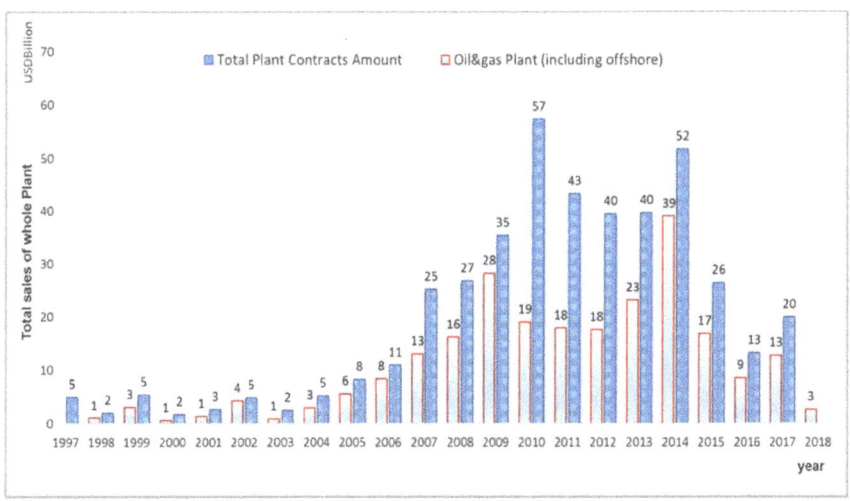

Figure 1. Total annual contracts amount by Korean EPC contractors on overseas projects [4].

2. Plant Piping Construction: Major Causes for Delays

In an onshore EPC project, piping construction comprises more than 40% of the work volume for the entire project, and welding is the major critical activity. From the lessons learned and experience of recently completed oil and gas EPC mega projects interviewed by the researchers, the main reasons for the delay of piping construction were determined to be resource constraints, lack of material management, and unrealistic work durations.

The critical resources for piping construction are those involved with welding: the welder, welding materials, and welding equipment. There are limited quantities of these resources, e.g. qualified welders, yet many simultaneous activities will require welding. The most common error for Korean mega-project EPC oil and gas contractors is not taking into consideration this limited resource. This is often caused by using the traditional Critical Path Method (CPM), ignoring resource constraints. To meet their developed schedules, welders must be mobilized without any constraint, which is impractical for most onshore EPC project sites.

The next main project management error is an insufficient, or lack of, material management plan. As stated, the piping activities are on the critical path, as are the procurement of piping materials [6]. Typical schedules include only milestones on the estimated time of arrival of piping materials. As such, the contractor is only notified of an issue once the material delivery is delayed. The contractor is not able to recognize the problem until a delay has already occurred [7]. Neither the PERT/CPM nor CCPM methods can sufficiently manage this process [8]. This paper presents a supplemental material management process, likely the paper's greatest addition to literature and the general body of knowledge of this paper.

Finally, during the estimation process, the schedulers often develop unrealistic work durations. This negatively impacts either efficiency and/or the ability for contractor's to meet their contractual schedule. This research has found the greatest error to be the use (or misuse) of schedule contingency/float. Often these contingencies are "hidden" within activities by schedulers, minimizing project management teams' ability to effectively manage the resultant float. The CCPM scheduling process, with the "buffer" concept, is proposed to mitigate this issue (explained in greater detail below).

CCPM Scheduling Process: The Use of a Buffer

A key part of estimating a project's cost and schedule is a contingency and scheduling float, called a buffer for CCPM and this paper. This buffer allows flexibility for unknown and/or unplanned occurrences without negatively impacting the schedule. There are two traditional methods used by Korean oil and gas contractors for estimating a buffer size: cut and paste and square root methods. In the cut/paste method a "safe estimate" (includes duration to account for uncertainty) is cut and half of the duration is attached to the end of all tasks. In the root square error method, uncertainty is calculated through the difference of the safe and average estimate. The uncertainty is then calculated through the square root of the sum of squares [9,10].

The CCPM buffer calculation differs to the traditional method. First, as opposed to CPM's critical path, CCPM has a resource-constrained critical, known as the critical chain [11,12]. The CCPM critical chain accounts for the resource competition (resource loaded) and includes duration buffers that are non-work schedule activities to manage uncertainty. CCPM's buffers fall into three categories [13]:

i. Project Buffer

   A. Placed at the end of the critical chain
   B. Protects target finish date from slippage along the critical chain

ii. Feeding Buffer

   A. Placed at each point where a chain of dependent activities is not on the critical chain
   B. Protects the critical chain from slippage along the feeding chains

iii. Resource Buffer

   A. The period of preparing resources to be used for critical chain activities
   B. Protects the delay of starting activities due to resource constraint

In CCPM, buffers are separate activities. The size of each buffer reflects the duration uncertainty of the dependent activities leading up to that buffer. Instead of managing the total float of network paths, the critical chain method focuses on managing the remaining buffer durations against the remaining duration of chains of activities. The CCPM buffer management process is typically split into three stages: OK, Watch and Plan, and Act. The OK stage is where approximately 100% to 67% of the buffer is left and no action is required. Next, in the Watch and Plan stage (67% to 33% buffer remaining) remaining buffers are continuously reduced. The trend needs to be monitored and the plan needs to be set for the action. Finally, in the Act stage, the buffers are exhausted, and it is thus time to execute the action (overtime work, additional resources, etc.) based on the plan.

3. Novelties and Contribution

Previous studies verified that CCPM is applicable to real projects [14–17] and can be used to manage project schedules [18,19]. These studies focused on architecture and civil industries, but did not consider plant construction. Previous studies have introduced, evaluated, and developed the CCPM concept, but lack CCPM implementation within a realistic environment [17]. In addition, no previous research has presented CCPM for robust schedules or sufficiently discussed how cost and quality fluctuate with schedule management.

This paper adds to the general body of knowledge by presenting a CCPM process applied to a robust EPC plant project schedule such that can be used by practitioners in a real-world scenario. However, the major uniqueness of this study is that it manages to integrate a material resource management process into the CCPM process. This proposed process will allow project managers to check the availability of the piping materials needed for installation in advance to avoid delays. Through this process, piping materials are seen as a resource and any unintended absences' schedule impacts are managed through a "resource buffer" flexibility. The proposed method's superior performance (compared to traditional CPM without a material management process) is verified through Monte Carlo Simulations of a project case-study.

This study focuses on producing a superior schedule management process for field installation of piping units for plant construction projects, validated through an executed Korean EPC project. This analysis has been isolated to the construction portion of the project and has ignored the engineering and procurement schedule and cost impacts. This isolation allows a focused discussion on the different construction management techniques (PERT/CPM vs. CCPM), removing variables that are not the focus of this paper. (Although the procurement schedule is also not included in the PERT/CPM vs. CCPM analysis, its management is discussed through a separate process.) Below, the four research steps (RS) taken are described. These research steps are also illustrated in a box-flow in Figure 2.

(RS1) Investigating existing PERT/CPM schedule limitations and Proposing the CCPM process as a way of mitigating limitations for PERT/CPM,

(RS2) CCPM Application on a sample, theoretical piping installation project,

(a) Propose the application step: CPM Schedule→Resource Allocation and Leveling→Creating Critical Chain Activities→Creating Buffers→Material Risk Management with Resource Buffer (details in Section 8).

(RS3) Propose a supplemental resource management process for piping material risk analysis,

(a) Identify the risk factors for project delay
(b) Propose the way to insert the resource buffer as a schedule milestone
(c) Propose the flow chart for material management process

(RS4) Apply CCPM and resource management process to two processes (Crude Pump Unit and Propylene Compressor Unit) within a previously executed Middle Eastern Refinery Expansion Project (performed by Korean EPC), verifying:

(a) Deterministic Schedule Analysis: CPM vs. CCPM applied resource management process using MS Project
(b) Probabilistic Schedule Analysis: CPM vs. CCPM applied resource management process using @Risk (schedule risk simulation software, Palisade, Sydney, Australia) on a previously executed Korean projects.

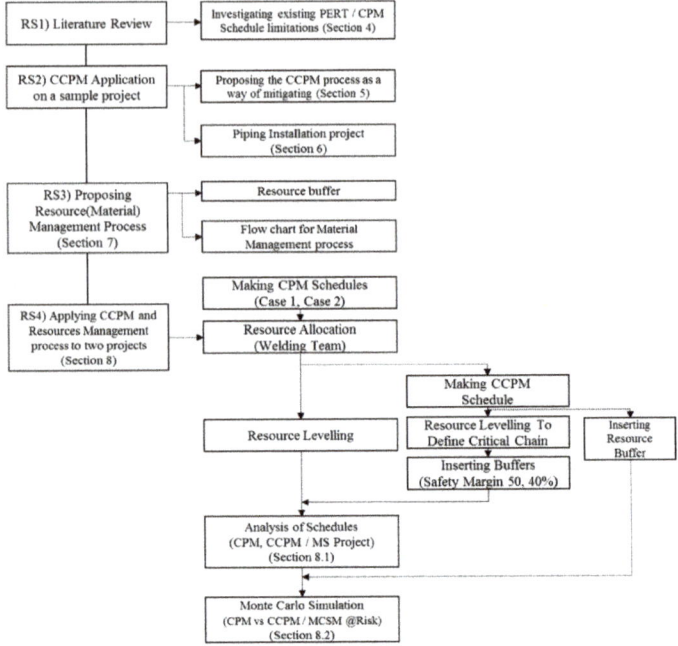

Figure 2. Research Methodology and Process.

4. Relevant Literature

Critical Chain Project Management is a method and collection of algorithms based on the Theory of Constraints (TOC). This theory is a management philosophy that proposes focusing on constraints is a better strategy to achieve the goal. The idea of CCPM was first introduced by Eliyahu M. Goldratt in 1997 [20]. After its introduction, many researchers have studied the efficacy of CCPM. Studies have included providing clarity on CCPM's use [21], presenting advantages and disadvantages of the CCPM approach [22], comparing traditional and TOC project management processes [23], and proposing an enhanced framework combining Supply Chain Management (SCM) and CCPM to manage EPC project uncertainty [24]. Wang et al. proposed an improved CCPM framework that addressed two major challenges in CCPM-based scheduling such as buffer sizing and multiple resources leveling to enhance the implementation of CCPM. This study's results displayed that performing multiple resources leveling in CCPM-based construction scheduling was both feasible and effective [25].

Although research in CCPM has been exhaustive, the systematic assessment of its performance within probabilistic schedules has been identified as a gap in the literature [26]. Specifically, no previous research has presented algorithms for robust schedules, sufficiently discussed how cost and quality fluctuate with schedule management, and/or were conceptual, lacking specific application. This paper seeks to bridge these gaps.

A subsection of CCPM, management of buffers, has also been studied. There exist general guidelines of managing buffer through the three layers (i.e., "OK", "Watch", and "Plan") [27]. In addition, Lee [28] experimented how the duration and cost of a project fluctuated depending on how the three levels of buffer management were set up. Alternatively, Lee [14] studied how to set the levels of buffer management according to the project's schedule margin. However, none reflect the characteristics of an EPC project.

As stated, a need was found in incorporating the material management procurement process into a scheduling process, currently missing in literature and industry. Several researchers have studied managing the procurement process through differing scheduling processes. Yeo and Ning proposed the enhanced framework for procurement by coupling the concepts of SCM and CCPM [24]. Bevilacqua developed a prioritization method for work packages, within the critical chain context to minimize the risks related to accidents in refinery plants [15]. Feng proposed a model with multi-resource constraints, which could be applied to the critical chain construction of the A-bid section in the South-to-North Water Diversion Project [18]. Dong attempted to apply the CCPM theory to the maintenance period arrangements of power plant electrical equipment, and a mathematical model of optimizing the maintenance period was established [19]. Jose Finocchio Jr. discussed the adequacy of the CCPM for scheduling projects involving shutdowns on an oil platform [29]. These findings were used as foundation for the scheduling portion of this research, but lack applicability to EPC mega-project procurement.

Previous studies, performed by practitioners, studied the relationship between PERT/CPM and CCPM in industry use. PERT/CPM and CCPM methodologies can coexist in project-based organizations [30] because "*CPM employs one time estimate and one cost estimate for each activity; PERT may utilize three time estimates (optimistic, expected, and pessimistic) and no costs for each activity. Although these are distinct differences, the term PERT is applied increasingly to all critical path scheduling*" [31]. Nowadays, PERT is commonly used in conjunction with the CPM, while PERT/CPM is the most widely used terminology for project management techniques. Even though PERT/CPM is the most widely used technique, many studies have shown it is lacking, containing fundamental limitations [32,33]. The most significant limitation of PERT/CPM is the inability to consider resource constraints and poor management of project uncertainty.

While PERT/CPM estimates an optimized schedule, the process only considers the work sequence and relations for planning the schedule, ignoring resources availability [11]. Resource optimization, such as resource leveling or resource smoothing, is a recommended scheduling practice to ensure resource availability is considered and is included in the CCPM process [11]. Alternatively, the traditional uncertainty management approach of PERT/CPM is also lacking. In this traditional process, schedulers create "safety time" or a "safety margin" to protect against project growth during execution caused by unknowns, often based on worst case scenarios [20]. This results in project uncertainties being "buried" and unrealistically long project durations [33]. The CCPM process includes separate buffer activities which are transparent safety margins allowing a more efficient schedule risk (safety margin) management process.

Researchers conclude that CCPM is superior in reliable planning and the execution process. CPM was found to offer no execution methodology, a fundamental limitation, and it is recommended that CPM should subordinate to CCPM during execution of the scheduling control [34]. CCPM also better manages limited essential resources (high-capacity lifting cranes, professional welders, piping fitters, etc.), experienced in recently EPC megaprojects in the middle east Asia. This is illustrated from the following quote, "*CPM is as scheduling method to identify the shortest time a project could be accomplished assuming resources are Infinite, whereas CCPM is a method of planning and managing projects that put the main emphasis on the resources required to execute project tasks*" This implies that CPM alone does not make sense for field resource allocations and supports this paper's use of CCPM as a more applicable scheduling analysis tool for EPC megaprojects [30]. Goldratt [20] and others pointed out with rectifiable reasons that the reliance on CPM as the sole schedule planning methodology frequently causes projects to miss deadlines and commitments to stakeholders. A major risk omitted from mitigation in CPM is the contention for scarce resources [34].

5. RS1: CCPM Application

CCPM differs from PERT/CPM mainly due to its inclusion of resource dependencies and fixed nature during the project period [12]. CCPM improves the project plan by ensuring that it is feasible and immune from reasonable common cause variation such as uncertainty or statistical fluctuations [33]. As stated, CCPM is based on the TOC, in which there is always at least one constraint. In TOC, a focusing process is used to identify constraints and reorganize around said constraint to minimize impacts [20]. CCPM uses logical relationships and resource availability and ensures activities durations do not include safety margins. It also uses statistically determined buffers, aggregated safety margins of activities placed at strategic points on the project schedule path. This accounts for limited resources and project uncertainty, limitations of PERT/CPM as discussed above.

In conclusion, CCPM mitigates the CPM limitations in the following ways [12]:

(1) *Risk Management*: Unexpected risks management and risk absorption with buffer management.
(2) *Focused Oversite*: Management attention remains centralized on critical chain which is fixed.
(3) *Resource contention*: The project duration is dependent on Resource Availability no later than the logical sequence of activities.

6. RS2: CCPM Application to Pipe Installation Works

The authors have chosen to apply CCPM on a sample piping installation project. To allow the reader a better understanding of the project, a brief description of activities, relations, durations, and basic scheduling follow. A sample piping field installation for plant construction project was developed with activities, relations, and general schedule seen below in Table 1. The case study in this paper basically uses "day" as the time unit for activity duration, according to a common industry practice for EPC projects. The critical path activity is activity Nos. 1→3→5→6 and the project duration is 18 days.

Table 1. Sample project—piping installation work.

Activity No.	Predecessor	Successor	Relations	Duration (Days)	Resource Name
1		2, 3	FS	2	A
2	1	4	FS	4	B
3	1	5	FS	6	B
4	2	6	FS	2	B
5	3	6	FS	4	C
6	4, 5		FS	6	A

The project was then resource loaded for each resource for the sample project. Resource competition occurs for activity numbers 2, 3 and 4. Resource B is allocated to both activity numbers 2 and 3 from Day 3 to Day 6, and both activity numbers 3 and 4 from Day 7 to Day 8. Figure 3 depicts the weakness of the traditional scheduling method: the schedule cannot be realistically completed in 18 days due to the resource constraints. If this were at true error, the project schedule would be delayed due to poor schedule estimating techniques.

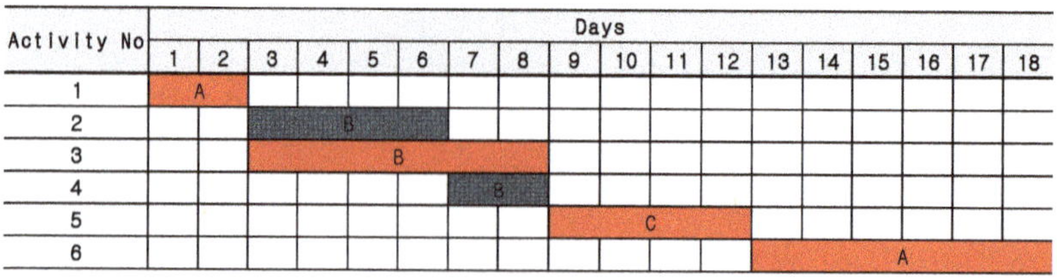

Figure 3. Critical path for sample project and resource allocation (labor). (A: Resource A, B: Resource B, C: Resource C).

To solve the resource competition, the activity sequence is rearranged based on resource B. This results in increasing the project duration from 18 days to 20 days. After resolving the resource constraint, the critical path activity is revised to a critical chain activity, activity numbers 1→3→2→4→6.

Next, safety margins, or buffers, are introduced to the schedule. The estimation of the buffer size is dependent on each activity and its safety margin. To reduce the behaviors and time wasting associated with having excessive embedded safety, CCPM recommends that the task estimates are reduced by half the length of a "normal" duration [20]. To develop the buffer to be used for this research, this paper relies on the findings of previous studies. Many researchers estimated the project buffer as the 50% of a total sum of the 50% safety margin [13,16,28,35–37]. However, Lee [38] suggested that it is not reasonable to estimate the buffer to 50% of work duration and proposes 20% of work duration be applied by Monte Carlo simulation. Liu [39] insisted that the buffer should be estimated according to the worker's ability and experience. Cho [40] argued that, when estimating the safety margin, the positive case in the 3-point estimation needs to be eliminated. Han [41] concluded that Many factors are considered when estimating a buffer.

For simplicity, and as this is common practice, the authors assume (as illustrates in Figure 4) that each activity has a 50% safety margin for the beginning, providing a project buffer of 50% of the total sum of the safety margin [20]. However, a project buffer with 40% or 30% of the total sum of the safety margin are compared when this process is applied to the project case study in Section 10. The critical chain activity is activity numbers 1→3→2→4→6 and the project duration is reduced to 15 days. The buffers include a feeding buffer (one day) and a project buffer (five days) (Figure 4).

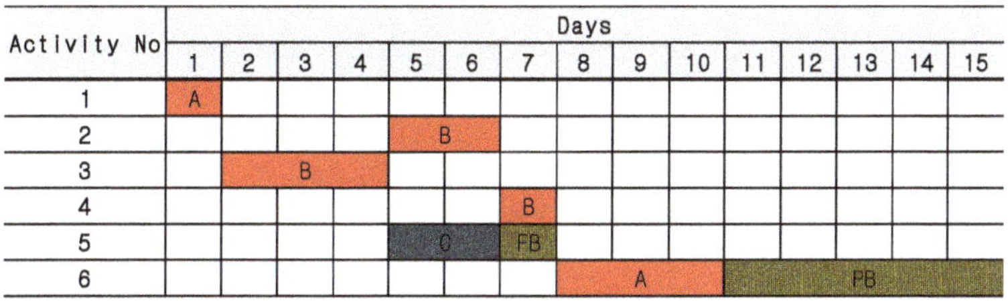

Figure 4. Critical chain activities (resource leveling) and inserting Buffers (project buffer and feeding Buffer). (A: Resource A, B: Resource B, C: Resource C, FB: Feeding Buffer, PB: Project Buffer).

7. RS3: CCPM Incorporating with Material Resource Management

Previous researchers concentrated on project buffer and feeding buffer size and there is a lack of investigations into the resource buffer [42]. Resource buffers have been treated only as a milestone activity notifying the start of successor activities. This leaves project managers at a disadvantage for managing this uncertainty and represents a gap in existing literature. Valikoniene proposed including resource buffers as time buffers with the assigned resources and cost, and concluded that *"application of the resource buffers on average shortens project duration"* [42]. In this study, resource buffers were created as a milestone activity for project schedule and it proposed the resource (material) management process is a flow chart. When simulating the projects for the case studies, the resource buffers are considered as time buffers with durations of 0–1.7 days. Concerning this gap in the literature, the checking and monitoring piping material availability is currently not supported by the CPM and CCPM processes used in practice. To fill this void, this study proposed a method to manage the materials related to the work activity from the concept of resource buffers. These resource buffers are inserted alongside the critical chain within the schedule to ensure that the appropriate people and skills are available to work as soon as needed [20]. Researchers suggested placing milestone activities in the CCPM schedule to indicate the timing for the delivery. However, supplementary to the schedule management, the practitioner should use the resource management process presented within this paper.

The resource management process proposed is depicted in Figure 5 as a flow chart. As can be seen, this process is complex and requires collaboration and efficient communication between the project manager (reviewing milestone schedules), material manager (monitoring piping availability and source of replenishment), and construction manager (controlling resource use and allocation within the work sequence).

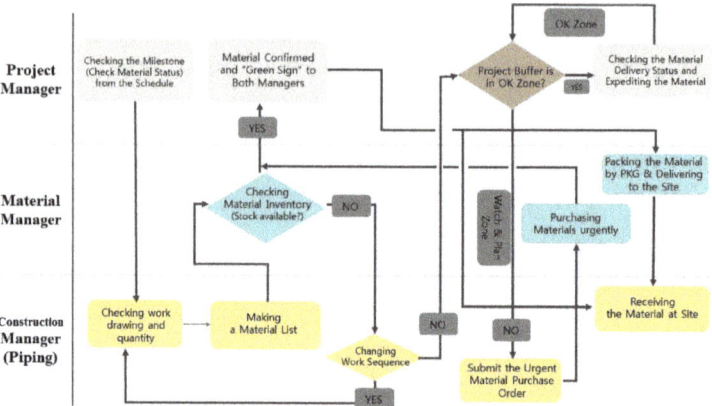

Figure 5. Flow Chart for Material Management Process.

Many factors can contribute to a delay in the availability of piping material, and the solution differs depending on the case. In a previous study, the risk factors contributing to the material issue for the plant construction project were presented. Kim, Yong-Su, Hwang, Moon-Hwan, Park, and Chan-Sik indicated the most common reason for delay is mobilization of workers, which accounts for 25% of project delays. They went on to account 13% of delays due to lack of project planning accounts, 11% of delays due to a delay in arrival of material, 2% of delays due to equipment mobilization and lack of qualified workers each [43]. Kim [44] insisted that the risks of design change and additional works have a significant impact on project duration. Kang [45] indicated that a delay of material procurement and logistics as well as a design change while construction work is in progress are the major factors for project delay.

8. RS4: Case Study of CCPM Application with Material Resource Management

The purpose of this section is to validate the application of the CCPM on the piping field installation work for a plant construction project. The following shows the project outline and piping unit information for the case study, which was recently completed. Table 2 depicts the work quantities for each case described below.

i. Project Name: Refinery Expansion Project (Middle East)
ii. Contractor: D Company
iii. Project Duration: 50 Months
iv. Unit: Crude Pump Unit (CPU)/Propylene Compressor Unit (PCU)
v. Discipline: Piping Field Installation

CPU (Case 1): The function of the Crude Pump Unit (CPU) is to transport the crude oil to the storage tank. It consists of three pumps that have suction, discharge, and min. flow lines. Diameter inch (DI) is the unit of measurement for pipe welds. If a particular spool has five joints of 6″ diameter, the total DI for the pipe welds is 30 DI (6″ × 5 joints = 30 DI).

PCU (Case 2): The function of the Propylene Compressor Unit (PCU) is to compress the propylene gas to transform it into a liquid state to minimize the propylene gas loss and increase the efficiency of the refinery. The unit consists of one compressor unit with three suction lines and one discharge line.

Table 2. Work Quantity for Both Cases.

Case	Line No.	Description	Piping Size (Inch)	Piping D/I * Field	PKG Q'ty ** (Test)
Crude Pump Unit (CPU-Case 1)	54″-P-006	Pump Suction Header	54	1689	1
	42″-P-268	G-001 Suction	42	326	1
	42″-P-271	G-002 Suction	42	182	1
	42″-P-274	G-003 Suction	42	452	1
	36″-P-021	Pump Discharge Header	36	955	2
	20″-P-023	Min. Flow Header	20	996	1
	24″-P-269	G-001 Discharge	24	570	1
	12″-P-270	G-001 Min. Flow	12	223	(Min. Flow Header)
	24″-P-272	G-002 Discharge	24	613	1
	12″-P-273	G-002 Min. Flow	12	186	(Min. Flow Header)
	24″-P-275	G-003 Discharge	24	651	1
	12″-P-277	G-003 Min. Flow	12	233	(Min. Flow Header)
Propylene Compressor Unit (PCU-Case 2)	20″-P-1968	K-001C Suction 1	20	380	2
	14″-P-	Return Line	14	476	(Discharge)
	14″-P-	K-001C Suction 2	14	218	2
	12″-P-	Return Line	12	96	(Discharge)
	16″-P-	K-001C Suction 3	16	122	2
	16″-P-	Return Line	16	240	(Discharge)
	18″-P-	K-001C Discharge	18	326	1

(Note: * Pipe D/I = pipe diameter—inch, which is typical unit of work-quantity for ping installation). ** PKG Q'ty = Pipe spool package quantity).

8.1. Deterministic Schedule Analysis: CPM vs. CCPM

To prepare a schedule, the following condition is assumed in this study:

(1) The schedule is limited to the field welding for piping work.
(2) The resources are welding teams (other resources such as cranes, loaders, operators, painting crews, insulation crews, etc. are not considered.)
(3) The day productivity is 8 DI/welder. Available resources are six welding teams for crude pump unit and four teams for propylene compressor unit.
(4) Piping welding is conducted by piping work package based on the isometric drawings.
(5) The work sequence requires the pump and header line to be welded first followed by the branch lines due to the space constraint.

Tables 3 and 4 indicate the package number, isometric drawing number, work sequence, work quantity, resource allocation (welding team), and work duration for the crude pump unit and propylene compressor unit.

As stated, in preparing the CCPM schedule, resource leveling and buffer creations are performed. Since resource leveling was performed when making the CPM schedule, the buffers can now be created and inserted into the CCPM schedule. In this study, the safety margins are assigned differently based on the activity, resource, and project, so the safety margin is set as three cases of 50%, 40%, and 30%. Tables 3 and 4 summarize the inputs setting for CCPM simulation. After creating the project buffer, feeding buffer, and resource buffer, they are incorporated into the CCPM schedule, as shown in Figures 6 and 7 for the crude pump unit and propylene compressor unit, respectively.

Table 3. Crude pump unit (CPU-Case 1)—activity sequence and resource allocation.

Activity No.	PKG No.	ISO DWG No.	Description	Predecessor	Piping D/I Field	Welding Team (8 DI/Day)	Duration (Day)	Duration (Day) SM 50%	Duration (Day) SM 40%	Duration (Day) SM 30%
1	PKG-01		Pump Suction Header		1512	3				
2		P-006-01		2	108		4.5	3.0	2.7	3.2
3		P-006-02		3	324		13.5	7.0	8.1	9.5
4		P-006-03		4	324		13.5	7.0	8.1	9.5
5		P-268-01		4	210		8.8	4.0	5.3	6.1
6		P-271-01		4	210		8.8	4.0	5.3	6.1
7		P-274-02		4	336		14.0	7.0	8.4	9.8
8	PKG-03		G-002 Suction		66	1				
9		P-271-02		7	66		8.0	4.0	4.8	5.6
10	PKG-02		G-001 Suction		108	1				
11		P-268-02		5	108		14.0	7.0	8.4	9.8
12	PKG-04		G-003 Suction		24	1				
13		P-274-01		7	24		3.0	2.0	1.8	2.1
14	PKG-05		Pump Discharge Header 1		252	2				
15		P-021-01			36		2.3	1.0	1.4	1.6
16		P-021-02		15	180		11.3	6.0	6.8	7.9
17		P-021-03		16	36		2.3	1.0	1.4	1.6
18	PKG-06		Pump Discharge Header 2		540	2				
19		P-021-03		11, 16, 39	180		11.3	6.0	6.8	7.9
20		P-269-03		4, 19	96		6.0	3.0	3.6	4.2
21		P-272-03		4, 19	144		9.0	5.0	5.4	6.3
22		P-275-03		4, 19	120		7.5	4.0	4.5	5.3
23	PKG-07		G-003 Discharge		370	2				
24		P-275-01		4	134		8.4	4.0	5.0	5.9
25		P-275-02		24	120		7.5	4.0	4.5	5.3
26		P-275-03		25	24		1.5	1.0	0.9	1.1
27		P-277-01		25	92		5.8	3.0	3.5	4.0
28	PKG-08		G-002 Discharge		248	2				
29		P-272-01		4, 17	120		7.5	4.0	4.5	5.3
30		P-272-02		29	72		4.5	3.0	2.7	3.2
31		P-272-03		29	24		1.5	1.0	0.9	1.1
32		P-273-01		31	32		2.0	1.0	1.2	1.4
33	PKG-09		G-001 Discharge		388	2				
34		P-269-01		4, 17	182		11.4	6.0	6.8	8.0
35		P-269-02		34	72		4.5	3.0	2.7	3.2
36		P-269-03		35	102		6.4	3.0	3.8	4.5
37		P-270-01		36	32		2.0	1.0	1.2	1.4
38	PKG-10		Min. Flow		850	2				

Table 3. Cont.

Activity No.	PKG No.	ISO DWG No.	Description	Predecessor	Piping D/I Field	Welding Team (8 DI/Day)	Duration (Day)	Duration (Day) SM 50%	Duration (Day) SM 40%	Duration (Day) SM 30%
39		P-023-01		3	60		3.8	2.0	2.3	2.6
40		P-023-01A		39	40		2.5	2.0	1.5	1.8
41		P-023-02		40	40		2.5	2.0	1.5	1.8
42		P-023-03		41	160		10.0	5.0	6.0	7.0
43		P-023-04		42	20		1.3	1.0	0.8	0.9
44		P-270-01		45	48		3.0	2.0	1.8	2.1
45		P-270-02		39	132		8.3	4.0	5.0	5.8
46		P-273-01		47	40		2.5	2.0	1.5	1.8
47		P-273-02		39	140		8.8	5.0	5.3	6.1
48		P-277-01		49	40		2.5	2.0	1.5	1.8
49		P-277-02		40	130		8.1	4.0	4.9	5.7
Total Piping D/I					4358					

Table 4. Propylene compressor unit (PCU-Case 2)—activity sequence and resource allocation.

Activity No.	PKG No.	ISO DWG No.	Description	Predecessor	Piping D/I Field	Welding Team (8 DI/Day)	Duration (Day)	Duration (Day) SM 50%	Duration (Day) SM 40%	Duration (Day) SM 30%
1	PKG-01		BL to K-001C Suction 1	17	192	2				
2		P-876-02			32		2.0	1.0	2.0	2.0
3		P-979-01A		2, 14	80		5.0	2.5	3.0	3.5
4		P-979-01		3, 23	80		5.0	2.5	3.0	3.5
5	PKG-02		K-001C Suction 1		90	1				
6		P-876-01		2	64		8.0	4.0	4.8	5.6
7		P-876-01A		6	26		3.3	1.6	2.0	2.3
8	PKG-03		BL to K-001C Suction 2		140	1				
9		P-878-02		9	72		9.0	4.5	5.4	6.3
10		P-980-01			68		8.5	4.3	5.1	6.0
11	PKG-04		K-001C Suction 2		90	1				
12		P-878-01		9	90		11.3	5.6	6.8	7.9
13	PKG-05		BL to K-001C Suction 3		440	2				
14		P-1968-02		14, 9, 20	160		10.0	5.0	6.0	7.0
15		P-978-01			280		17.5	8.8	10.5	12.3
16	PKG-06		K-001C Suction 3	14	160	1				
17		P-1968-01		9	160		20.0	10.0	12.0	14.0
18	PKG-07		K-001C Discharge		554	2				
19		P-892-01		20	92		5.8	2.9	3.5	4.0
20		P-892-02		21	126		7.9	3.9	4.7	5.5
21		P-892-03	To BL		36		2.3	1.1	2.0	1.6
22		P-979-01	Return Line (Suction 1)	20	80		5.0	2.5	3.0	3.5
23		P-978-01	Return Line (Suction 3)	20	196		12.3	6.1	7.4	8.6
24		P-980-01	Return Line (Suction 2)	20	24		1.5	0.8	2.0	2.0

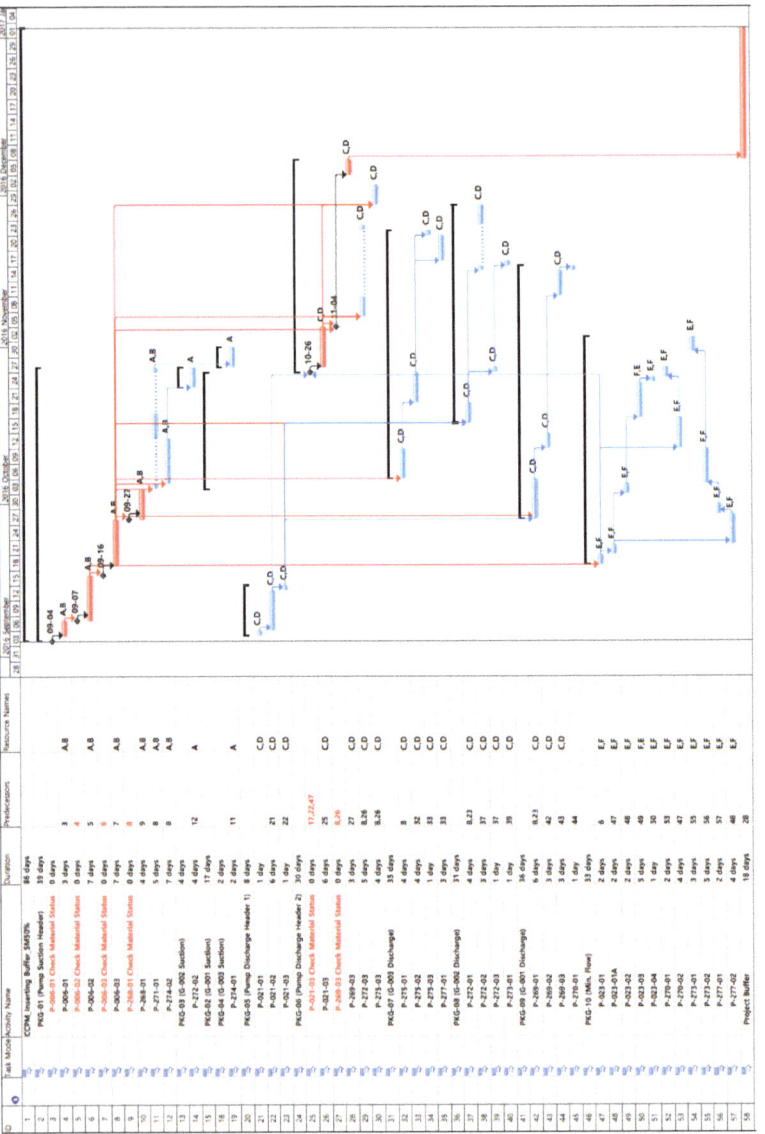

Figure 6. CCPM schedule for CPU-Case 1 (safety margin 50%).

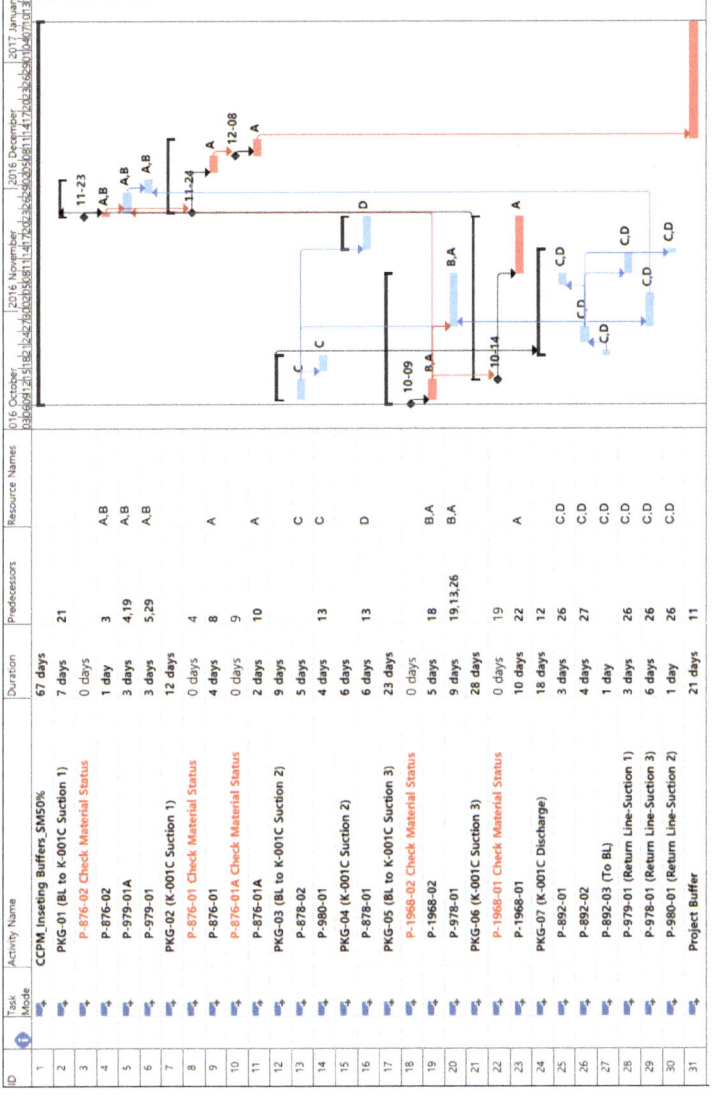

Figure 7. CCPM Schedule for PCU-Case 2 (safety margin 50%).

Table 5 shows a comparison of the CPM Schedule and the CCPM Schedule with a 50% safety margin. The units differ for the critical path (critical chain) activity and project duration. Critical Path for CPU-Case 1 is: Activity Nos. 2→3→4→5→11→19→21, while critical chain is: Activity Nos. 2→3→4→5→11→19→20. This is the result from the resource leveling. Critical chain for PCU-Case 2 is also different with critical path due to resource leveling. Applying CCPM to the project reduces the overall project duration by about 48 days from 134 days to 86 days for CPU-Case 1 and by about nine days from 76 days to 67 days for PCU-Case 2.

Table 5. Schedule comparison of CPM and CCPM (safety margin 50%).

Units	Description	Critical Path/Critical Chain	Project Duration
Crude Pump Unit (CPU-Case 1)	CPM	Activity Nos. 2→3→4→5→11→19→21	134 Days
	CCPM (Safety Margin 50%)	Activity Nos. 2→3→4→5→11→19→20	86 Days
Propylene Compressor Unit (PCU-Case 2)	CPM	Activity Nos. 14→16→2→6→7	76 Days
	CCPM (Safety Margin 50%)	Activity Nos. 14→17→2→6→7	67 Days

However, setting the safety margin at 50% for all projects would be impractical. To check the sensitivity due to the safety margin, a CCPM schedule was prepared in this study and 40% and 30% safety margins were applied for both units (Table 6).

Table 6. Deterministic schedule analysis: CPM vs. CCPM with different safety margins.

Project	(Unit: Day)	CPM	CCPM with Safety Margin		
			50%	40%	30%
Crude Pump Unit (Case 1)	Project Duration	134.0	86.0	93.0	97.0
	Project Buffer		18.0	15.0	12.0
	Feeding Buffer (1)		4.0	3.0	2.0
	Feeding Buffer (2)		1.0	1.0	1.0
Propylene Compressor Unit (Case 2)	Project Duration	76.0	67.0	71.0	76.0
	Project Buffer		21.0	16.0	12.0

With a smaller safety margin, the CCPM-predicted project duration increased, although it is still shorter than the CPM project duration. In the case of the crude pump unit, the project duration of CPM is 134 days. However, the project duration of CCPM with a 30% safety margin is 97 days including 15 days of buffers. For the propylene compressor unit, while the project duration in both CPM and CCPM with a 30% safety margin is 76 days, CCPM has a 12-day project buffer. For both projects, CCPM is more capable of reducing the project duration even when applying a 30% safety margin.

8.2. Case Studies for Probabilistic Simulation; CPM vs. CCPM

In this study, the Monte Carlo Simulation Model (MCSM) and @Risk Program (schedule risk simulation SW) are used for validating the project duration. The simulation procedure for the piping installation work is as follows:

i. *Developing Model*: Create the schedule using MS Project (network diagram).
ii. *Defining Parameter*: Define the piping installation work and input the project duration (parameter).
iii. *Defining Uncertainty*: Define the PERT distribution (three-point estimation) for each activity duration.
iv. *Simulation*: Check and analyze the total project duration.

The simulation is conducted 1000 times with random input and confidence of 85%. The project buffer is considered a fixed value as it is calculated from the project duration. Each activity is applied to PERT distribution (three-point estimating), a probabilistic distribution function, and each three have different values: minimum, most likely, and maximum. For the CCPM, the three cases of maximum value are 30%, 40%, and 50% followed by the safety margin proposition. For both CPM and CCPM, the minimum value is −10% and the most likely value is the project duration by the MS Project.

The resource buffer is presented as a milestone in the project schedule, so it has no duration and does not affect the schedule. In field application, the resources would be delayed for various reasons. The resource buffer is therefore considered a parameter. It could have a three-point distribution as a beta-distribution (i.e., most likely, pessimistic (min), and optimistic (max)). For example, each resource buffer has the most likely case of 1 day, a minimum value of 0 days (no delay), and a maximum value of 1.7 days, which is calculated based on the sum of all resource buffers being less than the sum of all project buffers. Figures 8 and 9 show the results of CCPM simulation.

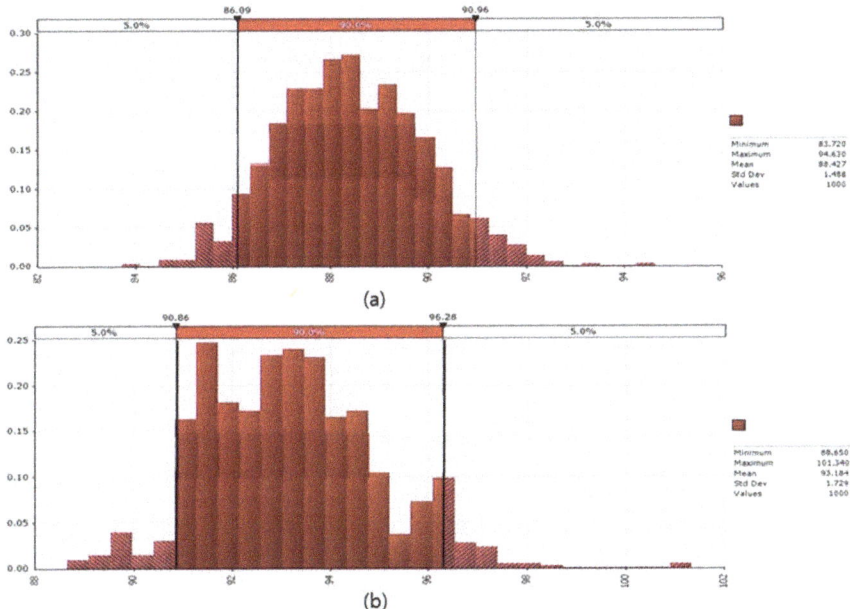

Figure 8. Project Duration for CCPM with: (**a**) Safety Margin of 50%; and (**b**) resource buffer applied (CPU-Case 1).

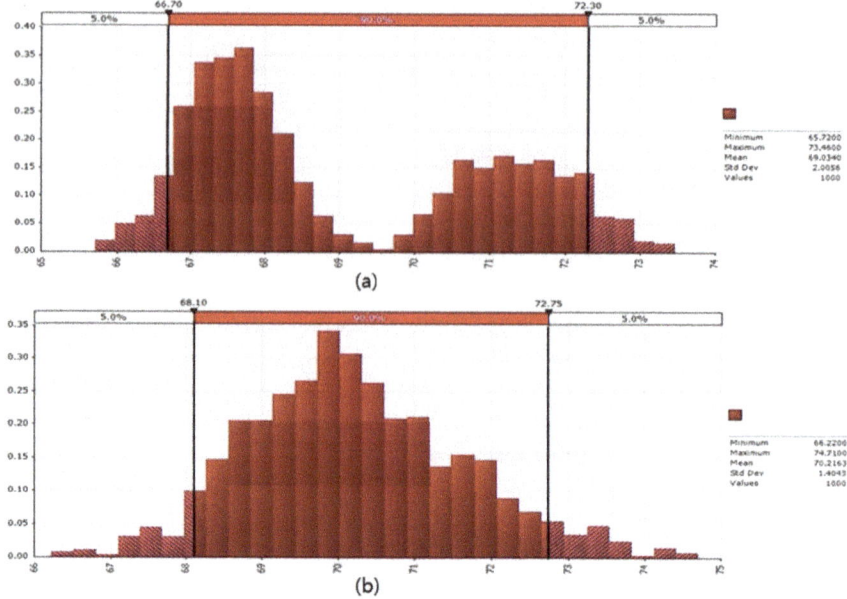

Figure 9. Project Duration for CCPM with: (**a**) safety margin of 50%; and (**b**) resource bugger applied (PCU-Case 2).

8.3. Comparison of Analysis Results—Probabilistic Simulations

8.3.1. CPM vs. CCPM without Resource Buffer

The summary of the simulation results between CPM and CCPM with various safety margins is shown in Table 7. In the case of the Crude Pump Unit (CPU-Case 1), the result indicates that the CPM project duration would have a range of 134 to 152 days. The project duration range of CCPM with a safety margin 30% is 95 days to 103 days. The CPM project duration from @Risk with 85% confidence is 146 days, while it is 101 days for CCPM (SM 30%). As confirmed at the deterministic schedule analysis (Section 8.1), CCPM is capable of shortening the project duration from 146 days to 101 days at 85% confidence in the probabilistic schedule analysis as well. CCPM, moreover, clearly indicates the project buffer to response to the project risk, while CPM could not present the project buffer. Comparing the project duration between deterministic and probabilistic analysis, the project duration of CPM increases from 134 days to 146 days (difference: 12 days) at 85% confidence, while CCPM (SM 30%) is from 97 days to 101 days (difference: four days only) at 85% confidence. These results ensure that CCPM is more stable than CPM and has the benefit of reducing the project duration. For applying the safety margin, as safety margin decreases from 50% to 30%, the project buffer is also reduced from 18 days to 12 days and the project durations are increased for both deterministic analysis (from 86 days to 97 days) and probabilistic analysis (from 91 days to 101 days). When it comes to the size of the project buffer, CCPM with SM 30% is still valid against CPM. In the case of Propylene Compressor Unit (PCU-Case 2), the same results are observed.

Table 7. Summary of simulation results.

Project	Description	Project Duration (Deterministic)	Project Buffer	MCSM (@Risk)			
				Min	ML	Max	Confidence 85%
Crude Pump Unit (Case 1)	CPM	134		134.0	141.2	151.8	146.3
	CCPM (SM 50%)	86	18	83.7	88.4	94.6	91.0
	CCPM (SM 40%)	93	15	91.5	95.4	99.8	98.4
	CCPM (SM 30%)	97	12	95.1	98.6	103.3	100.8
Propylene Compressor Unit (Case 2)	CPM	76		74.0	79.3	88.6	83.5
	CCPM (SM 50%)	67	21	65.7	69.0	73.5	72.3
	CCPM (SM 40%)	71	16	69.5	72.3	77.4	74.5
	CCPM (SM 30%)	76	12	74.3	77.2	81.3	80.4

(SM: Safety Margin, ML: Most Likely).

8.3.2. CPM vs. CCPM with Resource Buffer

This paper has claimed that the resource buffer is a greatly underutilized tool. To test its efficacy, a simulation is performed on the CCPM schedule in which the resource (material) buffer is considered. The findings are presented in Table 8. The duration of the resource buffers was determined to be 0–1.7 days for the PERT distribution. As can be seen, applying the resource buffer increases the project duration (comparing Table 7 to Table 8 findings). For example, Table 7 shows the CCPM (SM 50%) duration as 91 days (85% confidence), while Table 7 depicts 96 days for CCPM (SM 50%) with the resource buffer. Similar findings can be finding when comparing all SMs amongst the two Case Studies.

These findings illustrate that, without a resource buffer, projects are more likely to have extended schedules due to not adequately taking into account potential delivery delays. For example, when using a SM of 50%, the Crude Pump duration of 91 days includes no delivery delays. This means that when a delivery delay occurs, the project manager will likely have to pull from the project buffer "fund" which is likely allocated to other risks. Alternatively, the project manager using data from Table 8 will be better prepared to handle delivery delays and less likely to have schedule delays due to said delivery delays (as the project includes contingency for such an event).

It is also interesting to note that the project durations when applying CCPM with resource buffers are less than those of CPM. For example, in the case of CPU-Case 1, the project duration for CCPM (SM 30%) is 106 days at 85% confidence, while the project duration for CPM is 146 days. Although the CPM schedule will have ample float, it will likely be less efficient and/or result in the bidder not being competitive.

Table 8. Simulation result of CCPM schedule applying the resource (material) buffer.

Project	Description	Project Duration (Deterministic)	Project Buffer	MCSM (@Risk)				Remarks
				Min	ML	Max	Confidence 85%	
Crude Pump Unit (Case 1)	CCPM (SM 50%)	86	18	88.7	93.2	101.3	96.3	
	CCPM (SM 40%)	93	15	96.0	100.6	106.2	103.9	Resource Buffers are added from 0 to 1.7 days
	CCPM (SM 30%)	97	12	98.4	103.4	109.2	106.4	
Propylene Compressor Unit (Case 2)	CCPM (SM 50%)	67	21	66.2	70.2	74.7	72.8	
	CCPM (SM 40%)	71	16	70.2	74.1	78.5	76.7	
	CCPM (SM 30%)	76	12	74.8	78.5	85.0	81.5	

(SM: Safety Margin, ML: Most Likely).

9. Conclusions

Through an investigation of scheduling of a crude pump unit and propylene compressor unit, it was found that the traditional PERT/CPM scheduling method may be prone to overestimating project durations (~35% increase of duration when compared to the CCPM method in this study). Alternatively, excluding material uncertainty (represented by the resource buffer) in schedule development, a project has a greater potential for schedule growth due to improper risk preparedness (5% schedule growth potential found in this study). In summary, this paper finds that the CCPM scheduling technique, when coupled with adequate use of resource buffers and a procurement management process, is more apt at effectively and efficiently managing a project's risks.

The impacts of resource buffer on project duration are lacking in the literature. As such, the results of this paper represent an addition to the existing body of knowledge. Furthermore, this study provides practitioners a useful platform to manage piping installation work, eliminate project duration extension, and enable project teams to effectively manage limited human and material resources using the buffer management and material management processes.

10. Recommendations for Future Research

This study was conducted for piping field installation work for an EPC plant project. It was demonstrated that construction scheduling needs to be expanded to engineering and procurement scheduling. A method of estimating the buffer size needs to be developed and validated, such as project buffer, feeding buffer, and resource buffer, which is more reliable and realistic. While this study only considered the labor resource (welder), for the further study, additional resources such as equipment, scaffolding, etc. will be considered. The case study in this paper used "day" as the time unit basically (although it incorporated decimal points of project buffers), according to a common industry practice. Using "hour" time unit in the CCPM analysis might have some advantages (namely more precision to better present the sensitivity of resource allocations), which might be covered in future research.

Author Contributions: Sung-Hwan Jo developed the concept and drafted the manuscript based on the analysis. Kyoung-Youl Pyo verified the analysis and revised the manuscript. Eul-Bum Lee set up the study directions and supervised the overall work. All of the authors read and approved the final manuscript.

Acknowledgments: This work was supported by the Technology Innovation Program (10077606, To Develop an Intelligent Integrated Management Support System for Engineering Project) funded By the Ministry of Trade, Industry & Energy (MOTIE), Korea. The authors would like to thank D.S. Alleman (a Ph.D. candidate in University of Colorado—Boulder) for his industry feedback on this CCPM research.

Conflicts of Interest: The authors declare no conflict of interest.

Abbreviations

CCPM	Critical Chain Path Method
EPC	Engineering Procurement Construction
PERT	Program Evaluation and Review Technique
CPM	Critical Path Method
TOC	Theory of Constraints
ETA	Estimated Time of Arrival
P-I Matrix	Probability-Impact matrix
DI	Diameter Inch
BL	Branch Line

References

1. CERIK. *Construction Trend Monthly Briefing*; Korean Construction and Economy Research Institute of Korea (CERIK): Seoul, Korea, 2018.
2. Ahn, B. *Managing the Efficiency of Foreign Engineering Contracts: A Study of a Norwegian and South Korean Project Interface*; University of Stavanger: Stavanger, Norway, 2015.
3. Tanmay. *Oil & Gas EPC Market Perspective and Comprehensive Analysis to 2023*. 2018. Available online: https://theexpertconsulting.com/oil-gas-epc-market-perspective-and-comprehensive-analysis-to-2023/ (accessed on 20 May 2018).
4. National Contractors Association of Korea. *Statistics on Overseas Construction Contracts Amount*; National Contractors Association of Korea: Seoul, Korea, 2018; Available online: http://www.icak.or.kr/reg/login.php?s_url=L3BsdC9wbHRfc3RhMDQucGhwP2ZfZGF0ZT0xOTY5LzAxLzAxJnRfZGF0ZT0yMDE4LzA1LzIxJnJfY29kZT0= (accessed on 20 May 2018).

5. McNair, D. *EPC Contracts in the Process Plant Sector*; Asia Pacific Projects Update; DLA Piper: London, UK, 2011.
6. Blankenbaker, E.K. *Construction and Building Technology*; Goodheart-Willcox: Tinley Park, IL, USA, 2013.
7. Tommelein, I.D. Pull-driven scheduling for pipe-spool installation: Simulation of lean construction technique. *J. Constr. Eng. Manag.* **1998**, *124*, 279–288. [CrossRef]
8. Dilmaghani, F. *Critical Chain Project Management (CCPM) at Bosch Security Systems (CCTV) Eindhoven: A Survey to Explore Improvement Opportunities in the Scheduling and Monitoring of Product Development Projects*; University of Twente: Enschede, Netherlands, 2008.
9. Newbold, R.C. *Project Management in the Fast Lane: Applying the Theory of Constraints*; CRC Press: Boca Raton, FL, USA, 1998.
10. Tukel, O.I.; Rom, W.O.; Eksioglu, S.D. An investigation of buffer sizing techniques in critical chain scheduling. *Eur. J. Oper. Res.* **2006**, *172*, 401–416. [CrossRef]
11. PMI. *PMBOK® Guide*, 6th ed.; PMI (Project Management Institute): Newton Square, PA, USA, 2017.
12. PMI. *Practice Standard for Scheduling*, 2nd ed.; PMI (Project Management Institute): Newton Square, PA, USA, 2017.
13. Kim, K.H. *A Study on the Application of TOC Critical Chain Project Management*; National Contractors Association of Korea: Kwang-Ju, Korea, 2004.
14. Lee, K.C. *A Study of CCPM Improvement on Shipbuilding Industry*; Pusan National University: Busan, Korea, 2013.
15. Bevilacqua, M.; Ciarapica, F.; Giacchetta, G. Critical chain and risk analysis applied to high-risk industry maintenance: A case study. *Int. J. Proj. Manag.* **2009**, *27*, 419–432. [CrossRef]
16. Ha, B.G. *The Application of CCPM (Critical Chain Project Management) for Schedule Management in Construction Projects*; Dongwi-Univ.: Pusan, Korea, 2010.
17. Grant, K.P.; Marton, G. Critical chain project management: Under investigation or case closed? In Proceedings of the PMI® Research and Education Conference, Limerick, Ireland, 15–18 July 2012.
18. Feng, J.-C.; Li, L.; Yang, N.; Hong, Y.-Z.; Pang, M.; Yao, X.; Wang, L.-C. Critical chain construction with multi-resource constraints based on portfolio technology in South-to-North Water Diversion Project. *Water Sci. Eng.* **2011**, *4*, 225–236.
19. Dong, L.; Zhao, X.K.; Zhou, W.P.; Lv, Q. Application of Critical Chain Project Scheduling Theory in Power Plant Electrical Equipment Maintenance. In Proceedings of the Advanced Materials Research, Lulea, Sweden, 21–22 March 2013; pp. 1985–1989.
20. Goldratt, E. *Critical Chain*; The North River Press Publishing Corporation: Great Barrington, MA, USA, 1997.
21. Steyn, H. An investigation into the fundamentals of critical chain project scheduling. *Int. J. Proj. Manag.* **2001**, *19*, 363–369. [CrossRef]
22. Herroelen, W.; Leus, R. On the merits and pitfalls of critical chain scheduling. *J. Oper. Manag.* **2001**, *19*, 559–577. [CrossRef]
23. Wei, C.-C.; Liu, P.-H.; Tsai, Y.-C. Resource-constrained project management using enhanced theory of constraint. *Int. J. Proj. Manag.* **2002**, *20*, 561–567. [CrossRef]
24. Yeo, K.; Ning, J. Integrating supply chain and critical chain concepts in engineer-procure-construct (EPC) projects. *Int. J. Proj. Manag.* **2002**, *20*, 253–262. [CrossRef]
25. Ma, G.; Wang, A.; Li, N.; Gu, L.; Ai, Q. Improved critical chain project management framework for scheduling construction projects. *J. Constr. Eng. Manag.* **2014**, *140*. [CrossRef]
26. Su, Y.; Lucko, G.; Thompson, R.C., Jr. Evaluating performance of critical chain project management to mitigate delays based on different schedule network complexities. In Proceedings of the 2016 Winter Simulation Conference, Washington, DC, USA, 11–14 December 2016; pp. 3314–3324.
27. Patric, F. Critical Chain Scheduling and Buffer Management . . . . In Proceedings of the Project Management Institute, Philadelphia, PA, USA, 10–16 October 1999.
28. Lee, D.G. *A Study on the Buffer Management of Critical Chain Project Management*; Ulsan-University: Ulsan, Korea, 2002.
29. Junior, J.F.; Martins, M.R. Offshore Platform Turn around Using the Critical Chain Project Management Method (CCPM). In Proceedings of the ASME 2009 28th International Conference on Ocean, Offshore and Arctic Engineering, Honolulu, HI, USA, 31 May–5 June 2009; pp. 479–485.

30. Critical Chain and Critical Path, Can They Coexist? 2018. Available online: http://3escp33iuwsj485tugc1mb91.wpengine.netdna-cdn.com/wp-content/uploads/bsk-pdf-manager/2016/02/CPM-and-CCPM-a-Compare-and-Contrast-Pinnacle-Strategies.pdf (accessed on 20 May 2018).
31. Brennan, M. *PERT and CPM: A Selected Bibliography*; Council of Planning Librarians: Chicago, IL, USA, 1968.
32. Shin, K.Y. *A Study on the Method of Applying CCPM to Shipbuilding Industry*; Pusan University: Pusan, Korea, 2009.
33. Leach, L.P. Critical chain project management improves project performance. *Proj. Manag. J.* **1999**, *30*, 39–51.
34. Bagchi, T.P.; Sahu, K.; Jena, B.K. Why CPM Is Not Good Enough for Scheduling Projects. In Proceedings of the 2017 IEEE International Conference on Industrial Engineering and Engineering Management (IEEM), Singapore, 10–13 December 2017.
35. Im, H.M. *A Study on Schedule Management in Construction Projects by CCPM*; Pukyeong Univ.: Pusan, Korea, 2003.
36. Kim, Y.M. *Application of Critical Chain Technique in Project-Type Contract Production Environment*; Pukyeong-Univ.: Pusan, Korea, 2005.
37. Kim, O.S. *Optimal Schedule Selection by CCPM on Project Planning Phase*; Pukyeong-Univ.: Pusan, Korea, 2003.
38. Lee, M.R. *(A) Study on the Project Buffer Set-up Methodology of Critical Chain Project Management: Small Manufacturing Project*; Hanyang-University: Seoul, Korea, 2012.
39. Liu, J. *A Study on the Buffer Sizing Method of CCPM Technique*; Kachon University: Seongnam, Korea, 2012.
40. Cho, J.S. *A Study on Project Schedule Management System Modeling for Small Scale IT Companies*; Ulsan-University: Ulsan, Korea, 2009.
41. Han, H.D. *Study on Practical Application of Critical Chain Project Management*; Soongsil-University: Seoul, Korea, 2006.
42. Valikoniene, L. *Resource Buffers in Critical Chain Project Management*; University of Manchester: Manchester, UK, 2015.
43. Kim, Y.-S.; Hwange, M.-H.; Park, C.-S. A study on the present status and problem analysis of construction process management in domestic building construction sites. *J. Architect. Inst. Korea* **1996**, *12*, 253–264.
44. Kim, J.W.; Lee, J.-H. Identifying Risks of Power Plant EPC Business in the Middle East and Analyzing Their Priority by the AHP. *Plant J.* **2015**, *11*, 32–46.
45. W., K.H. *A Case Study on the Analysis of Maintenance Costs though the Estimation of Elemental Costs for Educational Buildings*; ChungAng-Universtiy: Seoul, Korea, 2009.

© 2018 by the authors. Licensee MDPI, Basel, Switzerland. This article is an open access article distributed under the terms and conditions of the Creative Commons Attribution (CC BY) license (http://creativecommons.org/licenses/by/4.0/).

Article

A Probabilistic Alternative Approach to Optimal Project Profitability Based on the Value-at-Risk

Yonggu Kim and Eul-Bum Lee *

Graduate Institute of Ferrous Technology & Graduate School of Engineering Mastership, Pohang University of Science and Technology, 77, Cheongam-ro, Nam-gu, Pohang-si, Gyeongsangbuk-do 37673, Korea; ch61park@postech.ac.kr
* Correspondence: dreblee@postech.ac.kr; Tel.: +82-10-5433-8940

Received: 4 February 2018; Accepted: 4 March 2018; Published: 8 March 2018

Abstract: This paper focuses on an investment decision-making process for sustainable development based on the profitability impact factors for overseas projects. Investors prefer to use the discounted cash-flow method. Although this method is simple and straightforward, its critical weakness is its inability to reflect the factor volatility associated with the project evaluation. To overcome this weakness, the Value-at-Risk method is used to apply the volatility of the profitability impact factors, thereby reflecting the risks and establishing decision-making criteria for risk-averse investors. Risk-averse investors can lose relatively acceptable investment opportunities to risk-neutral or risk-amenable investors due to strict investment decision-making criteria. To overcome this problem, critical factors are selected through a Monte Carlo simulation and a sensitivity analysis, and solutions to the critical-factor problems are then found by using the Theory of Inventive Problem Solving and a business version of the Project Definition Rating Index. This study examines the process of recovering investment opportunities with projects that are investment feasible and that have been rejected when applying the criterion of the Value-at-Risk method. To do this, a probabilistic alternative approach is taken. To validate this methodology, the proposed framework for an improved decision-making process is demonstrated using two actual overseas projects of a Korean steel-making company.

Keywords: Value-at-Risk; probabilistic alternative approach; Theory of Inventive Problem Solving; Project Definition Rating Index; optimal project profitability

1. Introduction

Since 2005, Korean steel-making companies have been attempting to establish overseas steel plant projects (SPPs). The experience of Korean steel-making companies in overseas SPPs is relatively low, whereas their experience in domestic SPPs is extensive. Korean steel-making companies have suffered numerous difficulties due to the uncertainty and risks of overseas SPPs [1]. Stakeholders are consistently exposed to risks when managing a project at any stage in the engineering, construction, procurement, or sustainment life-cycles [2–5]. The risks can lead to project failure [6]. Risk management of the project can be achieved through the economical application of resources to identify, assess, and prioritize risks as well as minimize, monitor, and control the likelihood or impact of unfortunate events, while maximizing the realization of opportunity [7]. The purpose of risk management is to ensure that uncertainties do not cause deviation from the project goals. Therefore, for sustainable project development, investors need to be aware of and deal with the risks associated with a project.

In particular, investors dealing with projects for sustainable development should be careful when making investment decisions. Investors have traditionally used the discounted cash flow (DCF) method to make decisions when investing in projects and it is widely used because it is simple to use and understand. The value of the DCF method is based on the cash flow of the investment project.

In this paper, the factors affecting cash flow are defined as profitability impact factors. The disadvantage of the DCF method is that it does not reflect risk, and so, it has difficulties in recognizing, preventing, and overcoming the risks of the project. A way to reflect these risks is the Value-at-Risk (VaR) method which can be applied as an improved project-profitability indicator to estimate the future uncertainty risks. The VaR method, introduced in the late 1980s by major financial institutions, is one of the risk-measurement methods that can be used to quantitatively predict the amount of loss due to risk. While the input factors used in the DCF method have fixed values, the input factors used in the VaR method have variable values depending on the risks. Although the VaR method has the advantage of reflecting the risk, its disadvantage is that the decision criterion is conservative due to excessive recognition of the risk, which often leads to loss of an investment opportunity for investors. In previous studies, the VaR method for various projects has been applied to evaluate the risk by reflecting the project [8–11]. For sustainable development, investors need to select good investment opportunities that are less risky and more profitable for their projects. Therefore, this paper proposes a method to support investors' through a probabilistic alternative approach that takes advantage of the VaR method. VaR has been used in previous studies to analyze risk factors through sensitivity analysis to select risk priorities [12–14]. This paper proposes a probabilistic approach to compare and analyze alternatives using the Theory of Inventive Problem Solving (TRIZ or TIPS) and the business version of the Project Definition Rating Index (PDRI) as a differentiating method from previous studies. Unlike previous studies, this method is not limited to presenting only alternatives, it also presents new input factors that reflect the risks and helps determine whether various alternatives are actually applicable for sustainable project development.

The objective of this paper is to propose an alternative approach for the optimization of project profitability through a quantitative evaluation of the various risks using the profitability indicators of sustainable investments. The contribution of this paper is the proposal of an alternative business perspective that can quantitatively analyze the project risk factors and improve the project value from a sustainability perspective in project development. This paper is organized as follows: Section 2 discusses the need for research from an investigation of the relevant literature; Section 3 introduces the project evaluation methodology for traditional decision-making processes and for improved decision-making processes; Section 4 identifies the factors that affect project value in terms of risk management and a method is suggested for quantitatively analyzing risk factors using a proposed probabilistic alternative approach employing TRIZ and PDRI; Section 5 validates this paper through a case study of two actual steel plant projects, and finally, a conclusion summarizes the paper and discusses limitations and future plans.

2. Related Work

A major indicator of project profitability is the results from the DCF method, which is a well-established valuation method for steel-plant projects (SPPs), for which cash flows are used [15]. The DCF method measures the future cash flows of a project from which the gains are converted into the present value (PV) [16]. The DCF method is typically represented by the net present value (NPV) and the internal rate of return (IRR), which are useful in the assessment of a reasonable value in terms of the difference between the present value and the future cash flow value [15]. However, SPPs can be inaccurately assessed when using traditional evaluation methods due to the large size of a project, its long-term operation period, the risk characteristics according to the uncertainty of the contract complexity, varying degrees of management flexibility, and the financial structure [15,17–19].

While making investment decisions for overseas SPP projects, investors analyze the project profitability according to the impact of various risks and they should estimate the realistic losses from the risks to ensure the sustainability of the development and their investments. This means that the NPV and IRR output variables must be calculated under uncertain input variables that vary within a certain range and shift with the occurrence of hazardous events; it is then possible to obtain NPV and IRR probability distributions [20]. In the VaR method, two risk definitions are applicable:

the potential-loss degree of the asset portfolio and the potential-profit standard deviation [21]. Risk can be understood as the potential-loss amount. The VaR is defined as the maximum amount of loss reserves in the time horizon of the portfolio at a given confidence level. The confidence level is defined by each company according to its standards and its financial condition [22]. The main purpose of the VaR is the quantification of the potential losses under normal market conditions [23]. Here, it is essential to pay attention to the term "normal." Fundamentally, the VaR does not make use of unusual market circumstances, such as the Great Depression of 1929 or the financial crisis of 2008. Therefore, to establish a risk-prevention plan with project-profitability forecasting, investors should focus on normal market conditions as well as abnormal conditions, as well as on the fluctuations in extreme situations.

Ye et al. [8] considered the VaR value as the NPV and carried out an infrastructure-project investment evaluation for which an NPV at-risk method was utilized. Habibi et al. [9] dealt with the conditional VaR of a cash-flow stream in the presence of an exchange-rate risk. Caron et al. [10] used the VaR to obtain an improvement in balancing the overall portfolio of power-plant projects for a company operating in the engineering and contracting industry. These studies attempted to consider the risk variability by using the NPV at-risk method from the perspective of the investor, the creditor, and the project constructor. The limitation of these studies, however, is their sole use of the NPV as a project-evaluation index derived from a Monte Carlo simulation. Monte Carlo simulations can be further utilized for the evaluation of potential risks with a sensitivity analysis, and Gatti et al. [12] used Monte Carlo simulations to calculate the VaR estimates for project-financing transactions, whereby suggestions were made regarding the ways that important issues can be discussed in the development of a model for the improvement of the project value.

Value engineering (VE) is used as an alternative approach in the engineering phase of a project. Miles [24] technically analyzed various cases using VE as a problem-solving system. Lbusuki [25] suggested the correct systematic approach of VE and a target-costing method for cost management. These studies mainly focus on approaches that improve the technical aspects of products. To increase the effectiveness of the VE practice, TRIZ can be applied at VE idea-gathering meetings [26]. The versatility of TRIZ means that it can be applied to business issues as well as the technical aspects of projects [27]. The TRIZ technique is used in this paper to solve the problem of major risk factors that are derived from an analysis of a Monte Carlo simulation.

TRIZ techniques are used in various disciplines and fields. Kim and Cochran [28] reviewed a number of TRIZ concepts from the perspective of the axiomatic-design framework. Yamashina et al. [29] proposed an effective integration of TRIZ and quality-function deployment, enabling technological innovations for the effectiveness and systematic operability of new products. John and Harrison [30] identified ways in which TRIZ tools and methodologies could be used to innovate the environment and then presented a way TRIZ could be applied as a sustainable design tool for specific purposes. Ilevbare et al. [31] moved away from the traditional TRIZ literature by exploring the challenges that are associated with the acquisition and application procedures for TRIZ beginners based on their practical experience as well as the benefits that are associated with the attainment of TRIZ knowledge. Souchkov [32] provided a brief overview of the manner in which TRIZ can benefit the business world, whereby business and management innovations are improved and implemented. This study evaluates project risks through the application of the business version of the PDRI, which is widely used by project managers of SPPs, and TRIZ-based solutions are proposed for which a project-risk evaluation result is employed.

3. Project-Evaluation Methodology

The methodology of this study consists of two tasks. The first task is to assess risk applied project evaluation. The second task is to select alternatives for optimal profitability by applying a probabilistic alternative approach to risk in project evaluation. The overall methodology framework of this paper is shown in Figure 1.

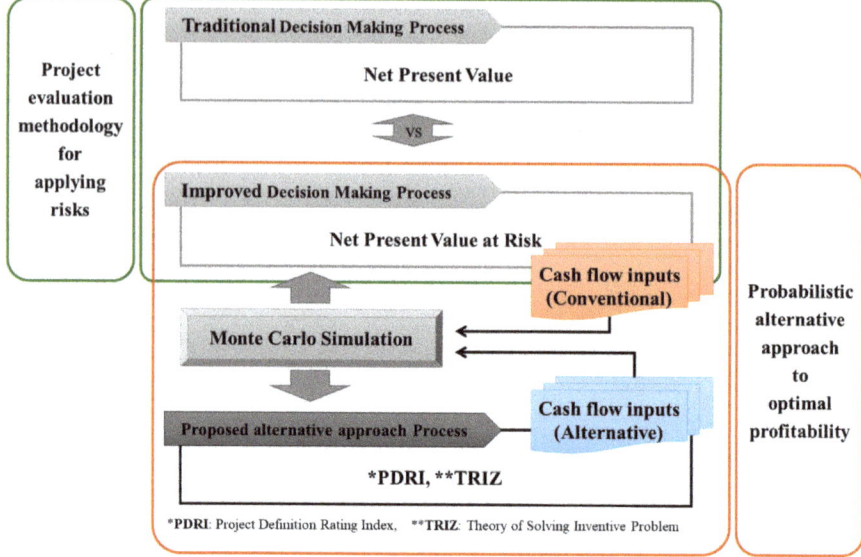

Figure 1. Overall project evaluation framework.

3.1. Traditional Decision-Making Process for the DCF Model

Project evaluation is based on a cash-flow model. Essentially, the cash-flow model defines the pro forma income-statement elements and the discount-rate elements as profitability impact factors. Figure 2 shows an example of a pro forma income statement.

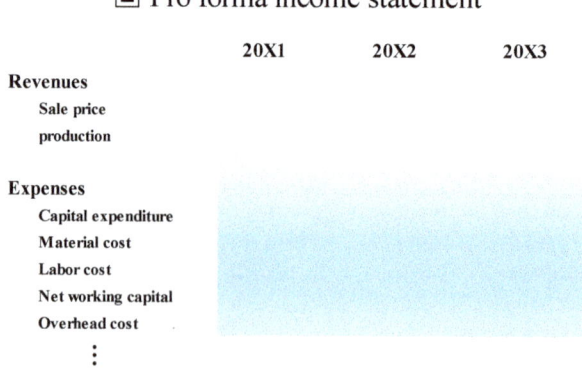

Figure 2. Example of a pro forma income statement.

To assess the project value, a reasonable cash-flow model that reflects the profitability impact factors that are the input variables of the DCF method should be established. This paper defines the following 15 profitability impact factors, (1) capital expenditure; (2) material cost; (3) labor cost; (4) net working capital; (5) overhead cost; (6) sale price; (7) production; (8) exchange rate; (9) corporate tax rate; (10) debt-to-equity ratio; (11) risk-free interest rate; (12) market risk premium; (13) beta; (14) cost of debt (COD) before tax; and (15) country risk. Further, this paper introduces the necessary calculation formulas. A capital cost that represents the weighted average cost of capital (WACC) is relevant here;

it consists of the cost of equity (COE), the COD, the debt-to-capital ratio (DC$_{ratio}$), and the corporate tax rate (TAX). In this case, the COE is derived using the capital-asset pricing model [33]. The risk-free interest rate (R$_{free}$), market risk premium, and beta also need to be considered. The cost of debt refers to the borrowing of the interest rate in project financing.

$$COE = R_{free} + beta \times (market\ risk\ premium - R_{free}), \quad (1)$$

$$WACC = (1 - DC_{ratio}) \times COE + DC_{ratio} \times COD \times (1 - TAX), \quad (2)$$

To calculate the project profitability, it is essential to obtain a discount rate which is an index of the project risks, as follows:

$$Discount\ rate = WACC + Country\ risk\ premium, \quad (3)$$

The project cash flow is divided into the construction-period component and the business-period component, and each part constitutes the cash inflows and outflows that are associated with the 15 profitability impact factors presented above.

Through the DCF method, a project is assessed using the NPV and the IRR. Further, the profitability of these two indicators is verified using the cash flow, as follows:

$$NPV = \sum_{t=1}^{N} \frac{CF_t}{(1+r)^{t-1}}, \quad (4)$$

where t is the year of the project period, N is the total project period, CF_t is the cash flow of the year, and r is the discount rate.

$$IRR = r\ value\ when\ the\ NPV = 0, \quad (5)$$

where r is derived using a trial-and-error method.

3.2. Improved Decision-Making Process for the VaR Model

During the planning stage of projects, investors consider the various risks. However, if traditional economic evaluation criteria such as the NPV and the IRR are being used, the investors are then at risk of overlooking the volatility of the project uncertainty risks [34]; therefore, it is necessary to consider the change in profitability according to the fluctuation of the risks. By developing a sophisticated probabilistic model of the future cash flows, investors can determine the project investment based on risk-based decision criteria. In this study, an improved DCF method is used, to which the profitability impact factors that reflect the project risks are applied. The improved DCF method is called VaR-based NPV at Risk (NPVaR) and is based on a cash-flow model [8]. The main difference between the traditional DCF method and the NPVaR method is the determination of whether each of the profitability impact factors accurately reflect the risks of the project. A Monte Carlo simulation shows the way that this distribution can be calculated for a given project, and how it can comprehensively measure the business risks of the project [10]. In fact, a Monte Carlo simulation can be easily applied to explain numerous types of practical assumptions regarding the probability distributions of the profitability impact factors of the cash-flow model. The probability distribution for each profitability impact factor of the NPVaR method is determined based on a number of literature researches [15,35,36].

In this paper, VaR refers to the NPV for the maximum loss that may occur during a relatively long operation period of a project, whereas the general VaR, which is used in financial sectors, refers to the amount of loss for a relatively short period of time. The NPVaR method is applied to the VaR in terms of the NPV as it applies a discount rate on the cash flow of the project during the project period. As shown in Figure 3, NPVα is defined as the NPV corresponding to the significance level of α; that is, it refers to the minimum NPV at the (1 − α) confidence level [8].

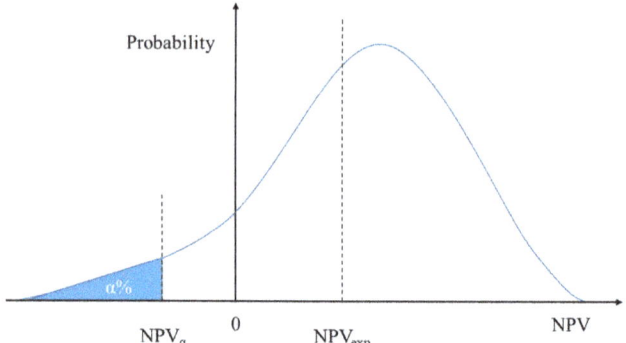

Figure 3. Probability distribution of the net present value (NPV).

Figure 3 shows two definitions of simple profitability forecasting for which the volatility of the risks is considered. These definitions will be used as the decision-making criteria in the following section. First, the expected NPV (NPV_{exp}) is defined as the mean NPV of a probability distribution. The NPV_{exp} is a level of profitability that can occur under normal circumstances. The second definition is the $NPV\alpha$, the lower cumulative $\alpha\%$ NPV of the probability distribution. The $NPV\alpha$ is a level of profitability that can occur under severe circumstances. Therefore, if the investors decide on whether to execute a project using the NPVaR, they will be able to receive assistance during the decision-making process by identifying the maximum number of losses that might occur during the operation period of the project. If the NPV, which does not consider the risk volatility, is not used as a traditional decision-making tool, the project profit may be less than the expected NPV, or a loss might result due to the unknown risks of an actual project. To prevent the uncertainty risks, this paper presents decision-making criteria that are determined by the nature of the investors according to the risk circumstances and in consideration of the risk volatility.

4. Probabilistic Alternative Approach

In terms of project evaluation, the NPVaR is presented as the decision-making criteria; although previous studies have been limited to the VaR decision-making process [8,10,20,35,36]. The authors of this current study noticed that the decision-making criterion for project investment is more conservative as it reflects the profitability impact factors of the project risks. When the investment is approved in the VaR process, the distinction from previous studies is not evident. The focus of this paper, however, is the actions that investors can take when their investment is rejected in the VaR process, which are discussed below.

First, in this study, the major risk factors are recognized based on previous studies of the many risks that arise for investors who are dealing with overseas construction projects, as well as on the database of a steel-making company. The major risk factors can be tabulated according to their associations with the profitability impact factors, as follows. A database of overseas constructions should be set and analyzed to identify the relevance between the associated major risk factors and the profitability impact factors. The analysis of the project profitability, for which the Monte Carlo simulation was employed, is a sensitivity analysis that finds the most influential factor among the profitability impact factors. Second, the business version of the PDRI, which is based on major risk factors, is helpful in determining the alternatives for achieving the optimal project profitability. The success regarding the optimal profitability requires identification of the most influential factors, analysis of the problems that are associated with the risks, and a TRIZ-based derivation of the optimal solution.

4.1. Identification of Overseas-Project Risk Factors

In this paper, the proposed method is used for the data analysis to make relationships between the profitability impact factors and the associated major risk factors, and the risk factors of overseas investment projects are determined based on the literature references that are associated with the overseas construction cases that are discussed in the remainder of this paper. The methods that have been widely used to evaluate the success of projects are the PDRI [37] and the front-end loading (FEL) index [38]. This study also includes an additional nine references in which other overseas and domestic plant projects have been assessed. It is advantageous that a variety of the risks that can occur in overseas construction-project cases can be recognized. Table 1 shows 66 overseas construction-project risk factors that are related to the contents of the literature references.

As shown in Table 1, the risk factors of overseas construction projects are largely classified into internal and external risk factors. The external risk factors are related to the project-investment environment and the internal risk factors are related to the managerial regulations of the project itself. For the classification, the risk factors have been organized into categories that are based on the literature references. The classification considers most of the risk factors, while also considering the investors during the investment-decision step.

When they are utilized in real-life projects, the major risk factors can be added and subtracted depending on the nature of the project. Additionally, it is more effective to build a database of risk factors so that investors can document a company's overseas experiences. In this study, a business version of the PDRI is analyzed based on the 66 previously mentioned risk factors to find the associated risk factors according to the profitability impact factors, as shown in Table 2.

Table 1. List of the risk factors for overseas construction projects based on the related literature.

	Classification of Risk Factors			Literature List										
L1	*L2	L3	Risk Factor	①[37]	②[38]	③[39]	④[40]	⑤[41]	⑥[42]	[43]	⑧[44]	⑨[45]	⑩[46]	⑪[47]
External Risk Factors	A.	A1.	Local administrative procedures and practices	✓		✓		✓			✓			✓
		A2.	Maturity of legal system	✓		✓		✓				✓	✓	✓
		A3.	Consistent local policy	✓		✓		✓				✓		
	B.	B1.	Economic condition of employer				✓		✓		✓		✓	✓
		B2.	Economic condition of local country			✓			✓					
	C.	C1.	Local social stability	✓		✓						✓		
		C2.	Local labor situation	✓	✓	✓					✓			
		C3.	Cultural traits	✓		✓			✓			✓	✓	
		C4.	Local reactions toward project			✓			✓			✓		✓
	D.	D1.	Climate feature	✓										
		D2.	Soil condition	✓	✓									
		D3.	Distance from Republic of Korea			✓			✓				✓	
	E.	E1.	Local standards of design	✓	✓	✓						✓		✓
		E2.	Licensing standards	✓	✓						✓	✓		✓
		E3.	Tariff standards						✓					✓
		E4.	Environmental regulations	✓	✓				✓					✓
		E5.	Profit-repatriation procedure			✓			✓		✓			
		E6.	Local-content requirement				✓							
	F.	F1.	Local infra and utility levels	✓	✓	✓						✓		✓
		F2.	Difficulty of infra- and utility-use contracts	✓	✓									
		F3.	Future additional expansion possibilities	✓										
	G.	G1.	Volatility of market demand				✓			✓	✓	✓		✓
		G2.	Local-competitor statuses									✓	✓	
		G3.	Local market share				✓					✓		
		G4.	Exchange-rate fluctuations					✓	✓	✓	✓	✓		✓
		G5.	Inflation fluctuations					✓	✓	✓				✓
		G6.	Interest-rate fluctuations					✓	✓	✓	✓	✓		✓
	H.	H1.	Additional potential projects in progress				✓			✓	✓			
		H2.	Cash-flow stability	✓						✓		✓	✓	
		H3.	Debt-equity ratio				✓			✓				
	I.	I1.	Joint-venture consideration				✓			✓		✓		✓
		I2.	Ability of employer		✓					✓		✓	✓	✓
		I3.	Ability of contractor			✓				✓	✓	✓	✓	✓
		I4.	Ability of local companies			✓				✓	✓	✓	✓	✓

Table 1. Cont.

L1	*L2	L3	Risk Factor	①[37]	②[38]	③[39]	④[40]	⑤[41]	⑥[42]	⑦[43]	⑧[44]	⑨[45]	⑩[46]	⑪[47]
	J.	J1.	Similar experiences	✓										
		J2.	Liquidated damages			✓					✓			
		J3.	Contract-terms changeability		✓					✓	✓		✓	✓
		J4.	Unclear contract terms			✓								✓
		J5.	Steel-purchase conditions						✓					✓
		J6.	Force majeure					✓	✓	✓				
	K.	K1.	Product characteristics	✓	✓									
		K2.	Construction schedule	✓	✓	✓				✓		✓	✓	
	L.	L1.	Initial investment	✓										
		L2.	Operation costs	✓								✓		
		L3.	Reasonable contingencies	✓								✓		
	M.	M1.	Conformity of new process	✓										
		M2.	Level of applied technology	✓								✓	✓	
	N.	N1.	Layout planning	✓	✓									
		N2.	Constructability/Complexity			✓				✓				
		N3.	Design for scalability	✓	✓									
		N4.	Major-equipment selection	✓						✓				
		N5.	Timeliness of design							✓				
		N6.	Value engineering	✓										
	O.	O1.	Work-safety management	✓	✓	✓								
		O2.	Construction-safety facilities	✓										
		O3.	Pollution-prevention plan	✓					✓					✓
	P.	P1.	Personnel-procurement plan	✓		✓	✓			✓	✓	✓		✓
		P2.	Equipment procurement	✓	✓		✓	✓		✓		✓		✓
	Q.	Q1.	Suitable working method	✓		✓				✓				
		Q2.	Equip./Material transport	✓										
		Q3.	Local-company collaboration	✓	✓									
		Q4.	Local-material quality			✓					✓			
		Q5.	Security & Safety		✓				✓			✓		
		Q6.	Commissioning & Acquisition requirements	✓										
	R.	R1.	Document management	✓										
		R2.	Performance requirements		✓									

* **A.** Credibility of Local Government, **B.** Economic Stability, **C.** Local Social & Cultural Characteristics, **D.** Geographical conditions, **E.** Legal Standard, **F.** Status of infrastructure, **G.** Market conditions, **H.** Financing Plan, **I.** Project Organization, **J.** Contract Condition, **K.** Scope of Work, **L.** Expenses, **M.** Process & Technology, **N.** Engineering Period, **O.** Health, Safety, & Environment, **P.** Procurements, **Q.** Construction Period, **R.** Completion Requirements.

Table 2. Relationships between the profitability impact factors and the associated major risk factors.

Profitability Impact Factors	Associated Major Risk Factors							
	External Risk Factors				Internal Risk Factors			
Capital Expenditure	C.	D.	F.		I.	J.	L.	N.	Q.
Material cost	G.				J.	L.	M.	N.
Labor cost	C.				L.	O.	P.	
Net working capital					L.	M.		
Overhead cost					L.	M.		
Sale price	B.	E.	G.		J.	K.		
Production	B.	G.			K.	R.		
Exchange rate	G.							
Corporate tax rate	A.	B.	C.	E.				
Debt/(equity + debt)	H.							
Cost of equity								
Risk-free interest rate	H.							
Market risk premium	B.				I.	J.		
Beta	B.							
Cost of debt	H.				I.	J.		
Country risk	A.	B.	C.					

A. Credibility of Local Government, **B.** Economic Stability, **C.** Local Social & Cultural Characteristics, **D.** Geographical conditions, **E.** Legal Standard, **F.** Status of infrastructure, **G.** Market conditions, **H.** Financing Plan, **I.** Project Organization, **J.** Contract Condition, **K.** Scope of Work, **L.** Expenses, **M.** Process & Technology, **N.** Engineering Period, **O.** Health, Safety & Environment, **P.** Procurements, **Q.** Construction Period, **R.** Completion Requirements.

The level of risk associated with overseas projects is greater than that of domestic projects. Therefore, as the role of the investors is the planning of projects and the provision of project investments, it is important for the investors to recognize and respond to such risks in advance. Also, it is essential to set the cash-flow model to predict the project profitability; and to reflect the cash-flow model for each of these risks, it is important to analyze the relationship between the profitability impact factors and the risk factors in advance. This study defines the major factors that affect the project profitability according to 15 factors based on the pro forma income statement shown in Figure 2 of the risk factors of overseas projects, as shown in Table 2. The relationships between the profitability impact factors and the associated risk factors is linked to the references on overseas projects [48–51]; furthermore, these relationships are utilized in the problem-solving of this paper.

This paper presents a method which is applicable to an increase in the NPV_α, which is an important indicator of investment decisions based on the relationships between the profitability factors and the associated major risk factors. It is important to identify the major contributors among the profitability impact factors; to find these factors, the use of a sensitivity analysis is recommended. A sensitivity analysis refers to the impact on the input variables in terms of the value of the results [52]. In a profitability-forecasting model that utilizes the NPVaR through an analysis of the major profitability impact factors affecting the NPV_α, the investors may establish a risk-prevention plan; if the investors can analyze the critical profitability impact factors influencing the NPV and manage the volatility of the controllable risk factors, the probability of the retention of the NPV_{exp}, the initial expected project profit, will be further increased.

4.2. Developing Alternatives and the Alternative Selection Process

In order to develop alternatives, the major risk factors are first identified by recognizing the risks given in Table 1 and assessing the priorities of these risks. The most vulnerable risk factor can be selected by analyzing the relationship between the profitability impact factors and the associated major risk factors given in Table 2 and then checking the pre-assessed PDRI scores. The most vulnerable risk factors are used in the idea meetings to discover alternatives to TRIZ. In this study, additional project evaluation is performed using probabilistic alternatives as the input factors in order to select an optimal alternative. The schematic process of this methodology is shown in Figure 4.

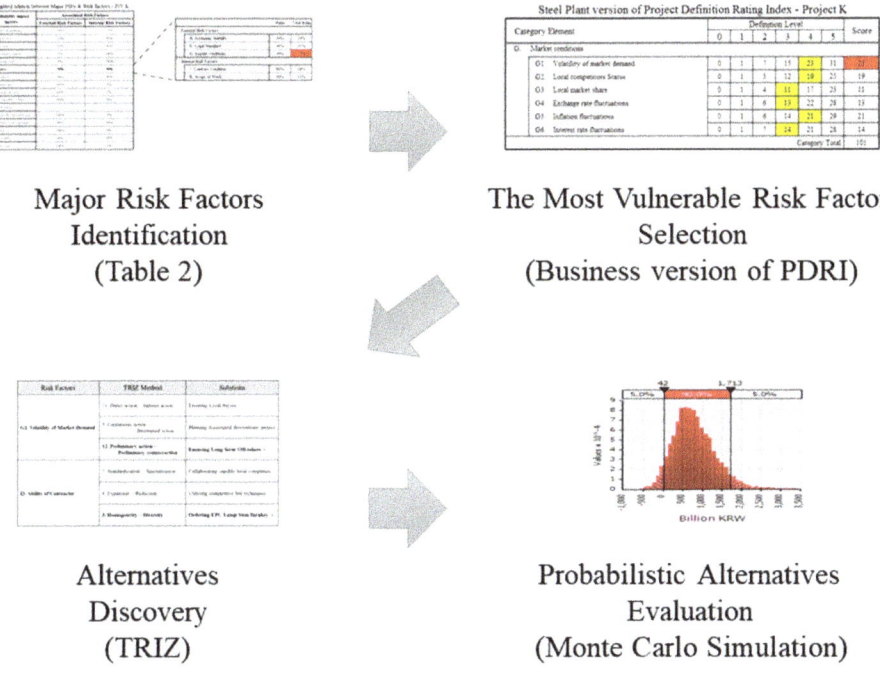

Figure 4. Flow diagram for alternative development and alternative selection.

4.2.1. Business-Version of PDRI

As shown in Figure 4, the investors first determined the weight of each risk factor of the project. This activity is an early part of the process of creating the business-version PDRI on a project-by-project basis. The major risk factors are identified in Table 2 using the most influential profitability impact factors given in the previous section. The list of 66 elements of the business-version PDRI was originally identified and categorized according to 10 references [37–47]. The elements were weighted in order of importance using the inputs from 15 experienced project managers and estimators who each have between 10 and 20 years of experience. An example of the business version of PDRI is presented in Appendix A. These employees used the TRIZ to seek out alternatives to overcome the most vulnerable risk factors that were applied for the weighted evaluation table.

4.2.2. TRIZ

The TRIZ offers a systematic approach to gain an understanding and definition of difficult problems. Difficult problems typically require unique solutions, and the TRIZ offers a variety of strategies and tools to facilitate the formulation of creative solutions. One of the earliest theory-based discoveries of large-scale projects is that most of the problems that require creative solutions typically reflect a need to overcome the dilemma or the tradeoff between two contradictory factors. A key goal of the TRIZ-based analysis is the systematic application of strategies and tools to find superior solutions that overcome the need to enact tradeoffs or compromises between the two contradictory elements. Twelve TRIZ principles are introduced in this current study to find alternatives, as shown in Appendix B.

4.2.3. Probabilistic Alternative Selection

The profitability impact factor of alternatives based on TRIZ is first transformed into probabilistic values through a literature review and the company's internal database. Based on the previously derived alternatives, the probability-distribution fit is used to formulate probability distributions. The transformed probabilistic values are used as the input variables to perform the Monte Carlo simulation using the VaR method. Among the found alternatives, the alternative with the largest NPV_α is salient to solve the risk problem. The largest NPV_α is the optimal project value.

5. Case Study

The leading Korean steel-making company, P, encountered many difficulties during their participation in three overseas projects. The Indonesian Krakatau project (Project K) of 2009, experienced difficulties such as a sales decrease due to a change in the local-market conditions and high steep-price volatility in the local market. The Brazilian Companhia Siderúrgica do Pecém project (Project C) of 2011, was hindered by difficulties such as a construction delay due to local-union strikes, an increase in the investment cost that was caused by inflation and political issues, and local policing and environmental issues. The current statuses of these projects considering the corresponding difficulties are as follows. A long-term demand in the market had not been established for Project K, so the sale price slowly decreased while profitability was even worse. In Project C, the investment cost became much larger than expected due to the frequent design changes and decreased workability. Overseas SPPs can pose serious risks and various detrimental environmental conditions in accordance with the investment uncertainties in the target countries. Investors should be more cautious in terms of investment decisions, as the risks of overseas SPPs are greater than those of local projects.

In this section, first, the profitability of two projects is analyzed with the traditional DCF-based decision-making process. Second, the profitability of the projects with the improved decision-making process is analyzed based on the NPVaR method. The difference between the two results is then analyzed and discussed. Finally, the selection of alternatives for the optimal profitability of the two projects is validated through a probabilistic alternative comparison of the major risk factors derived from the improved decision process.

5.1. Traditional DCF-Based Decision-Making Process

The case models for Project K (Indonesia, 2009) and Project C (Brazil, 2011) are shown in Table 3, based on the profitability impact factors that were formulated from the pro forma financial-statement and the discount-rate elements. Values that are as close to the real values as possible serve as the basis for the modelling of both cases. The profitability impact factors are estimated based on the company's financial and accounting disclosure documents that are stored in the Republic of Korea (ROK)'s electronic disclosure system. The cash-flow data for Project K and Project C, are presented in Tables 4 and 5, respectively.

Table 3. Case information for a cash-flow model.

Profitability Impact Factors	Unit	Project K (Indonesia, 2009)	Project C (Brazil, 2011)
Construction/Operation period	Year	3/15	4/20
* Capital expenditure	Million USD	3000	3800
Material cost	Million USD per year	480	460
Labor cost	Million USD per year	300	280
Net working capital	Million USD per year	40	30
Overhead cost	Million USD per year	50	30
Sale price	Million USD/Million tons	680	640
Production	Million USD per year	3	3
Revenue	Million USD per year	2040	1920
Exchange rate	KRW/USD	1065.92	1122.10
Corporate-tax rate	%	28	34
Debt-to-equity ratio		1.5 (60%/40%)	1.0 (50%/50%)
Cost of equity	%	7.758	6.901
Risk-free interest rate	%	4.341	3.691
Market risk premium	%	7.392	7.219
Beta		1.12	0.91
Cost of debt after tax	%	3.622	2.845
Cost of debt before tax	%	5.031	4.311
WACC	%	5.277	4.873
Country risk premium	%	3.400	3.400
Discount rate	%	8.677	8.273

* Capital Expenditure (CAPEX) is amortized on a straight-line basis calculation.

Table 4. Cash-flow data sheet for Project K (Indonesia, 2009).

Period	Construction			Operation						
Year	1	2	3	4	5	...	15	16	17	18
Cash outflow	*(1000)	*(1000)	*(1040)	*(40)	*(40)	...	*(40)	*(40)	*(40)	-
Capital Expenditure	*(1000)	*(1000)	*(1000)	-	-	...	-	-	-	-
Net working capital	-	-	*(40)	*(40)	*(40)	...	*(40)	*(40)	*(40)	-
Cash Inflow	-	-	-	*718	*718	...	*628	*634	*641	*647
Profit after tax	-	-	-	*622	*622	...	*572	*578	*585	*591
Depreciation saving	-	-	-	*56	*56	...	*56	*56	*56	*56
Net cash flow	*(1000)	*(1000)	*(1040)	*678	*678	...	*588	*594	*601	*647

* Unit: Million USD.

Table 5. Cash-flow data sheet for Project C (Brazil, 2011).

Period	Construction				Operation					
Year	1	2	3	4	5	...	21	22	23	24
Cash outflow	*(950)	*(950)	*(950)	*(980)	*(30)	...	*(30)	*(30)	*(30)	-
Capital Expenditure	*(950)	*(950)	*(950)	*(950)	-	...	-	-	-	-
Net working capital	-	-	-	*(30)	*(30)	...	*(30)	*(30)	*(30)	-
Cash Inflow	-	-	-	-	*644	...	*600	*604	*607	*611
Profit after tax	-	-	-	-	*580	...	*536	*539	*543	*546
Depreciation saving	-	-	-	-	*64	...	*64	*64	*64	*64
Net cash flow	*(950)	*(950)	*(950)	*(950)	*614	...	*570	*574	*577	*611

* Unit: Million USD.

Table 4 shows the spending of the capital expenditure (CAPEX) from equity to debt; the CAPEX sequence uses the debt after the spending from the equity. Loan repayment is a method of fully amortizing loans, and it starts five years after the beginning of the operation period. The principal is repaid evenly during the remaining operation period. It is assumed here that the entirety of the CAPEX is a depreciable asset and is amortized on a straight-line basis calculation throughout the operation period. All products are sold out every year, thereby negating the need for a goods inventory. The net working capital is a plant-operation cost that is addressed one year prior to the start of the operation period and is then recovered in the final year of the operation period. This cost is used in the additional equity because debt is strictly defined by a term sheet for the CAPEX calculation.

Table 6 shows the calculation results of four of the profitability indicators, including the NPV and the IRR, using two cash-flow tables.

Table 6. Profitability-calculation results.

Profitability Indicators	Project K (Indonesia, 2009)	Project C (Brazil, 2011)
NPV (Million USD/Billion KRW)	1432/1525	901/1011
IRR (%)	15.25%	11.08%
Decision Making Result	Investment approval	Investment approval

All profitability indices are excellent and indicate that the implementation of the two projects should be approved. Both projects show NPVs that are greater than zero and IRRs that are greater than the discount rate (Project K is 8.677% and Project C is 8.273%), which is the minimum acceptable rate of return (MARR). Traditionally, investors would not hesitate in making the decision to invest in these projects. In practice, however, each of the profitability impact factors has a risk volatility. These profitability indices are prone to change depending on the potential economic situations. In the next section, the investment decision-making process is described using a new profitability index that reflects the risk variation.

5.2. Improved Decision-Making Process Based on the NPVaR Method

This case study is also based on the cash-flow model in Table 3. In traditional methods, the profitability impact factors are all fixed. If the profitability impact factors of each project are variable due to uncertainty risks, the decision-making results regarding the investment can be varied. In the improved decision-making process, the probability distribution of each profitability impact factor is set, as shown in Table 7, and these can be applied to a Monte Carlo simulation to forecast the project profitability considering the risk volatility.

Table 7. Probability-distribution-fitting results for major profitability impact factors.

Profitability Impact Factors	Unit	Project K (Indonesia, 2009)		Project C (Brazil, 2011)	
		Probability Distribution	Variation	Probability Distribution	Variation
Capital expenditure	Million USD	Triangular	Min: 2900 Mode: 3000 Max: 3200	Triangle	Min: 3600 Mode: 3800 Max: 4600
Material cost	Million USD/Year	Triangular	Min: 470 Mode: 490 Max: 530	Triangle	Min: 440 Mode: 460 Max: 490
Labor cost	Million USD/Year	Uniform	Min: 280 Max: 310	Uniform	Min: 260 Max: 290
Net working capital	Million USD/Year	Uniform	Min: 30 Max: 50	Uniform	Min: 25 Max: 35
Overhead cost	Million USD/Year	Uniform	Min: 40 Max: 60	Uniform	Min: 25 Max: 35

Table 7. Cont.

Profitability Impact Factors		Unit	Project K (Indonesia, 2009)		Project C (Brazil, 2011)	
			Probability Distribution	Variation	Probability Distribution	Variation
Sale price		Million USD/Million tons	Exponential	β: 26.7 Min: 620 Mean: 680	Triangle	Min: 600 Mode: 630 Max: 650
Production		Million USD/Year	Uniform	Min: 2.8 Max: 3.1	Uniform	Min: 2.95 Max: 3.1
Exchange rate		KRW/USD	Beta	α_1: 1.1427, α_2: 3.4881 Min: 755.75 Max: 1973.80	Normal	μ: 1122.10 σ: 175.92
Corporate tax rate		%	Triangular	Min: 25 Mode: 28 Max: 30	Uniform	Min: 34 Max: 34
Debt/(equity + debt)			Uniform	Min: 0.55 Max: 0.70	Uniform	Min: 0.45 Max: 0.60
Cost of equity	Risk-free interest rate	%	Uniform	Min: 1.0783 Max: 5.8217	Uniform	Min: 0.9135 Max: 5.1265
	Market risk premium	%	Triangular	Min: 5.6759 Mode: 5.6759 Max: 10.4031	Uniform	Min: 5.2502 Max: 8.8857
	Beta		Uniform	Min: 1.07 Max: 1.17	Uniform	Min: 0.89 Max: 0.91
Cost of debt before tax		%	Uniform	Min: 2.90 Max: 6.70	Triangular	Min: 3.128 Mode: 3.128 Max: 6.106
Country risk		%	Uniform	Min: 2.4 Max: 4.4	Uniform	Min: 3.0 Max: 3.8

A probability-distribution fitting produces probability distributions that are used for the fitting of a set of data that has been accumulated for more than 10 years with respect to the variable profitability impact factors. The probability distributions that present a similar fit are assumed to lead to an excellent profitability estimation. A variety of probability distributions can be produced, some of which can be adapted more easily to the gathered data than others, depending on the characteristics of the profitability impact factors [53]. A probability-distribution fitting and Monte Carlo simulation are presented in this study, using the commercial statistical software @Risk for Excel Version 6.3.1 (Palisade Corporation, Ithaca, NY, USA).

A Monte Carlo simulation is presented here by using the cash-flow data from Tables 4 and 5 and the probability distributions of the major profitability impact factors of Project K (Indonesia, 2009) and Project C (Brazil, 2011) from Table 7. The number of simulation repetitions is set to 10,000 to ensure the reliability of the simulation using the @Risk software. The derived simulation results are as close to reality as possible, and it is assumed that the investors are risk-averse. Two of the NPV probability distributions of the two projects were derived from the Monte Carlo simulation, as shown in Table 7. The detailed simulation results are also given in Table 8.

In the DCF methods, the two profitability indices are the NPV and the IRR, as can be seen in Table 6. Usually, the two investment cases would be approved where the NPV is larger than zero and the IRR is larger than the discount rate (MARR). The two projects are ideally set up to receive investments. However, according to the improved method, both projects are rejected. This investor attribute means that the decision-making criterion is based on the decision criterion $NPV_\alpha > 0$. The significance level that serves as the reference for the estimation of the NPV_α is typically set at 5% (confidence level of 95%) [8,10]. Because the $NPV_{0.05}$ values of these projects, which represent the profitability index of the improved decision-making method, are negative, the risk-averse investors would reject both projects based on the decision criterion ($NPV_{0.05} > 0$). Therefore such investors might miss suitable investment opportunities compared to relatively risk-neutral or risk-amenable investors. To resolve this problem, a method is suggested for enabling investors to find probabilistic alternatives to the risks for target projects.

Table 8. Simulation summary statistics for the net present value (NPV).

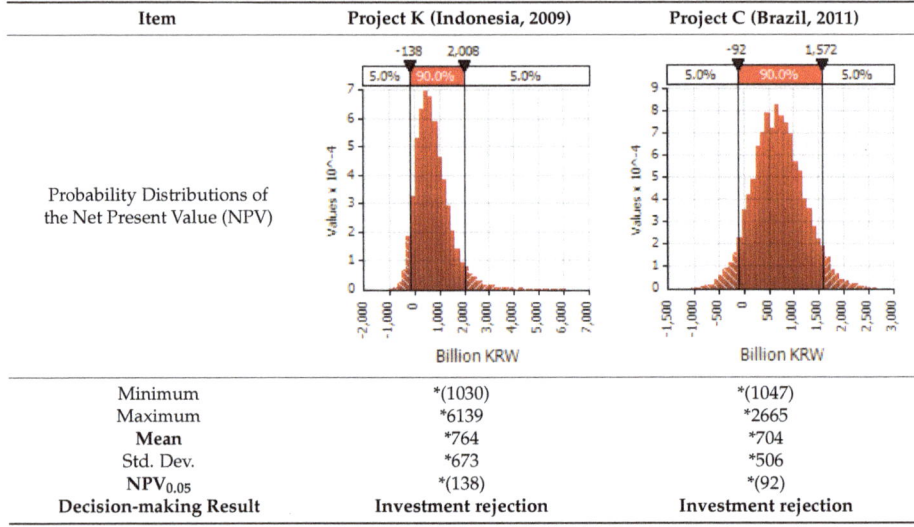

Item	Project K (Indonesia, 2009)	Project C (Brazil, 2011)
Probability Distributions of the Net Present Value (NPV)		
Minimum	*(1030)	*(1047)
Maximum	*6139	*2665
Mean	*764	*704
Std. Dev.	*673	*506
$NPV_{0.05}$	*(138)	*(92)
Decision-making Result	Investment rejection	Investment rejection

* Unit: Billion KRW.

5.3. Probabilisic Alternative Approach for Optimal Profitability

In this section, a method is proposed for the optimal increase of the $NPV_{0.05}$. First, it is important to find the major contributors among the profitability impact factors, as this will identify the most influential profitability impact factor. To find this factor, investors should use a sensitivity analysis, which investigates the impact of the input variables on the results [52]. In a profitability-forecasting model that utilizes the NPVaR method, through an analysis of the major profitability impact factors affecting the $NPV_{0.05}$, risk-prevention alternatives can be identified. Thus, if investors can find the critical profitability impact factor that has the most influence on the NPV through a sensitivity analysis, they can connect the associated risks, as shown in Table 2, by using a backward tracing method to seek the requisite solutions.

In Table 9, the most influential profitability impact factors are the sale price and the CAPEX in Project K and Project C, respectively. This paper utilizes the risk information in Tables 1 and 2 to find solutions to reduce the risk-factor volatility by using the business-version PDRI and the proposed TRIZ problem-solving method [27,54].

As shown in Table 10, in Project K, the sale price has five associated risk factors that are weighted in advance by overseas SPPs experts. According to this method, the risk factor "G. Market Condition" is the most influential factor, while the risk factor "G1. Volatility of market demand" is the most vulnerable PDRI score. In Project C, the CAPEX has eight associated risk factors that are also weighted. The risk factor "I. Project Organization" is the most influential factor, while the risk factor "I3. Ability of contractor" is the most vulnerable PDRI score.

Table 9. Sensitivity analysis of the net present value (NVP) for the two projects.

Project K (Indonesia, 2009)

Project C (Brazil, 2011)

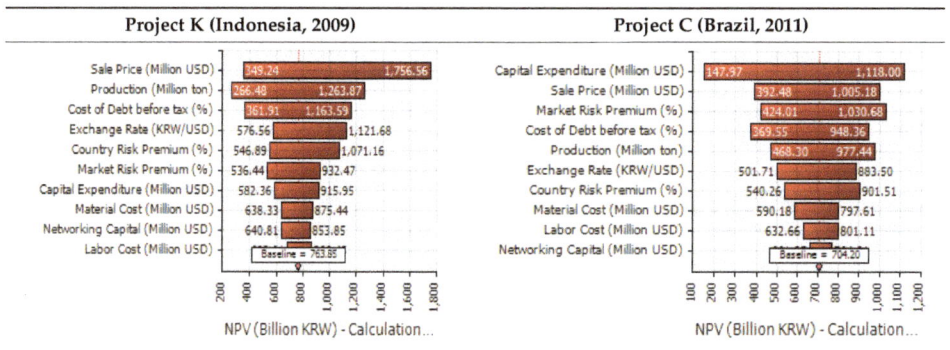

Table 10. Most vulnerable risk-factor selection.

Project K

Weighted Relation between Major Profitability Impact Factors & Risk Factors

Profitability impact factors	Associated Risk Factors: External Risk Factors	Associated Risk Factors: Internal Risk Factors
Capital Expenditure	25%	75%
Material cost	15%	85%
Labor cost	20%	80%
Net working capital	0%	100%
Overhead cost	0%	100%
Sale price	70%	30%
Production	40%	60%
Exchange rate	100%	0%
Corporate tax rate	100%	0%
Debt / (equity + debt)	100%	0%
Cost of equity		
Risk-free interest rate	100%	0%
Market risk premium	60%	40%
beta	100%	0%
Cost of debt	60%	40%
Country risk	100%	0%

	Ratio	Total Ratio
External Risk Factors		
B. Economic Stability	20%	14%
E. Legal Standard	30%	21%
G. Market conditions	**50%**	**35%**
Internal Risk Factors		
J. Contract Condition	60%	18%
K. Scope of Work	40%	12%

Business-Version PDRI

Category Element	Definition Level 0	1	2	3	4	5	Score
G. Market conditions							
G1. Volatility of market demand	0	1	7	15	23	31	23
G2. Local competitors Status	0	1	5	12	19	25	19
G3. Local market share	0	1	4	11	17	23	11
G4. Exchange rate fluctuations	0	1	6	13	22	28	13
G5. Inflation fluctuations	0	1	6	14	21	29	21
G6. Interest rate fluctuations	0	1	7	14	21	28	14
						Category Total	101

Project C

Weighted Relation between Major Profitability Impact Factors & Risk Factors

Profitability impact factors	Associated Risk Factors: External Risk Factors	Associated Risk Factors: Internal Risk Factors
Capital Expenditure	25%	75%
Material cost	15%	85%
Labor cost	20%	80%
Net working capital	0%	100%
Overhead cost	0%	100%
Sale price	70%	30%
Production	40%	60%
Exchange rate	100%	0%
Corporate tax rate	100%	0%
Debt / (equity + debt)	100%	0%
Cost of equity		
Risk-free interest rate	100%	0%
Market risk premium	60%	40%
beta	100%	0%
Cost of debt	60%	40%
Country risk	100%	0%

	Ratio	Total Ratio
External Risk Factors		
C. Local Social & Cultural Characteristics	40%	10%
D. Geographical conditions	30%	8%
F. Status of infrastructure	30%	8%
Internal Risk Factors		
I. Project Organization	**40%**	**30%**
J. Contract Condition	30%	23%
L. Expenses	10%	8%
N. Engineering Period	5%	4%
Q. Construction Period	15%	11%

Business-Version PDRI

Category Element	Definition Level 0	1	2	3	4	5	Score
I. Project Organization							
I1. Considering joint venture	0	1	4	8	13	19	8
I2. Ability of employer	0	1	6	12	19	26	12
I3. Ability of contractor	0	1	8	17	25	31	25
I4. Ability of local companies	0	1	7	15	22	29	22
						Category Total	67

The three hypothetical alternatives that are presented for each project were identified through a process of long discussion. The alternatives were derived from the application of the TRIZ method. Further, the probability distribution of each alternative is weighted with the assumption of the situation of each alternative, where it is assumed that the alternatives only affect the corresponding risk factor, and the probability distribution of each alternative can be found by using the distribution fit. The Monte Carlo simulation derived the best alternative using the $NPV_{0.05}$, for which a 10,000-time repetition was performed for each alternative. The simulation results are shown in Table 11.

Table 11. Investment decisions for the net present value ($NPV_{0.05}$) for the identification of the alternatives from an improved process.

Risk Factor	TRIZ Method Alternative Solution	Probability Distribution	Variation	$NPV_{0.05}$ (Billion KRW)	Proposed Decision Making
G1. Volatility of Market Demand (Project K)	9. Continuous action—Interrupted action → Planning associated downstream project	Uniform	Min: 630 Max: 630	(204)	Rejection
	11. Direct action—Indirect action → Ensuring local buyers	Exponential	β: 26.7 Min: 626.2 Mean: 686.8	(40)	Rejection
	12. Preliminary action—Preliminary counteraction → Ensuring long-term off-takers ✓	Triangular	Min: 626 Mode: 656 Max: 676	65	Approval
I3. Ability of Contractor (Project C)	3. Homogeneity—Diversity → Ordering EPC Lump-sum Turnkey ✓	Triangular	Min: 3900 Mode: 4000 Max: 4200	26	Approval
	4. Expansion—Reduction → Utilizing competitive bid techniques	Triangular	Min: 3400 Mode: 3400 Max: 4800	(147)	Rejection
	7. Standardization—Specialization → Collaboration-capable local companies	Triangular	Min: 3500 Mode: 3600 Max: 4700	(126)	Rejection

In Project K, the sale price decreased due to the provision of discounts to the long-term off-takers. In Project C, the CAPEX increased because the EPC Lump-Sum Turnkey contract method transfers the construction risks to a contractor who then spends more capital. Viewed through the DCF indicators such as the NPV and the IRR, the results calculated for both projects worsened because of the cost of the risk hedges; however, an alternative in each project is the positive $NPV_{0.05}$ in Table 10, whereby the decision criterion is fulfilled. Eventually, both projects could be approved by the investors according to the selected alternatives. Rather than looking for ways to overcome the uncertainty risks by listing the risk items when the risks are being resolved, it is preferable for the investors to quantify the weighted risk items for each project in advance and find the solutions using the TRIZ method.

The two cases in this study are projects currently in progress. Project K was completed in December 2013 and has been operating for three years; however, the company has experienced financial difficulties for three consecutive years. As the number of fixed sale points is low, the selling prices of the products are continuously falling. Project C was completed in June 2016 and has been operating for six months. Unlike the CAPEX of 3800 MUSD that was expected at the feasibility-study stage, the final CAPEX is 4500 MUSD, which is an increase of 18.42%. Although the investors decided to approve both projects based on traditional decision-making methods, the final results of these projects still need to be monitored, since the interim results are not promising. To prepare for this situation, the improved method recommended in this study provides an additional opportunity to protect the projects against the related risks. The effectiveness of the proposed method was verified by reviewing the actual project results from two overseas SPPs, thereby allowing for a comparison of the results from the traditional and improved methods. However, this study is limited because the solutions that are provided by the alternative approach cannot be verified with respect to the two presented cases.

6. Conclusions

Korean steel-making companies need a new decision-making process for selecting profitable projects that result in successful overseas investments for sustainable development. In traditional decision-making process, it is difficult to reflect risks, and improved decision-making processes have thus been proposed in several studies. The improved decision-making process used in this paper is the NPVaR method, for which the investment decision-making criteria regarding overseas SPPs are derived according to a project-profitability estimation. Unlike the traditional decision-making process, the improved method reflect the risks of the profitability impact factors that affect SPP projects and this supports better decision-making by Korean steel-making companies. In particular, if the prospective investment project is rejected after the initial decision-making process, the subsequent actions that the investors should take are described in this study. Notably, he "probabilistic alternative approach" described in this study is not covered in detail in other studies and it is a significant contribution. This approach entails financial arbitrage so that decision makers can be presented with several choices to avoid risky situations. Further, this approach can be extended to the field of options-valuation research.

However, this study has two limitations. First, the methodology of this paper considers only the macro business elements to evaluate the project value and does not consider the micro technical aspects. Technical project evaluation is made primarily on the basis of practical experience and engineering knowledge. Technical project evaluation is different from business project evaluation because it is difficult to appreciate its value. The proposed methodology does not describe the technical aspects of the project, so the results may be less realistic. To overcome this problem, this study proposed that the project be evaluated based solely on a cash flow model. Therefore, future research needs to present a model for evaluation of micro-technology in project value evaluation. Second, it is assumed in this study that the selected alternative influences only one input variable of the probabilistic alternative. However, it is more realistic to consider that selected alternatives can affect various profitability impact factors in a project. A simple methodology is assumed in this study and simulation is performed to verify the effectiveness of the probabilistic alternative approach. In the future, the authors propose analyzing the various effects of profitability impact factors by thoroughly analyzing the alternatives.

Acknowledgments: This work was supported by the Technology Innovation Program (10077606, To Develop an Intelligent Integrated Management Support System for Engineering Project) funded by the Ministry of Trade, Industry & Energy (MOTIE, Korea).

Author Contributions: Yonggu Kim developed the concept and drafted the manuscript. Eul-Bum Lee revised the manuscript and supervised the overall work. All authors read and approved the final manuscript.

Conflicts of Interest: The authors declare no conflicts of interest.

Appendix

Category Element			Definition Level						Score
			0	1	2	3	4	5	
A.		Credibility of Local Government							
	A1.	Local administrative procedures and practices							
	A2.	Maturity of legal system							
	A3.	Consistent local policy							
								Category Total	0
B.		Economic Stability							
	B1.	Economic condition of employer							
	B2.	Economic condition of local country							

								Category Total	0
C.	\multicolumn{2}{l}{Local Social & Cultural Characteristics}							
	C1.	Local social stability							
	C2.	Local labor situation							
	C3.	Cultural traits							
	C4.	Local reactions of project							
								Category Total	0
D.	\multicolumn{2}{l}{Geographical conditions}							
	D1.	Climate feature							
	D2.	Soil condition							
	D3.	Distance from Korea							
								Category Total	0
E.	\multicolumn{2}{l}{Legal Standard}							
	E1.	Local standards of design							
	E2.	Licensing standards							
	E3.	Tariff standards							
	E4.	Environmental regulations							
	E5.	Repatriation of profits procedure							
	E6.	Local content requirement							
								Category Total	0
F.	\multicolumn{2}{l}{Status of infrastructure}							
	F1.	Local infra and utility level							
	F2.	Difficulty of infra and utilities use contracts							
	F3.	Future additional expansion possibilities							
								Category Total	0
G.	\multicolumn{2}{l}{Market conditions}							
	G1.	Volatility of market demand							
	G2.	Local competitors Status							
	G3.	Local market share							
	G4.	Exchange rate fluctuations							
	G5.	Inflation fluctuations							
	G6.	Interest rate fluctuations							
								Category Total	0
H.	\multicolumn{2}{l}{Financing Plan}							
	H1.	Additional potential projects in progress							
	H2.	Cash flow stability							
	H3.	Debt equity ratio							
								Category Total	0
I.	\multicolumn{2}{l}{Project Organization}							
	I1.	Considering joint venture							
	I2.	Ability of employer							
	I3.	Ability of contractor							
	I4.	Ability of local companies							
								Category Total	0
J.	\multicolumn{2}{l}{Contract Condition}							
	J1.	Similar experiences							
	J2.	Liquidated damages							
	J3.	Contract terms changeability							
	J4.	Unclear contract terms							
	J5.	Steel purchase conditions							
	J6.	Force majeure							

								Category Total	0
	Scope of Work								
K.	K1.	Characteristics of product							
	K2.	Construction schedule							
								Category Total	0
	Expenses								
L.	L1.	Initial investment							
	L2.	Operation costs							
	L3.	Reasonable Contingencies							
								Category Total	0
	Process & Technology								
M.	M1.	Conformity of new process							
	M2.	Level of applied technology							
								Category Total	0
	Engineering Period								
	N1.	Layout planning							
	N2.	Constructability/Complexity							
N.	N3.	Design for Scalability							
	N4.	Major equipment selection							
	N5.	Timeliness of design							
	N6.	Value engineering							
								Category Total	0
	Health, Safety & Environment								
O.	O1.	Work safety management							
	O2.	Construction safety facilities							
	O3.	Pollution prevention Plan							
								Category Total	0
	Procurements								
P.	P1.	Personnel procurement plan							
	P2.	Equipment procurement							
								Category Total	0
	Construction Period								
	Q1.	Suitable working method							
	Q2.	Equip./Material transport							
Q.	Q3.	Local company collaboration							
	Q4.	Local materials quality							
	Q1.	Security & Safety							
	Q2.	Commissioning & Acquisition requirements							
								Category Total	0
	Completion Requirements								
R.	R1.	Document Management							
	R2.	Performance requirements							
								Category Total	0
								Total Score	0

Appendix

12 TRIZ Principles for alternatives

Principle 1: Combination–Separation

Part of an object or process step is combined to form a uniform object or process. Separate uniform objects or uniform processes to form independent parts or phases.

Principle 2: Symmetry–Asymmetry

Change the symmetrical shape or property to an asymmetrical shape or property or change the asymmetrical shape or property to a symmetrical shape or property.

Principle 3: Homogeneity–Diversity

Change from a homogeneous structure, system or environment to a complex structure, another system, or environment. Reduce the variety of structures, systems, or environments.

Principle 4: Expansion–Reduction

Increase or decrease the number of functions in the system or process. Increase or decrease the amount, duration, cost, speed, or other attribute of the process.

Principle 5: Mobility–Immovability (Dynamic–Static)

The fixed part of the system or environment is movable, and vice versa.

Principle 6: Consumption–Regeneration

Elements that are consumed by a system or process are regenerated within the same system or process. The consumed and accomplished elements are removed or modified in other applications.

Principle 7: Standardization–Specialization

Use more standardized processes, procedures, methods, and products. Take advantage of special processes, products, or methods.

Principle 8: Action–Reaction

Action–Reaction boosts the effect you desire. Acquire the opposite effect and amplify.

Principle 9: Continuous Action–Interrupted Action

Critical processes must be performed without interruption or idle time. They should be carried out on a constant load and constantly monitored. Hinder continuous action; prepare to pause in a continuous process.

Principle 10: Partial Action–Excessive Action

Use surplus or excessive action to achieve maximum or optimum effect. Protect sensitive areas from undesired behavior. Focus on the essential tasks to achieve maximum or optimal results. Strengthen your activity in areas that yield optimal results.

Principle 11: Direct Action–Indirect Action

Replace indirect action with direct action immediately. Or replace direct action with indirect action immediately.

Principle 12: Preliminary Action–Preliminary Counteraction

References

1. Kyeong-Chan, K. The reorganization of global steel industry and the implications for POSCO. *SERI Q.* **2009**, *2*, 78.
2. Bal, M.; Bryde, D.; Fearon, D.; Ochieng, E. Stakeholder Engagement: Achieving Sustainability in the Construction Sector. *Sustainability* **2013**, *5*, 695–710. [CrossRef]
3. Jillella, S.; Matan, A.; Newman, P. Participatory Sustainability Approach to Value Capture-Based Urban Rail Financing in India through Deliberated Stakeholder Engagement. *Sustainability* **2015**, *7*, 8091–8115. [CrossRef]
4. Liang, X.; Yu, T.; Guo, L. Understanding Stakeholders' Influence on Project Success with a New SNA Method: A Case Study of the Green Retrofit in China. *Sustainability* **2017**, *9*, 1927. [CrossRef]
5. Tseng, Y.-C.; Lee, Y.-M.; Liao, S.-J. An Integrated Assessment Framework of Offshore Wind Power Projects Applying Equator Principles and Social Life Cycle Assessment. *Sustainability* **2017**, *9*, 1822. [CrossRef]
6. *Risk Management–Principles and Guidelines*; ISO 31000: 2009; International Organization for Standardization: Geneva, Switzerland, 2009.
7. Hubbard, D.W. *The Failure of Risk Management: Why it's Broken and How to Fix it*; John Wiley & Sons: Hoboken, NJ, USA, 2009.
8. Ye, S.; Tiong, R.L.K. NPV-at-Risk Method in Infrastructure Project Investment Evaluation. *J. Construct. Eng. Manag.* **2000**, *126*, 227–233. [CrossRef]
9. Habibi, H.; Habibi, R. Applications of Simulation-Based Methods in Finance: The Use of ModelRisk Software. *J. Adv. Stud. Financ.* **2016**, *7*, 82.
10. Caron, F.; Fumagalli, M.; Rigamonti, A. Engineering and contracting projects: A value at risk based approach to portfolio balancing. *Int. J. Proj. Manag.* **2007**, *25*, 569–578. [CrossRef]
11. Shi, J.; Wang, Y.; Fu, R.; Zhang, J. Operating Strategy for Local-Area Energy Systems Integration Considering Uncertainty of Supply-Side and Demand-Side under Conditional Value-At-Risk Assessment. *Sustainability* **2017**, *9*, 1655. [CrossRef]
12. Gatti, S.; Rigamonti, A.; Saita, F.; Senati, M. Measuring Value-at-Risk in Project Finance Transactions. *European Financial Manag.* **2007**, *13*, 135–158. [CrossRef]

13. Zhang, C.; Pu, Z.; Zhou, Q. Sustainable Energy Consumption in Northeast Asia: A Case from China's Fuel Oil Futures Market. *Sustainability* **2018**, *10*, 261. [CrossRef]
14. Zhu, L.; Ren, X.; Lee, C.; Zhang, Y. Coordination Contracts in a Dual-Channel Supply Chain with a Risk-Averse Retailer. *Sustainability* **2017**, *9*, 2148. [CrossRef]
15. Kim, Y.; Shin, K.; Ahn, J.; Lee, E.-B. Probabilistic Cash Flow-Based Optimal Investment Timing Using Two-Color Rainbow Options Valuation for Economic Sustainability Appraisement. *Sustainability* **2017**, *9*, 1781. [CrossRef]
16. Damodaran, A. *Investment Valuation: Tools and Techniques for Determining the Value of any Asset*; John Wiley & Sons: Hoboken, NJ, USA, 2012; Volume 666.
17. Hacura, A.; Jadamus-Hacura, M.; Kocot, A. Risk analysis in investment appraisal based on the Monte Carlo simulation technique. *Eur. Phys. J. B Condens. Matter Complex Syst.* **2001**, *20*, 551–553. [CrossRef]
18. Rezaie, K.; Amalnik, M.S.; Gereie, A.; Ostadi, B.; Shakhseniaee, M. Using extended Monte Carlo simulation method for the improvement of risk management: Consideration of relationships between uncertainties. *Appl. Math. Comput.* **2007**, *190*, 1492–1501. [CrossRef]
19. Suslick, S.B.; Schiozer, D.; Rodriguez, M.R. Uncertainty and risk analysis in petroleum exploration and production. *Terrae* **2009**, *6*, 2009.
20. Liu, J.; Jin, F.; Xie, Q.; Skitmore, M. Improving risk assessment in financial feasibility of international engineering projects: A risk driver perspective. *Int. J. Proj. Manag.* **2017**, *35*, 204–211. [CrossRef]
21. Dempster, M.A.H. *Risk Management: Value at Risk and Beyond*; Cambridge University Press: Cambridge, UK, 2002.
22. Hull, J.C. *Options, Futures, and Other Derivatives*, 8th ed.; Pearson Education: New York, NY, USA, 2012.
23. Jorion, P. *Value at Risk*; McGraw-Hill: New York, NY, USA, 1997.
24. Miles, L.D. *Techniques of Value Analysis and Engineering*; Miles Value Foundation: Portland, OR, USA, 2015.
25. Ibusuki, U.; Kaminski, P.C. Product development process with focus on value engineering and target-costing: A case study in an automotive company. *Int. J. Prod. Econ.* **2007**, *105*, 459–474. [CrossRef]
26. Mao, X.; Zhang, X.; AbouRizk, S.M. Enhancing value engineering process by incorporating inventive problem-solving techniques. *J. Construct. Eng. Manag.* **2009**, *135*, 416–424. [CrossRef]
27. Ruchti, B.; Livotov, P. TRIZ-based innovation principles and a process for problem solving in business and management. *TRIZ J.* **2001**, *1*, 677–687.
28. Kim, Y.-S.; Cochran, D.S. Reviewing TRIZ from the perspective of axiomatic design. *J. Eng. Des.* **2000**, *11*, 79–94. [CrossRef]
29. Yamashina, H.; Ito, T.; Kawada, H. Innovative product development process by integrating QFD and TRIZ. *Int. J. Prod. Res.* **2002**, *40*, 1031–1050. [CrossRef]
30. Jones, E.; Harrison, D. Investigating the use of TRIZ in Eco-innovation. *TRIZ J.* 2000. Available online: https://www.researchgate.net/profile/David_Harrison10/publication/49401085_Investigating_the_use_of_TRIZ_in_eco-innovation/links/54bccdbb0cf253b50e2d6200/investating-the-use-of-TRIZ-in-eco-innovation.pdf (accessed on 19 January 2015).
31. Ilevbare, I.M.; Probert, D.; Phaal, R. A review of TRIZ, and its benefits and challenges in practice. *Technovation* **2013**, *33*, 30–37. [CrossRef]
32. Souchkov, V. Breakthrough thinking with TRIZ for business and management: An overview. *ICG Train. Consult.* 2007, pp. 3–12. Available online: http://scinnovation.cn/wp-content/uploads/soft/100910/BreakthroughThinkingwithTRIZforBusinessandManagementAnOverview.pdf (accessed on 19 March 2007).
33. Bodie, Z.; Kane, A.; Marcus, A.J. *Investments, 10e*; McGraw-Hill Education: New York, NY, USA, 2014.
34. Trigeorgis, L. *Real Options: Managerial Flexibility and Strategy in Resource Allocation*; MIT Press: Cambridge, MA, USA, 1996.
35. Wibowo, A.; Kochendörfer, B. Financial risk analysis of project finance in Indonesian toll roads. *J. Constr. Eng. Manag.* **2005**, *131*, 963–972. [CrossRef]
36. Yun, S.; Han, S.H.; Kim, H.; Ock, J.H. Capital structure optimization for build–operate–transfer (BOT) projects using a stochastic and multi-objective approach. *Can. J. Civ. Eng.* **2009**, *36*, 777–790. [CrossRef]
37. Bingham, E.; Gibson, G.E., Jr. Infrastructure Project Scope Definition Using Project Definition Rating Index. *J. Manag. Eng.* **2016**, *32*, 04016037. [CrossRef]
38. Sindhu, J.R.N. *Investigating the Effect of Front-End Planning in Fast-Track Delivery Systems for Industrial Projects*; Texas A&M University: College Station, TX, USA, 2016.

39. Yildiz, A.E.; Dikmen, I.; Birgonul, M.T.; Ercoskun, K.; Alten, S. A knowledge-based risk mapping tool for cost estimation of international construction projects. *Autom. Constr.* **2014**, *43*, 144–155. [CrossRef]
40. Deng, X.; Low, S.P.; Li, Q.; Zhao, X. Developing competitive advantages in political risk management for international construction enterprises. *J. Constr. Eng. Manag.* **2014**, *140*, 04014040. [CrossRef]
41. Lee, N.; Schaufelberger, J.E. Risk management strategies for privatized infrastructure projects: Study of the build–operate–transfer approach in east Asia and the Pacific. *J. Manag. Eng.* **2013**, *30*, 05014001. [CrossRef]
42. Zhao, X.; Hwang, B.-G.; Yu, G.S. Identifying the critical risks in underground rail international construction joint ventures: Case study of Singapore. *Int. J. Proj. Manag.* **2013**, *31*, 554–566. [CrossRef]
43. Subramanyan, H.; Sawant, P.H.; Bhatt, V. Construction project risk assessment: Development of model based on investigation of opinion of construction project experts from India. *J. Constr. Eng. Manag.* **2012**, *138*, 409–421. [CrossRef]
44. Cheng, M.-Y.; Tsai, H.-C.; Chuang, K.-H. Supporting international entry decisions for construction firms using fuzzy preference relations and cumulative prospect theory. *Expert Syst. Appl.* **2011**, *38*, 15151–15158. [CrossRef]
45. Han, S.H.; Kim, D.Y.; Kim, H. Predicting profit performance for selecting candidate international construction projects. *J. Constr. Eng. Manag.* **2007**, *133*, 425–436. [CrossRef]
46. Ozorhon, B.; Dikmen, I.; Birgonul, M.T. Case-based reasoning model for international market selection. *J. Constr. Eng. Manag.* **2006**, *132*, 940–948. [CrossRef]
47. Wang, S.Q.; Tiong, R.L.; Ting, S.; Ashley, D. Evaluation and management of political risks in China's BOT projects. *J. Constr. Eng. Manag.* **2000**, *126*, 242–250. [CrossRef]
48. Zou, P.X.; Zhang, G.; Wang, J. Understanding the key risks in construction projects in China. *Int. J. Proj. Manag.* **2007**, *25*, 601–614. [CrossRef]
49. Kim, D.Y.; Han, S.H.; Kim, H.; Park, H. Structuring the prediction model of project performance for international construction projects: A comparative analysis. *Expert Syst. Appl.* **2009**, *36*, 1961–1971. [CrossRef]
50. Chen, H.L.; Chen, C.-I.; Liu, C.-H.; Wei, N.-C. Estimating a project's profitability: A longitudinal approach. *Int. J. Proj. Manag.* **2013**, *31*, 400–410. [CrossRef]
51. Qazi, A.; Quigley, J.; Dickson, A.; Kirytopoulos, K. Project Complexity and Risk Management (ProCRiM): Towards modelling project complexity driven risk paths in construction projects. *Int. J. Proj. Manag.* **2016**, *34*, 1183–1198. [CrossRef]
52. Saltelli, A.; Chan, K.; Scott, E.M. *Sensitivity Analysis*; Wiley: New York, NY, USA, 2000; Volume 1.
53. Ritzema, H. *Drainage Principles and Applications*; International Institute for Land Reclamation and Improvement (ILRI): Wageningen, The Netherlands, 1994.
54. Altshuller, G.; Shulyak, L.; Rodman, S. *40 Principles: TRIZ keys to Innovation*; Technical Innovation Center, Inc.: Worcester, MA, USA, 2002; Volume 1.

 © 2018 by the authors. Licensee MDPI, Basel, Switzerland. This article is an open access article distributed under the terms and conditions of the Creative Commons Attribution (CC BY) license (http://creativecommons.org/licenses/by/4.0/).

Article

Towards a Fair and More Transparent Rule-Based Valuation of Travel Time Savings

Kingsley Adjenughwure * and Basil Papadopoulos

Department of Civil Engineering, Democritus University of Thrace, 67100 Xanthi, Greece; papadob@civil.duth.gr
* Correspondence: kadjenug@civil.duth.gr; Tel.: +30-69934-99389

Received: 15 January 2019; Accepted: 6 February 2019; Published: 13 February 2019

Abstract: The value of travel time savings (VOTTS) is one of the most important variables for calculating the benefits of transportation projects. However, the way it is currently calculated (usually via discrete choice models) is complex, tedious and subject to a reasonable level of uncertainty. Furthermore, the method is not easily understood by government officials who use the VOTTS for appraisal and the citizens are not fully aware how such values are calculated. This lack of understanding and transparency in methodology may lead to misuse of the VOTTS during transport project appraisals which in turn can result in unfair transport decisions for citizens, government and the environment. To solve these problems, a fuzzy logic rule-based approach is proposed. With this approach, the rules can be made based on economic and behavioral theories by experts, government officials and citizens (via surveys). This approach makes it understandable to everyone how values are calculated. To test the applicability of the approach, a simple numerical example is presented by estimating the VOTTS of various countries using their gross domestic product-purchasing power parity (GDP-PPP) and the traffic congestion level. Results are then compared to values obtained from a recent metanalysis on VOTTS in Europe and some official VOTTS.

Keywords: value of travel time savings; fuzzy logic; rule-based systems; transport project evaluation; cost benefit analysis; traffic congestion; transport planning

1. Introduction

The value of travel time savings (VOTTS) is a very important component of most cost-benefit analyses (CBA) of transportation projects. It is estimated that travel time gains make up about 60 to 80% of the benefits of transportation infrastructure projects [1]. It is well-known that transportation infrastructure projects have both socio-economic and environmental effects which means that any mistakes in the project evaluation could result to socio-economic and environmental problems in the long run. Such an important evaluation variable like VOTTS could have serious consequences for the goals of sustainability of transportation projects if not calculated properly. Given the importance of this variable, it is no surprise that most developed countries like the Netherlands, UK, Denmark and Sweden have established recommended values which are used for project evaluation [2–4]. These recommended values are usually calculated via stated preference surveys where respondents are given choices to make a trade-off between cost and time. Discrete choice models are then applied to the responses to calculate the average VOTTS. The methods used in the survey design and model estimations are well-established (standardised) and are mathematically rigorous, thus making the results generally acceptable. Although this process sounds straightforward, the actual effort needed to carry out the surveys, estimate the values and arrive at the recommended VOTTS is very enormous and time consuming. For this and other reasons, most national value of time studies are carried out every 5 to 10 years.

Despite the effort, time, money and expert knowledge required for such studies the results are still very much subject to uncertainties. The main sources of uncertainty in such studies are from the design of the survey and the model estimation. Sometimes there is ambiguity in the attribute levels to use for time and cost as these also affect the estimated VOTTS [5,6]. It is also possible that respondents do not answer as expected (non-trading habits) or get tired because of the many trade-offs they make (usually between 6 and 12 trade-offs) [6]. Furthermore, the chosen model may not fully capture how respondents make their choices. For instance, there is uncertainty in the type of theory respondents use to select their response as most models currently use utility maximisation whereas regret minimisation could also be used [7,8]. In most studies, the uncertainties in the design and model estimation are normally just discussed and some solutions offered to mitigate them. There are usually not many discussions on alternative approaches for calculating the VOTTS. The main reason is that the models used are based on economics and behavioural theories which have long been established and there is available expertise and commercial software for both survey design and model estimations.

The complexity of the survey design, and model estimation techniques make most studies understandable only to experts. Additionally, the sample of respondents used in the survey may not be a good representative of the whole population. To solve this issue, estimated values must be converted to recommended VOTTS after a weighting by distance, income and trip purpose [2–4]. These recommended VOTTS are then presented to the government. To give further credibility to the method of calculating values, the whole process is normally subjected to external audit by various experts in the field [3]. Obviously, it is difficult to explain such designs and models to government officials who use these values for project evaluation. Therefore, such recommended VOTTS are usually taken at face value, albeit with a bit of scepticism. There is, however, not much that can be done about it since government officials are usually not experts in the topic.

Very few studies have proposed a different, transparent and practical approach for the estimation of VOTTS. Most studies are usually focused on developing more complex methods to capture better human behaviour [5]. Other studies focus on improving the survey design techniques to make it easier for people to respond and improve the plausibility of results [9,10]. One important reason for the lack of different approaches for estimating VOTTS is that it is not observable but rather derived from a trade-off involving cost and time. These models are generally grouped under the name of discrete choice modelling [11,12]. In this modelling framework, travellers are faced with a series of trade-offs between a longer journey at cheaper cost and a shorter journey with a higher cost. The VOTTS is then calculated and interpreted as a traveller's willingness to pay to save an hour of travel time [11,12]. Other studies have also used such trade-off techniques to estimate the distribution of VOTTS in a non-parametric way [13,14].

Generally, the VOTTS of an individual cannot be directly calculated using a classical regression technique. Some studies have used meta-analysis to create regression models to estimate VOTTS, but they must assume a model [15,16]. What is normally done is to use the income to infer the VOTTS for example by saying VOTTS is about some percentage of the hourly income, the so-called cost saving approach (CSA) [16]. Apart from income, studies have shown that variables like comfort of the travel, trip length etc. also affect VOTTS [16]. However, even though all these variables are known to affect VOTTS, it still not certain to what extent they do so.

To tackle the problem of the uncertainty and understandability of VOTTS results, a new estimation approach is proposed based on fuzzy logic and expert knowledge. Fuzzy logic is a well-known technique for handling uncertainty [17]. The main advantage is that it is close to the way humans think and express uncertainty. So, a variable maybe considered high, low, very high etc. Also, humans often make simple rules to help them in decision making. Fuzzy logic has been applied successfully to solve many transportation problems ranging from traffic control to transportation planning. A review of the applications of fuzzy logic in transport can be found in [18]. In contrast to previous studies, this paper does not focus on using fuzzy logic for choice modelling (route choice, mode choice) [19–22] but rather it proposes to use a fuzzy inference system (FIS) to directly estimate VOTTS of an individual, country (or at least of a group of people).

The proposed approach is chosen for two reasons. First, the rules can be implemented based on economic and behavioural theories, by experts, government officials, citizens (via surveys) etc. The rules can be updated or modified periodically according to current trends. This makes it understandable to everyone how VOTTS are calculated.

The second reason is that the model outputs and the uncertainties of the current methods can be incorporated in the model input and output of the fuzzy model. For example, findings from various VOTTS studies using choice models and other related techniques can be used to define membership functions for inputs, rules and the final output.

Given the nature of the proposed method, equating the VOTTS calculated by the fuzzy method with that estimated from standard discrete choice and related economic models is not justified because the values are calculated from rules rather than by economic theory. The main purpose of this paper is to put forward a research agenda for a fair, more transparent rules-based approach for the future calculation of VOTTS which will be used for evaluating transportation projects. The proposed method is easy to implement and can be used as an alternative means of calculating VOTTS or as a supporting tool for VOTTS estimations from discrete choice or other related econometric models.

The rest of the paper is organised as follows. In the next section a brief description of the current method for estimating VOTTS using discrete choice is presented. This is followed by a description of fuzzy logic and the fuzzy inference system in general. Thereafter, a conceptual model of the proposed approach is presented and discussed. Then a simple numerical example is presented using the proposed approach to illustrate the method and check validity of outputs. Finally, a discussion of how the proposed fuzzy method can be improved and incorporated into current VOTTS estimation methods is presented.

2. Theoretical Background

2.1. Value of Travel Time Savings (VOTTS) from Discrete Choice Model

The VOTTS is derived from choices made by individuals in a stated choice experiment. The simplest experiment consists of a series of choices between time and cost attributes. The choice experiments are carefully designed using efficient designs such that the VOTTS can be estimated [9]. Discrete choice models use the theory of utility maximization which postulates that when faced with choices rational individuals choose the one that maximises their utility [23]. In the context of VOTTS, the systematic utility function (U) can be expressed as a linear function of both cost and time defined as follows [3,23]:

$$U = \beta_c C + \beta_t T \qquad (1)$$

where β_c and β_t are the marginal utilities of the cost of travel (C) and travel time (T), respectively. The VOTTS savings is derived as the ratio $\frac{\beta_t}{\beta_c}$. This is the marginal rate of substitution between time and money [3]. Note that the complete utility function contains an unobserved error term which is assumed to follow a certain distribution usually distributed independently, identically extreme value [23]. This assumption leads to classical logit models. Using the choices made by respondents, the parameters β_c and β_t can be estimated using standard techniques like maximum likelihood estimation.

2.2. Fuzzy Logic and Fuzzy Inference System (FIS)

In this section, a brief description of fuzzy sets and fuzzy inference systems is given. The goal is not to go into the details of this method as it is well established; rather, the focus is on the elements relevant for this paper. A more concise introduction to fuzzy logic and fuzzy inference systems can be found in [24].

2.2.1. Fuzzy Sets

Fuzzy sets introduced by Lofti Zadeh [17] is an alternative way of representing uncertainty. In classic set theory, an object either belongs to a set or it does not. Fuzzy sets help to represent the uncertainty (or lack of complete information) whether an object belongs to the set or not [25]. This is achieved by assuming that the object belongs to the set with a certain degree in the closed interval [0 1]. Formally, a fuzzy set is defined as: Let X be a universe of discourse. The fuzzy set A is characterised by its membership function [25,26]:

$$\mu_A(x) : X \rightarrow [0,1] \text{ for } x \in X \tag{2}$$

The membership functions usually represent "linguistic" variables such that they are easily understandable. For example, the cost of travel may be described by the linguistic variables, high, medium and low each with their corresponding membership function. Using these membership functions, membership values can then be assigned for each cost of travel in a specified range of values. This membership value indicates how low, medium or high a given travel cost is. The most common membership function is the triangular membership function defined as [25,26]:

$$A(x) = \begin{cases} \frac{x-a}{b-a} & \text{for } a \leq x \leq b \\ \frac{c-x}{c-b} & \text{for } b \leq x \leq c \\ 0 & \text{otherwise} \end{cases} \tag{3}$$

The fuzzy number is usually represented by $A = (a, b, c)$ its centre value b, the left and right values a and c.

2.2.2. Fuzzy Rules

These are rules made using the linguistic variables like low, medium, high. They enable decision making in a simple and human-like manner. Fuzzy rules are usually of the form IF A then B [27]. Where A and B are linguistic variables representing input and output respectively. For example, a simple transportation rule could be IF travel time of a mode is HIGH, THEN mode share is expected to be LOW. It is then left for experts to decide which range of travel time is considered HIGH and which range of mode share is considered LOW. Once this is known, these rules can then be used to make decisions concerning expected mode shares given the travel time of the mode. One advantage of using fuzzy rules for decision making is that we do not have to know the exact value of the variables (input or output) but just a range. This is very close to the how humans think since it is more likely that they use simple rules to make decisions with approximate values rather than complex mathematical equations and exact values [27].

2.2.3. FIS

This is a decision-making system which uses a set of fuzzy rules. The FIS uses a set of input and a knowledge base (rules and data) to make inference about a set of outputs [24,25]. The rules and data are usually supplied by experts with experience in that subject. A fuzzy inference system is made up of 5 components [24] (Figure 1):

- A **fuzzification** unit: this transforms the crisp value into the interval [0 1] using a specified membership function.
- A **rule-base**: this contains IF-THEN rules used to represent a link between the input and output variables.
- A **database**: this contains the membership functions of the linguistic variables used in the fuzzy rules.
- **Decision unit**: this is where an inference is made by using the specified rules to get a fuzzy output.

- **Defuzzification unit**: In this unit, the fuzzy output from the decision unit is converted to a final crisp output.

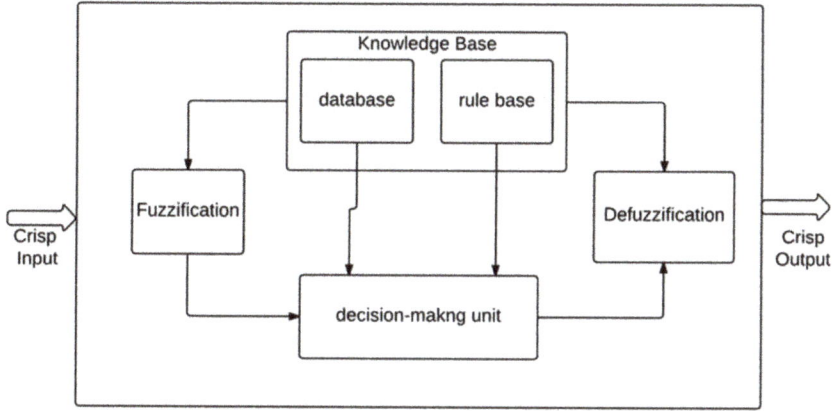

Figure 1. Fuzzy inference system (FIS). Adapted from [24].

In contrast to most artificial intelligence (AI)-based techniques which are black-box models, all inputs, rules and outputs of a fuzzy inference system can be easily scrutinised and modified accordingly. This makes it very suitable for evaluating variables like VOTTS which are subject to debate and uncertainty.

In the next section, we present a conceptual model using FIS which can be used to support the estimation of VOTTS. The conceptual mode is general and can be used in any a VOTTS studies.

3. Methodology

3.1. Proposed Conceptual Model

A generalised rule-based model for estimating the VOTTS is proposed. The model is basically a FIS comprising of rules defined by a consensus of experts, citizens and government officials (Figure 2). This is to increase the transparency and understandability of the process used to calculate the VOTTS. The output of the model is the VOTTS. The input for the model can be various variables which have significant effects on the VOTTS. The proposed model may be constructed based on values from previous VOTTS or independently. The structure of the model is explained below.

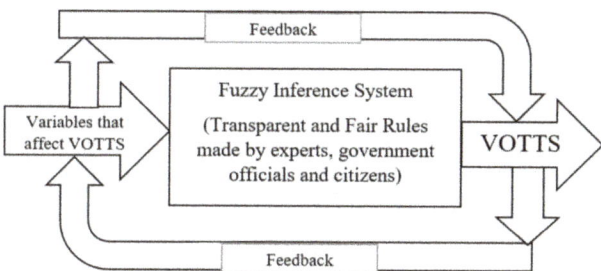

Figure 2. Conceptual model.

3.1.1. Input Variables

The chosen input variables should be those that are known to directly affect the VOTTS. Generally, many variables affect the VOTTS like the transport mode [28], the region [29], temporal differences [30], individual differences [31]. However, for appraisal purposes, the VOTSS is usually not differentiated by all these variables except by the transport mode [32]. Various studies have shown that income has a significant effect on the VOTTS [4,15]. However, care should be taken when using this variable since it automatically gives higher VOTTS to those with high income and lower VOTTS to low income earners. For the purpose of the model, an average income can be used, or income can be excluded entirely. This is to ensure that everyone is considered equal during the calculation of the VOTTS.

Another important variable is the comfort of travel. It has been shown in various studies that travellers' VOTTS are higher when a trip is made in a crowded or congested condition [33,34]. This comfort variable can be easily incorporated in the proposed model through surveys on the travel conditions or other directly thorough calculated delays. Other variables like trip distance can also be used but it is debatable how much effects they have and whether it is appropriate to differentiate VOTTS based on trip distance [4,29,32].

The choice of input variables to use should be made via consensus between, experts and government officials considering evidence from previous studies.

3.1.2. Membership Functions

The membership functions and linguistic variables can be easily defined by experts and government officials or by citizens (via surveys). For example, government can classify income levels as low, medium high, experts can classify trip distance as short, medium and long distance and citizens can classify transport comfort levels as low, medium and high comfort through a survey.

The shape of the membership functions (triangular, trapezoidal, Gaussian etc) can be chosen based on expert advise). This also applies to the defuzzification techniques to use. Note that optimizing membership functions can also be performed automatically without expert intervention. Most techniques rely on genetic algorithms [35–38]. These methods can also be applied in this context. However, if such methods are used, they should be subject to evaluation before the final results are accepted.

3.1.3. Rule Base

The rule base should be defined by experts, government officials and through surveys from citizens. The rules should be understandable, plausible and fair. An example of a rule could be:

IF Comfort Level is low THEN VOTTS is HIGH. This rule is easy to understand for experts, government officials and citizens. The rule is also plausible because, when people travel under uncomfortable conditions they would want to arrive at their destination as quickly as possible (i.e., higher VOTTS). Finally, this rule is fair because it does not distinguish between the rich and poor, or type of travel mode etc. This is true for all modes and for all groups of travel and travel purposes.

3.1.4. Output Variable

The output is the VOTTS. Since the true value is not known, experts together with government officials can agree on values. The advantage of this model is that the exact values do not need to be known but just a range of values. These range of values can be supplied by experts or from previous studies like those in [15,16]. Also, linguistic variables like Low, Medium, High, Very High can be used to describe VOTTS. This can depend on different factors such as previous studies, GDP of the country, availability of transport funds. For example, VOTTS of 15 $/h may be considered high in one country and medium in another country.

3.1.5. Feedback Loop

To improve the model accuracy, a feedback loop between input and output should be maintained. This can be done by periodically updating rules, membership functions to reflect current trends. These periodic updates can be done based on recent surveys or value of time studies or based on economic, social or environmental trends. This will make the proposed method compatible with existing discrete choice models. The feedback loop can also be used as a means of validating output from the model and comparing it with those of discrete choice models. For example, if recent studies show that a new variable is now important in calculating the VOTTS, this variable can be included when building the rule-base. This can also be done if a variable becomes less important for instance due to technological advances. A typical example is the use of automated vehicles instead of a normal car. The idea of wasting time in traffic becomes questionable because an automated vehicle could provide comfort like an office and allow productive use of the time spent in congestion.

3.2. Numerical Example

To illustrate the proposed approach, a simple numerical example is presented using two input variables, economic prosperity of the country and travel comfort level. These two variables are chosen because the way in which they affect VOTTS is easily understandable, plausible and fair to all (within the country). The economic prosperity of a country determines the wage-rate which in turn determines how much money travellers can trade-off for shorter travel time. According to previous studies, VOTTS increases with increase in GDP with an elasticity close to 1.0 [16].

The variable comfort of travel is normally determined by the level of crowding in case of public transport or by the traffic congestion level for car travel. Studies have shown that the lower the comfort level, the higher the VOTTS. The way this is represented is through crowding and congestion multipliers which have been estimated depending on the level of comfort [33,34].

In this example, the gross domestic product-purchasing power parity (GDP-PPP) is used as input variable instead of the nominal GDP. This is because it reflects both the economic prosperity and standard of living of the country [16].

For the comfort level of travel, the average number of hours spent in traffic congestion is used. Of course, this applies to car travel but it is possible to illustrate the method with it. Other variables like public transport crowding can also be used if available. The GDP-PPP data is from the World Bank 2017 ranking [39]. The congestion data is from the INRIX traffic scorecard report 2017 [40]. The values for the VOTTS are based on ranges from previous studies [16].

Linguistic Variables

Five Linguistic variables are used for the GDP and 4 variables for the traffic congestion level. The output variable (VOTTS) has 4 linguistic variables (Table 1).

Table 1. Linguistic variables.

	Linguistic Variables				
Gross domestic product (GDP) ($k)	Low [0 20]	Lower Middle [10 40]	Middle [30 60]	Upper Middle [50 80]	High [60 130]
Traffic (h)	Low [0 20]	Moderate [10 40]	Congested [30 60]	Highly Congested [50 120]	
Value of travel time savings (VOTTS) ($/h)	Below Average [0 10]	Average [5 20]	Above average [15 30]	Very High [25 40]	

The values chosen for the membership functions of the linguistic variables are those that reflect the names of the variable (Figure 3). The approach is to first determine how many linguistic variables are needed for each input and output, then the most representative value is assigned as the middle value of each linguistic variable. The value for the middle linguistic variable is chosen close to the overall average value for each variable. For example, the average GDP for all countries in the list is $42.64k so the linguistic variable "Middle" is in the range [30 60] with GDP values between $40k and $50k surely "Middle". After that, the next criteria is that there should be a reasonable overlap between the variables [24]. Starting from the middle linguistic variable the other linguistic variables are created such that the overlap gives a reasonable membership function value [24]. Similar approach is applied to both the traffic congestion and VOTTS variables (Figures A1 and A2 Appendix A). There are various methods of constructing membership functions for linguistic variables [41,42] but this approach is adopted for simplicity. In practical applications, it is expected that membership functions and linguistic variables will be selected based on expert opinion with contribution from both the government and citizens or by optimization [35–38].

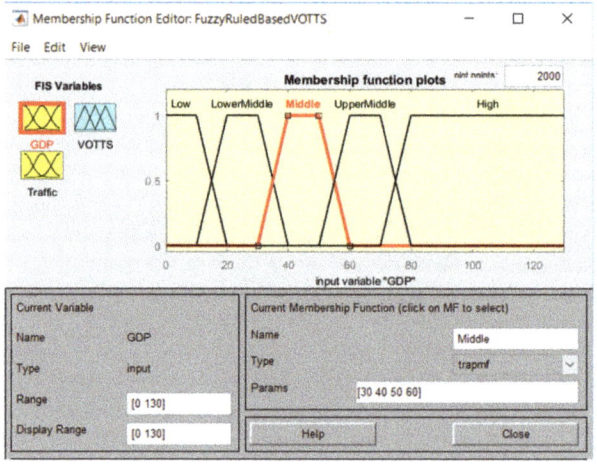

Figure 3. Membership function and linguistic variable for gross domestic product (GDP).

3.3. Rules

A total of 26 rules are generated. The rules are made based on how the two variables (GDP and Traffic) are expected to affect the VOTTS (see Section 3.2). Some selected subsets of the rules are shown in Table 2 below (see Appendix A, Figure A3, for all rules).

Table 2. Example Rules.

No	Rule
3	IF GDP is Low AND Traffic is Congested THEN VOTTS is Average
10	IF Traffic is Highly Congested then VOTTS is Above Average
11	IF GDP is Lower Middle AND Traffic is Low THEN VOTTS is Below Average
17	IF GDP is Middle AND Traffic is Congested THEN VOTTS is Average
18	IF GDP is Middle AND Traffic is Highly Congested THEN VOTTS is Above Average
20	IF GDP is Upper Middle AND Traffic is Moderate THEN VOTTS is Average
26	IF GDP is High AND Traffic is Highly Congested THEN VOTTS is Very High

For most rules, it is assumed that the GDP plays a more important role than congestion. Income is the most important variable that affects VOTTS [15,16]. The income level of a country is determined by its GDP. Countries with higher GDP generally tend to have higher average income and thus the VOTTS is expected to be high. The more money individuals earn, the more they are willing to pay to save travel time. For example, if GDP is High, VOTTS should be above average regardless of the congestion level since there is enough money to pay to save time. On the other hand if GDP is Low, then VOTTS should be Average. This is expected because even if traffic is congested and you do not have enough disposable income, you may not have enough to pay to save travel time. Nevertheless, some rules are also inserted to consider fairness, for example if Traffic is highly congested then the VOTTS should be above average. This is regardless of the GDP. The idea is that people should not be allowed to suffer in highly congested conditions just because the country or region has a low GDP. This type of rule could be used by organisation such as the World Bank for evaluating projects especially in low-income countries who suffer from severe traffic congestion.

Membership functions: the trapezoidal membership functions are used for all three variables. The main reason is that the range of values for the linguistic variables are more realistic. Trapezoidal membership functions are chosen for simplicity because they help represent more values in the core. This is more flexible than using a single value [43]. This is particularly suited for this application since it is difficult to assign just one value as being High, Medium or Low for the variables used like traffic congestion, GDP and VOTTS.

The triangular membership functions or any other membership function can also be used. The purpose of the paper is not to show which choice of membership shape is better but rather to show that a transparent and rule-based approach is better in estimating VOTTS because every decision made like rules, type of variable, membership functions can be easily scrutinized and modified in a fair and transparent way. Obviously for the model to be used for appraisal, the choices made will be based on consensus between experts, government officials and citizens.

The complete model was built using MATLAB (2017a) Fuzzy Logic Designer using the default settings. The structure of the model is shown below (Figure 4).

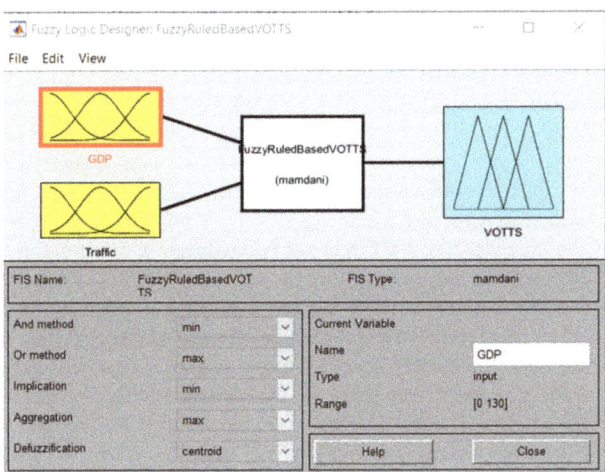

Figure 4. MATLAB model.

4. Results and Discussion

The proposed FIS is used to calculate VOTTS for various countries in the INRIX traffic report card and is shown in Figure 5 below (See Appendix A, Table A1, for exact estimates for all countries).

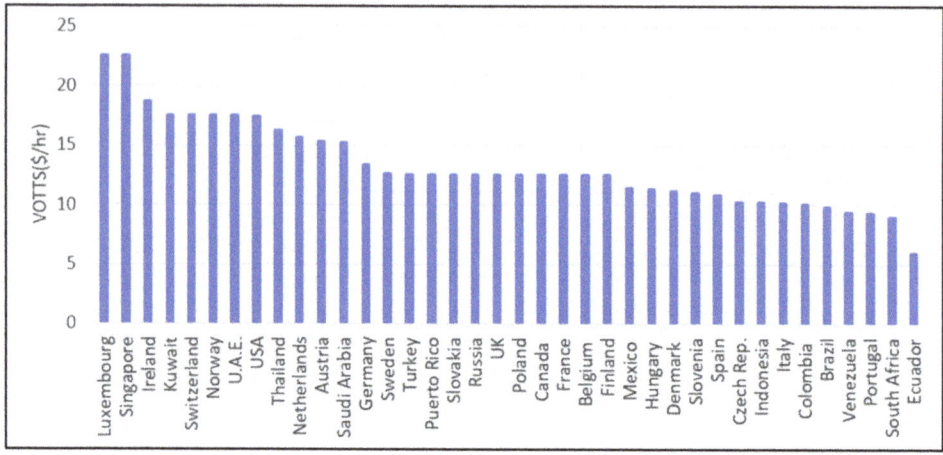

Figure 5. Estimated VOTTS for all countries in the INRIX traffic score card.

As expected, countries with lower GDP have low VOTTS while those with high GDP have high VOTTS with the exception of Thailand which has relatively high VOTTS despite its low GDP. This is due to its highly congested traffic situation which should justify a high VOTTS. In normal appraisals the VOTTS is expected to be lower than the one estimated here but this will be unfair to its citizens given the stress they have to go through daily. If the project is handled by an external body such as the World Bank, then the higher estimated value should be used. If the project is sponsored by the country alone, then government can decide whether a VOTTS is considered high or not and they may use lower values. For fairness, it is better that the government's willingness to pay to save a citizen one hour of travel should be higher than what the citizen is willing to pay especially for lower-income countries. The ability to include fairness rules in the proposed model is an extra advantage over discrete choice models. Discrete choice models will normally produce higher VOTTS for countries with high GDP regardless of the level of congestion in the country. This does not favour poor countries.

To check the validity of estimated results, a comparison is made with a recent metanalysis of VOTTS in European countries [16]. The comparison is made for the average value for car traffic with congestion for both commuting and leisure [16]. The metanalysis values were in 2010 equivalent euros. For a fair comparison, the values are all converted to 2017-dollar equivalent by taking into account average euro inflation (consumer price index (CPI) ratio of approximately 1.09) from 2010 to 2017 [44]. The average euro to dollar exchange rate is taken as 1 euro to 1.2 dollars [45]. The original values can be found in [16]. Note that the estimated VOTTS are not directly comparable to the ones estimated from discrete choice and econometric models since we use rules to infer VOTTS. However, the estimated VOTTS are compared to those estimated from discrete choice and econometric models as a benchmark.

Figure 6 shows the difference between the metanalysis estimate and the estimated values. As expected, the proposed model calculates lower values compared to metanalysis values for high GDP countries such as Luxembourg and Switzerland. On the other hand, the model estimates higher values for lower GDP countries such as Poland, Hungary and Slovakia. This is because of the fairness rules introduced in the model. For projects sponsored by the European Union, it will be wise to use such augmented values for lower GDP regions with moderate traffic congestion and use reduced values for high GDP regions with slightly better traffic conditions. For most countries, the metanalysis and the estimated values do not differ by more than 10%. The average VOTTS from the estimated model is almost the same as the model from metanalysis (see Appendix A, Table A2). This shows that the proposed model gives plausible results.

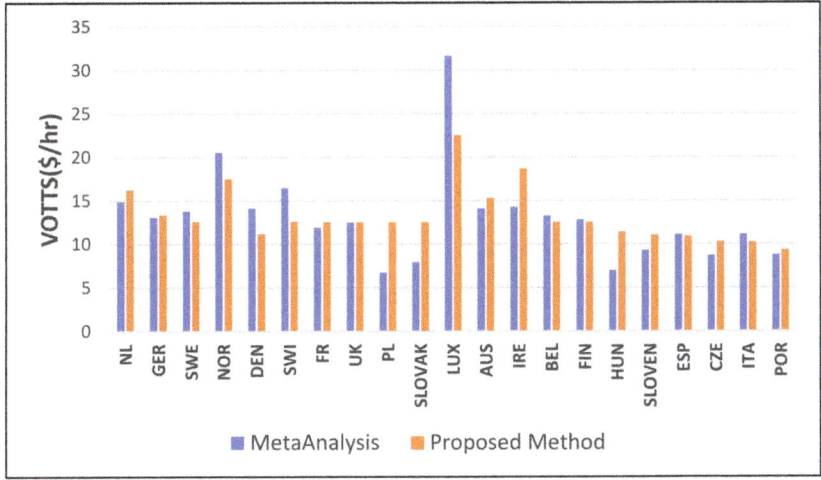

Figure 6. Comparison of estimated VOTTS for some European countries in the INRIX traffic score card with metanalysis values in [14].

Figure 7 below shows the estimated values compared to official recommended VOTTS for a number of countries (see Appendix A Table A3 for exact estimates). The official values are those for car travel except for UK which is for all modes. All values have been converted to 2017 dollars as stated before. Again, the average estimated VOTTS from the model is close to the average official values for the VOTTS further supporting the view that the proposed model can be used to estimate or recommend official VOTTS for countries and regions. However, there are noticeable differences. For instance, the official values for UK and Denmark are much higher than the estimated values. The model values are much closer to the metanalysis estimates for both countries. The average congestion in UK is much higher than in Denmark so the high official values for UK is justifiable but that's not true for Denmark. The Danish government could consider lowering their official VOTTS for car travel because the congestion level is not very high.

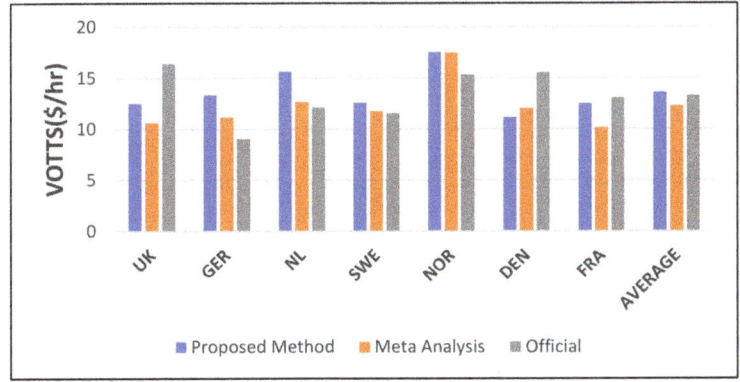

Figure 7. Estimated VOTTS for some European countries compared to official values in [14].

On the other hand, the official values for Germany and the Netherlands are lower than the estimated values. The average congestion in the Netherlands is lower than in Germany, so the lower official values for Netherlands is justifiable but that's not true for Germany. The German government could consider increasing their official VOTTS for car travel to compensate for the congestion level.

4.1. Sensitivity Analysis (Membership Function type)

As with most fuzzy inference systems, the membership function type used can affect the results. Although this is not the focus of the paper, we test the scenario where all trapezoidal membership functions (except the extreme values, see Appendix A Figure A4 are replaced by triangular membership functions [46]. The same set of rules were used for the FIS. The Figure 8 below shows the estimated VOTTS for all countries using the triangular membership functions compared to trapezoidal membership functions. The changes in VOTTS were below 2 $/h for most countries. However, some countries like Germany, Sweden and Denmark have noticeable changes.

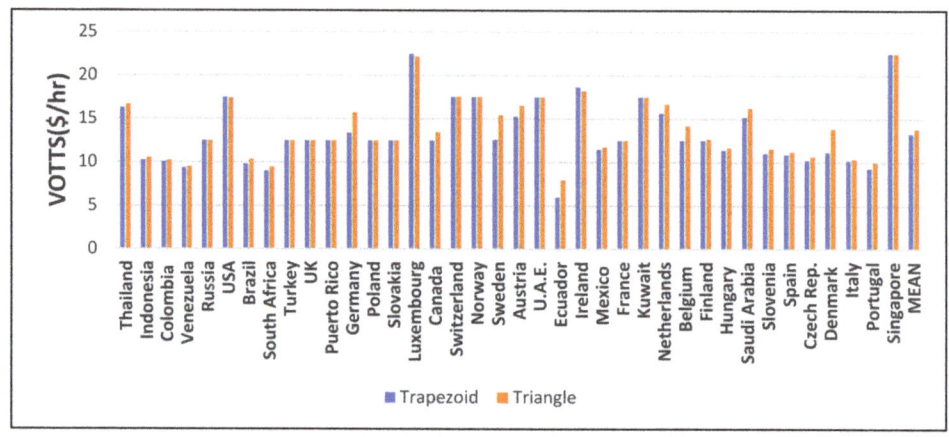

Figure 8. Estimated VOTTS using trapezoidal and triangular membership functions.

This shows that indeed choices like membership types and values can affect the model result. However, as stated before, the purpose of the paper is not to show which choice of membership shape is better but rather to show that a transparent and rule-based approach is better for estimating VOTTS because every decision made like rules, types of variables, membership functions can be easily scrutinized and modified in a fair and transparent way. The final model (rules, membership function types and values) to be used will depend on consensus between expert, government and citizens. This is an advantage of the proposed method when compared to the current approach of using discrete choice.

4.2. Sensitivity Analysis of Fairness Rules

It is expected that the fairness rules used in the model when reduced or modified can affect model results. The case of rule reduction is tested. The original model uses 26 rules to make inferences. Twenty rules make use of the two variables GDP and traffic while six of those rules where introduced for fairness which make use of either traffic or GDP. A simple sensitivity analysis is performed by reducing the number of rules from 26 to 20. The six rules which use only one variable were removed. Figure 9 shows the difference between the VOTTS predicted for all countries under both scenarios.

The FIS with 20 rules gave generally lower values than the one with 26 rules. This is because the rules concerning GDP and traffic congestion gives higher values when the GDP of the country is high or when the traffic is congested. Countries like Switzerland, Norway, Kuwait and Saudi Arabia with high GDP but with a relatively good traffic condition now have reduced VOTTS. In contrast, countries like Indonesia, Colombia, Venezuela with lower GDP but heavy traffic conditions now have a higher VOTTS. This clearly shows the fairness issue discussed before. The results show that the proposed method can use rules to reduce the effect of estimating high VOTTS for high GDP countries and low

VOTTS for low GDP countries. This makes it suitable for use by bodies such as the World Bank in determining the VOTTS to be used for the appraisal of projects in a fair and transparent manner.

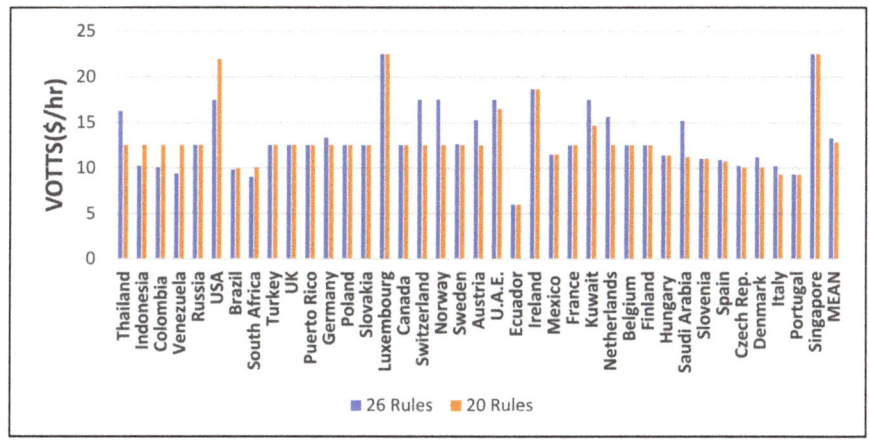

Figure 9. Estimated VOTTS for all countries using 20 and 26 rules.

5. Conclusions

A rule-based approach for estimating VOTTS has been proposed. The results show that the proposed model can give plausible values despite the simple rules used. The advantage of the model is the simplicity, understandability and transparency. All the processes and rules used to derive results can be easily scrutinised. The rules can be subject to a vote by governments, experts and citizens. This high level of flexibility and transparency is not available in current discrete choice models. The model can be applied for estimating VOTTS for urban regions, countries or even for the estimation of individual VOTTS. Another advantage is that fairness can be introduced in the rules, through input and output membership functions. For example, the government can decide which VOTTS is considered High, Medium or Low. The citizens can decide which comfort level is considered High, Medium or Low. This can be achieved through a survey. In this way, values that are considered fair can be used for the linguistic variables. The linguistic variables and their values can be appropriately chosen in a transparent way such that the model is completely understandable to the government, experts and the public.

Obviously, this model is not a replacement for econometric and discrete choice models but rather can serve as complementary estimation tool for such models. VOTTS estimated from discrete choice models are based on sound mathematical and economic theory whereas the proposed model is based on human experience and fairness principles. However, the model outputs and the uncertainties of the current discrete choice methods can be incorporated in the model input and output of the rule-based model. For example, findings from various VOTTS studies using choice models and other related techniques can be used to create membership functions for inputs, rules and the final output. Further research is needed to verify whether rules are enough to fully and correctly estimate VOTTS.

A limitation of the model is that the flexibility of the method makes it also vulnerable to misuse. For example, who decides what is a high VOTTS during projects: government, citizens or experts? Some variables can be more important than others: should weights apply to rules? This could pose a problem during implementation.

Another limitation of the method is that the variables used should be continuous. Categorical variables like gender or transport modes are difficult to fuzzify. If the proposed method is to be used for categorical variables, then they must be converted to continuous variables. Discrete choice models can handle both categorical and continuous variables. In future research, the proposed model will be applied for country wide estimation of VOTTS of individuals and various trip purposes and then compare values to those obtained from discrete choice models.

Author Contributions: The two authors contributed equally to most aspects of the paper.

Funding: This research received no external funding.

Conflicts of Interest: The authors declare no conflict of interest.

Appendix A

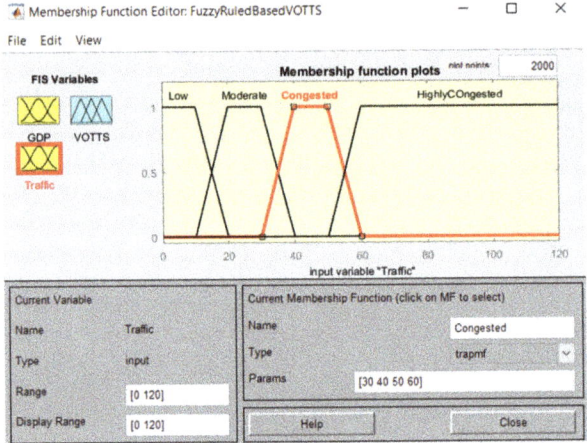

Figure A1. Membership function and linguistic variable for traffic congestion.

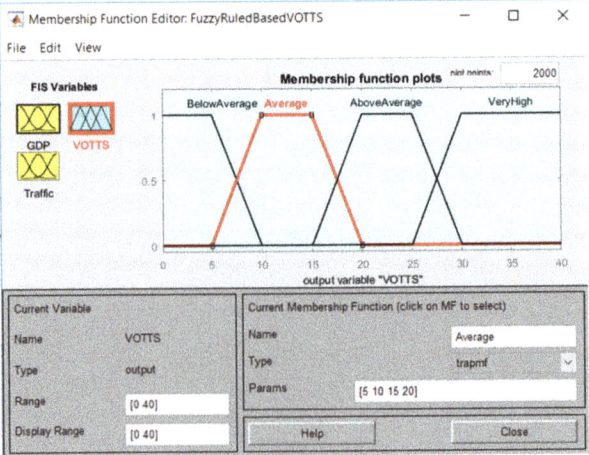

Figure A2. Membership function and linguistic variable for VOTTS.

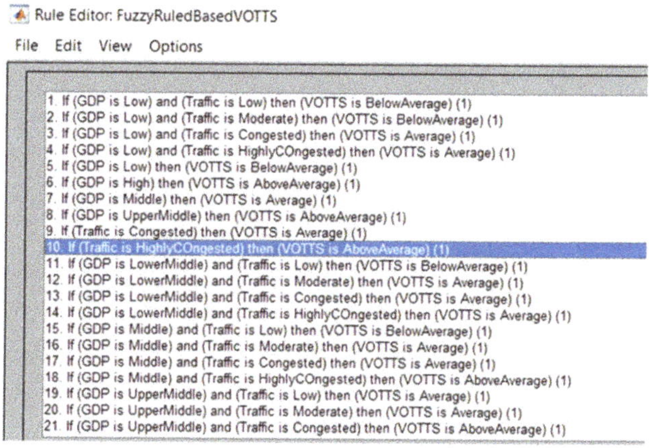

Rule Editor: FuzzyRuledBasedVOTTS
File Edit View Options
1. If (GDP is Low) and (Traffic is Low) then (VOTTS is BelowAverage) (1)
2. If (GDP is Low) and (Traffic is Moderate) then (VOTTS is BelowAverage) (1)
3. If (GDP is Low) and (Traffic is Congested) then (VOTTS is Average) (1)
4. If (GDP is Low) and (Traffic is HighlyCOngested) then (VOTTS is Average) (1)
5. If (GDP is Low) then (VOTTS is BelowAverage) (1)
6. If (GDP is High) then (VOTTS is AboveAverage) (1)
7. If (GDP is Middle) then (VOTTS is Average) (1)
8. If (GDP is UpperMiddle) then (VOTTS is AboveAverage) (1)
9. If (Traffic is Congested) then (VOTTS is Average) (1)
10. If (Traffic is HighlyCOngested) then (VOTTS is AboveAverage) (1)
11. If (GDP is LowerMiddle) and (Traffic is Low) then (VOTTS is BelowAverage) (1)
12. If (GDP is LowerMiddle) and (Traffic is Moderate) then (VOTTS is Average) (1)
13. If (GDP is LowerMiddle) and (Traffic is Congested) then (VOTTS is Average) (1)
14. If (GDP is LowerMiddle) and (Traffic is HighlyCOngested) then (VOTTS is Average) (1)
15. If (GDP is Middle) and (Traffic is Low) then (VOTTS is BelowAverage) (1)
16. If (GDP is Middle) and (Traffic is Moderate) then (VOTTS is Average) (1)
17. If (GDP is Middle) and (Traffic is Congested) then (VOTTS is Average) (1)
18. If (GDP is Middle) and (Traffic is HighlyCOngested) then (VOTTS is AboveAverage) (1)
19. If (GDP is UpperMiddle) and (Traffic is Low) then (VOTTS is Average) (1)
20. If (GDP is UpperMiddle) and (Traffic is Moderate) then (VOTTS is Average) (1)
21. If (GDP is UpperMiddle) and (Traffic is Congested) then (VOTTS is AboveAverage) (1)

(a)

Rule Editor: FuzzyRuledBasedVOTTS
File Edit View Options
6. If (GDP is High) then (VOTTS is AboveAverage) (1)
7. If (GDP is Middle) then (VOTTS is Average) (1)
8. If (GDP is UpperMiddle) then (VOTTS is AboveAverage) (1)
9. If (Traffic is Congested) then (VOTTS is Average) (1)
10. If (Traffic is HighlyCOngested) then (VOTTS is AboveAverage) (1)
11. If (GDP is LowerMiddle) and (Traffic is Low) then (VOTTS is BelowAverage) (1)
12. If (GDP is LowerMiddle) and (Traffic is Moderate) then (VOTTS is Average) (1)
13. If (GDP is LowerMiddle) and (Traffic is Congested) then (VOTTS is Average) (1)
14. If (GDP is LowerMiddle) and (Traffic is HighlyCOngested) then (VOTTS is Average) (1)
15. If (GDP is Middle) and (Traffic is Low) then (VOTTS is BelowAverage) (1)
16. If (GDP is Middle) and (Traffic is Moderate) then (VOTTS is Average) (1)
17. If (GDP is Middle) and (Traffic is Congested) then (VOTTS is Average) (1)
18. If (GDP is Middle) and (Traffic is HighlyCOngested) then (VOTTS is AboveAverage) (1)
19. If (GDP is UpperMiddle) and (Traffic is Low) then (VOTTS is Average) (1)
20. If (GDP is UpperMiddle) and (Traffic is Moderate) then (VOTTS is Average) (1)
21. If (GDP is UpperMiddle) and (Traffic is Congested) then (VOTTS is AboveAverage) (1)
22. If (GDP is UpperMiddle) and (Traffic is HighlyCOngested) then (VOTTS is AboveAverage) (1)
23. If (GDP is High) and (Traffic is Low) then (VOTTS is AboveAverage) (1)
24. If (GDP is High) and (Traffic is Moderate) then (VOTTS is AboveAverage) (1)
25. If (GDP is High) and (Traffic is Congested) then (VOTTS is VeryHigh) (1)
26. If (GDP is High) and (Traffic is HighlyCOngested) then (VOTTS is VeryHigh) (1)

(b)

Figure A3. All rules.

FIS Variables
GDP
VOTTS
Traffic
Membership function plots
2000
Low
Moderate
Congested
HighlyCOngested
1
0.5
0
0
20
40
60
80
100
120
input variable "Traffic"
Current Variable
Name
Traffic
Type
input
Range
[0 120]
Display Range
[0 120]
Current Membership Function (click on MF to select)
Name
Congested
Type
trimf
Params
[25 45 65]
Help
Close
Selected variable "Traffic"

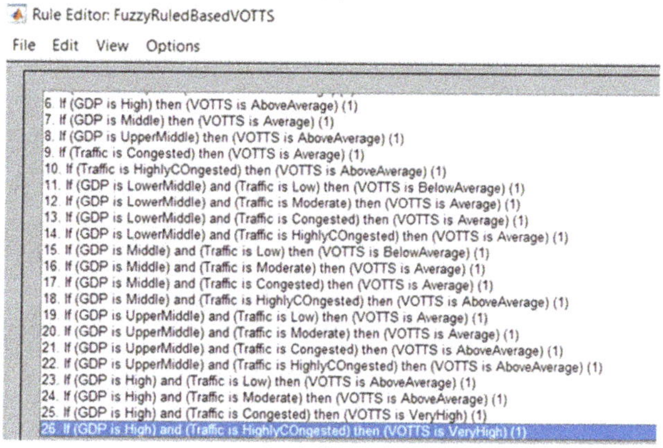

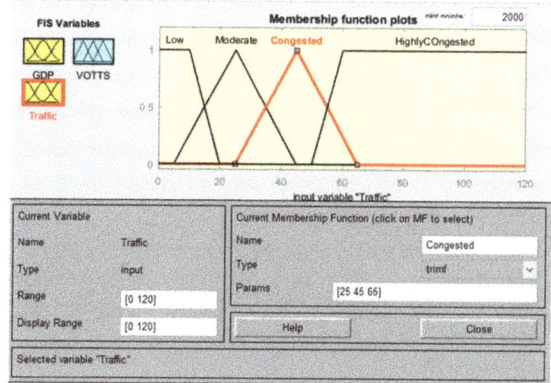

Figure A4. Triangular membership function and linguistic variable for traffic congestion.

Table A1. Estimated values of the proposed model compared value, from metanalysis values in [14].

Country	Estimated ($/h)	Meta-Analysis ($/h)	Difference (×100%)	Abs (×100%)
LUX	22.50	31.62	−0.41	0.41
SWI	12.58	16.45	−0.31	0.31
DEN	11.16	14.12	−0.27	0.27
NOR	17.50	20.54	−0.17	0.17
SWE	12.58	13.81	−0.10	0.10
ITA	10.18	11.07	−0.09	0.09
BEL	12.50	13.20	−0.06	0.06
FIN	12.50	12.76	−0.02	0.02
ESP	10.87	11.07	−0.02	0.02
UK	12.50	12.47	0.00	0.00
GER	13.35	13.10	0.02	0.02
FR	12.50	11.88	0.05	0.05
POR	9.29	8.72	0.06	0.06
AUS	15.27	14.07	0.08	0.08
NL	16.24	14.87	0.08	0.08
CZE	10.23	8.65	0.15	0.15
SLOVEN	11.02	9.26	0.16	0.16
IRE	18.66	14.25	0.24	0.24
SLOVAK	12.50	7.93	0.37	0.37
HUN	11.37	6.96	0.39	0.39
PL	12.50	6.73	0.46	0.46
MEAN	**13.23**	**13.03**	0.03	0.17

Table A2. Estimated and official values.

Country	Estimated ($/h)	Meta-Analysis ($/h)	Official ($/h)
UK	12.50	10.62	16.38
Germany	13.35	11.16	9.02
Netherlands	15.63	12.66	12.11
Sweden	12.58	11.76	11.56
Norway	17.50	17.49	15.32
Denmark	11.16	12.03	15.54
France	12.50	10.12	13.09
AVG	13.60	12.26	13.29

Table A3. Estimated VOTTS for various countries.

Country	GDP ($×1000)	Hours Lost in Traffic (h)	Estimated VOTTS ($/h)
Thailand	17.871	56	16.24
Indonesia	12.284	51	10.21
Colombia	14.552	49	10.03
Venezuela	12.400	42	9.36
Russia	25.533	41	12.50
USA	59.532	41	17.43
Brazil	15.484	36	9.80
South Africa	13.498	36	8.99
Turkey	27.916	32	12.50
UK	43.877	31	12.50
Puerto Rico	37.895	31	12.50
Germany	50.715	30	13.35
Poland	29.291	29	12.50
Slovakia	32.111	29	12.50
Luxembourg	103.662	28	22.50
Canada	46.378	27	12.50
Switzerland	65.006	27	17.50
Norway	60.978	26	17.50
Sweden	50.07	26	12.58
Austria	52.558	25	15.27
U.A.E.	73.879	24	17.50
Ecuador	11.617	23	5.99
Ireland	76.305	23	18.66
Mexico	18.149	23	11.46
France	42.779	22	12.50
Kuwait	71.943	22	17.50
Netherlands	52.941	22	15.63
Belgium	47.561	21	12.50
Finland	45.192	21	12.50
Hungary	28.375	18	11.37
Saudi Arabia	53.845	18	15.19
Slovenia	34.802	18	11.02
Spain	38.091	17	10.87
Czech Republic	36.916	16	10.23
Denmark	50.541	16	11.16
Italy	39.817	15	10.18
Portugal	32.199	15	9.29
Singapore	93.905	10	22.50

References

1. Mackie, P.J.; Jara-Diaz, S.; Fowkes, A.S. The value of travel time savings in evaluation. *Transp. Res. Part E Logist. Transp. Rev.* **2001**, *37*, 91–106. [CrossRef]
2. Kouwenhoven, M.; de Jong, G.C.; Koster, P.; van den Berg, V.A.C.; Verhoef, E.T.; Bates, J.; Warffemius, P.M.J. New values of time and reliability in passenger transport in The Netherlands. *Res. Transp. Econ.* **2014**, *47*, 37–49. [CrossRef]
3. Börjesson, M.; Eliasson, J. Experiences from the Swedish Value of Time study. *Transp. Res. Part A Policy Pract.* **2014**, *59*, 144–158. [CrossRef]
4. Abrantes, P.A.L.; Wardman, M. Meta-analysis of the UK values of time: An update. *Transp. Res. Part A Policy Pract.* **2011**, *45*, 1–17. [CrossRef]
5. Hess, S.; Daly, A.; Dekker, T.; Cabral, M.O.; Batley, R. A framework for capturing heterogeneity, heteroskedasticity, non-linearity, reference dependence and design artefacts in value of time research. *Transp. Res. Part B Methodol.* **2017**, *96*, 126–149. [CrossRef]
6. Hess, S.; Hensher, D.A.; Daly, A. Not bored yet—Revisiting respondent fatigue in stated choice experiments. *Transp. Res. Part A Policy Pract.* **2012**, *46*, 626–644. [CrossRef]
7. Ramos, G.M.; Daamen, W.; Hoogendoorn, S. A state-of-the-art review: Developments in utility theory, prospect theory and regret theory to investigate travellers' behaviour in situations involving travel time uncertainty. *Transp. Rev.* **2014**, *34*, 46–67. [CrossRef]
8. Haghani, M.; Sarvi, M. Hypothetical bias and decision-rule effect in modelling discrete directional choices. *Transp. Res. Part A Policy Pract.* **2018**, *116*, 361–388.
9. Bliemer, M.C.; Rose, J.M.; Hensher, D.A. Efficient stated choice experiments for estimating nested logit models. *Transp. Res. Part B Methodol.* **2009**, *43*, 19–35. [CrossRef]
10. Ojeda-Cabral, M.; Hess, S.; Batley, R. Understanding valuation of travel time changes: Are preferences different under different stated choice design settings? *Transportation* **2018**, *45*, 1–21. [CrossRef]
11. De Borger, B.; Fosgerau, M. The trade-off between money and travel time: A test of the theory of reference dependent preferences. *J. Urban Econ.* **2008**, *64*, 101–115. [CrossRef]
12. Fowkes, T.; Wardman, M. The design of stated preference travel choice experiments. *J. Transp. Econ. Policy* **1988**, *22*, 27–44.
13. Fosgerau, M. Using non-parametrics to specify a model to measure the value of travel time. *Transp. Res. Part A Policy Pract.* **2007**, *41*, 842–856. [CrossRef]
14. Fosgerau, M. Investigating the distribution of the value of travel time savings. *Transp. Res. Part B Methodol.* **2006**, *40*, 688–707. [CrossRef]
15. Shires, J.D.; de Jong, G.C. An international meta analysis of values of travel time savings. *Eval. Prog. Plan.* **2009**, *32*, 315–325. [CrossRef] [PubMed]
16. Wardman, M.; Chintakayala, V.P.K.; de Jong, G. Values of travel time in Europe: Review and meta-analysis. *Transp. Res. Part A Policy Pract.* **2016**, *94*, 93–111. [CrossRef]
17. Zadeh, L. Outline of a new approach to the analysis of complex systems and decision processes. *IEEE Trans. Syst. Man Cybern.* **1973**, *3*, 28–44. [CrossRef]
18. Teodorovic, D. Fuzzy logic systems for transportation engineering: The state of the art. *Transp. Res. Part A Policy Pract.* **1999**, *33*, 337–364. [CrossRef]
19. Vythoulkas, P.; Koutsopoulos, H. Modeling discrete choice behavior using concepts from fuzzy set theory, approximate reasoning and neural networks. *Transp. Res. Part C Emerg. Technol.* **2003**, *11*, 51–73. [CrossRef]
20. Kedia, S.; Bhuneshwar, K.; Katti, B. Fuzzy logic approach in mode choice modelling for education trips: A case study of Indian metropolitan city. *Transport* **2015**, *30*, 286–293. [CrossRef]
21. Olaru, D.; Smith, B. Modelling Daily Activity Scheduleswith Fuzzy Logic. In Proceedings of the 10th International Conference on Travel Behaviour Research, Lucerne, Switzerland, 10–15 August 2003.
22. Ming LU, IVT, BAUG, ETHZ, Exploring discrete choice model with fuzzy control theory. In Proceedings of the Conference STRC, Monte Verità, Ascona, Switzerland, 2–4 May 2012.
23. Train, K. *Discrete Choice Methods with Simulation*; Cambridge University Press: Cambridge, UK, 2003.
24. Mendel, J. Fuzzy logic systems for engineering, A tutorial. *Proc. IEEE* **1995**, *83*, 345–377. [CrossRef]
25. Zimmermann, H. *Fuzzy Set Theory and Its Applications*; Kluwer-Nijhoff: Boston, MA, USA, 1991.

26. Klir, G.J.; Yuan, B. *Fuzzy Sets and Fuzzy Logic: Theory and Applications*; Prentice Hall: Upper Saddle River, NJ, USA, 1995.
27. Dubois, D.; Prade, H. What are fuzzy rules and how to use them. *Fuzzy Sets Syst.* **1996**, *84*, 169–185. [CrossRef]
28. Fosgerau, M.; Hjorth, K.; Lyk-Jensen, S.V. Between-mode-differences in the value of travel time: Self-selection or strategic behaviour? *Transp. Res. Part D Transp. Environ.* **2010**, *15*, 370–381. [CrossRef]
29. Axhausen, K.W.; Hess, S.; König, A.; Abay, G.; Bates, J.J.; Bierlaire, M. Income and distance elasticities of values of travel time savings: New Swiss results. *Transp. Policy* **2008**, *15*, 173–185. [CrossRef]
30. Börjesson, M. Inter-temporal variation in the marginal utility of travel time and travel cost. *Transportation* **2014**, *41*, 377–396. [CrossRef]
31. Börjesson, M.; Cherchi, E.; Bierlaire, M. Within-individual variation in preferences. *Transp. Res. Rec. J. Transp. Res. Board* **2013**, *2382*, 92–101. [CrossRef]
32. Börjesson, M.; Eliasson, J. Should values of time be differentiated? *Transp. Rev.* **2018**. [CrossRef]
33. OECD. *International Transport Forum: Valuing Convenience in Public Transport*; Roundtable Report 156; OECD: Paris, France, 2015.
34. Wardman, M.; Nicolás, J. The congestion multiplier: Variations in motorists' valuations of travel time with traffic conditions. *Transp. Res. Part A Policy Pract.* **2012**, *46*, 213–225. [CrossRef]
35. Homaifar, A.; McCormick, E. Simultaneous design of membership functions and rule sets for fuzzy controllers using genetic algorithms. *IEEE Trans. Fuzzy Syst.* **1995**, *3*, 129–139. [CrossRef]
36. Gourdazi, P.; Hassanzadeh, R. A GA-Based Fuzzy Rate Allocation Algorithm. In Proceedings of the IEEE International Conference on Communications, Instabul, Turkey, 11–15 June 2006.
37. Permana, K.; Hashim, S. Fuzzy Mem-bership Function Generation using Particle Swarm Optimization. *Int. J. Open Probl. Compt. Math.* **2010**, *3*, 27–41.
38. Omizegba, E.; Adebayo, G. Optimizing fuzzy membership functions using particle swarm algorithm. In Proceedings of the IEEE International Conference on Systems, Man and Cybernetics, San Antonio, TX, USA, 11–14 October 2009; pp. 3866–3870.
39. Worldbank GDP PPP Report. Available online: https://data.worldbank.org/indicator/NY.GDP.PCAP.PP.CD?view=chart%2C&year_high_desc=true (accessed on 14 January 2019).
40. INRIX Traffic Report. Available online: http://inrix.com/scorecard/ (accessed on 14 January 2019).
41. Medasani, S.; Kim, J.; Krishnapuram, R. An overview of membership function generation techniques for pattern recognition. *Int. J. Approx. Reason.* **1998**, *19*, 391–417. [CrossRef]
42. Wijayasekara, D.; Manic, M. Data Driven Fuzzy Membership Function Generation for Increased Understandability. In Proceedings of the IEEE International Conference on Fuzzy Systems (FUZZ-IEEE), Beijing, China, 6–11 July 2014.
43. Barua, A.; Mudunuri, L.S.; Kosheleva, O. Why Trapezoidal and Triangular Membership Functions Work So Well: Towards a Theoretical Explanation, Departmental Technical Reports (CS), Paper 783. 2013. Available online: http://digitalcommons.utep.edu/cs_techrep/783 (accessed on 4 February 2019).
44. Inflation Website. Available online: http://www.in2013dollars.com/2010-euro-in-2017?amount=100 (accessed on 14 January 2019).
45. Euro to Dollar Exchange. Available online: https://www.statista.com/statistics/412794/euro-to-u-s-dollar-annual-average-exchange-rate/ (accessed on 14 January 2019).
46. Pedrycz, W. Why Triangular Membership Functions? *Fuzzy Sets Syst.* **1994**, *64*, 21–30. [CrossRef]

© 2019 by the authors. Licensee MDPI, Basel, Switzerland. This article is an open access article distributed under the terms and conditions of the Creative Commons Attribution (CC BY) license (http://creativecommons.org/licenses/by/4.0/).

MDPI
St. Alban-Anlage 66
4052 Basel
Switzerland
Tel. +41 61 683 77 34
Fax +41 61 302 89 18
www.mdpi.com

Sustainability Editorial Office
E-mail: sustainability@mdpi.com
www.mdpi.com/journal/sustainability